SYSTEMS ANALYSIS AND DESIGN FOR SAFETY

Safety Systems Engineering

DAVID B. BROWN

Auburn University

PRENTICE-HALL, INC., *Englewood Cliffs, New Jersey*

Library of Congress Cataloging in Publication Data

Brown, David B
   Systems analysis and design for safety.

   Bibliography: p.
   Includes index.
   1. Industrial safety. 2. System analysis.
I. Title.
T55.B75   614.8'52   75–44020
ISBN 0–13–881177– 6

To Dad, Carolyn, and Lesley

10 9 8 7 6 5 4 3

Printed in the United States of America

PRENTICE-HALL INTERNATIONAL, INC., *London*
PRENTICE-HALL OF AUSTRALIA, PTY. LTD., *Sydney*
PRENTICE-HALL OF CANADA, LTD., *Toronto*
PRENTICE-HALL OF INDIA PRIVATE LIMITED, *New Delhi*
PRENTICE-HALL OF JAPAN, INC., *Tokyo*
PRENTICE-HALL OF SOUTHEAST ASIA (PTE.) LTD., *Singapore*

CONTENTS

5   Fault-Tree Analysis   152

6   Statistical Analysis   196

PREFACE

The safety director is often faced with multi-dimensional problems which must be understood and solved *before* they lead to accidents. Both the complexity of the man-machine environment in which he functions and the probabilistic nature of accident cause-effect relationships prevent him from having the flexibility in design and execution afforded the typical manager. The safety director and his staff must be right, not by a process of trial and error, but by an a priori decision based upon facts. As such they must use every bit of pertinent information at their command to accomplish their objective.

Recent articles on the systems approach to safety have given new hope to safety practitioners. Quantitative methods for evaluating information and accident situations have been proposed. Also, the "systems" approach to design and evaluation has been discussed. These tools are essential if the science of safety is to continue to progress. Accident statistics prove that the "common sense" required for safety is not very common at all.

The intent of this work is twofold: (1) To provide a conceptual understanding of systems engineering and its application to safety and (2) To provide an analytic structure through which safety decision making can be performed. It is based upon basic principles of systems analysis, probability and statistics. The concept of tradeoffs and optimization is also stressed as a philosophy; although one very general technique, dynamic programming, is presented for those who have the necessary data for its use.

Chapter I begins with a series of definitions designed to set the stage for the discussion of systems engineering to follow. The concept of control is presented in order to establish a framework from which an effective accident control system can be designed. Chapter II is a stepping stone from the purely qualitative considerations of Chapter I to the quantitative techniques presented in the remainder of the book. The methods of hazard analysis and cost effectiveness in Chapter II are of particular value where quantitative data are unavailable.

Chapter III begins to present the analytical tools necessary for a logical approach to accident investigation. More importantly, however, this chapter provides an understanding of the concepts of Boolean Algebra necessary for the use of the probabilistic concepts of Chapter V. Here fault trees are presented in their classical usage and then extended to cost/benefit evaluations.

At this point in the book, supportive material is presented for the practical application of the techniques presented. Chapter VI presents statistical techniques useful for evaluating data and formulating estimates for fault tree analysis. Chapter VII is even more basic in demonstrating methods for data collection and preprocessing.

Chapter VIII goes beyond the pure cost/benefit approach and considers the discrete nature of alternative investments. Chapter IX applies this technique to roadway improvements in highway and traffic safety, using the case study method of presentation. These applications are oriented toward the viewpoint of traffic and highway maintenance engineers in all levels of government. However, the principles exemplified by these applications are equally applicable to occupational accident prevention. Traffic and highway safety applications were chosen for two reasons: (1) The author's expertise in these areas and (2) The general familiarity of all readers with roadway systems.

Chapter X is an extension of the systems approach to a broader look at related problems involving safety, health, and the environment in general. Legal and practical considerations are brought into the picture at this point to emphasize the author's view of a reasonable approach to safety.

Appendix I supports Chapter VII in supplying the coding scheme for data gathering. Appendix II provides the computer programs and their documentation for implementation of the concepts presented in the book and exemplified by the applications. These programs are the results of actual applications, and they have been rigidly tested in practice. As described in Chapter IX, they are directed at specifying solutions for implementation, and not just for the generation of more "interesting" information.

The intent of this book is to provide a reference for the practicing safety professional. However, it can be used as a text for a typical three credit hour course in systems analysis for safety. It is recommended that a small project be performed as a requirement of the course. This project could take the form of a design of an overall safety system. Alternatively, the student may be required to rigorously apply one of the quantitative techniques, such as fault tree analysis, to a particular industrial situation. In this case it would be recommended to cover those techniques first which are to form the basis for the project.

A final note to those skeptical of quantitative techniques. It is true that most quantitative techniques assume a knowledge of the future which is not available. Past probabilities, measures of effectiveness, and even costs are subject to future change. But in condemning a quantitative technique because some judgment of input data is required, consider the alternatives. All decisions require an estimate of some future outcome which is to some extent uncertain. So the choice is not between an imperfect quantitative technique and an alternative technique which is perfect. At best the choice is between two imperfect techniques; and usually the choice is between the quantitative technique and no technique at all. The question then resolves itself to one of deciding which technique will lead to the most beneficial decision given the imperfect inputs.

The arguments for quantitative models are threefold. First, the human estimates of future performance can be made on detailed characteristics of the system as opposed to overall performance. Estimates at this level are more accurate, and even if they are not perfect, a minor error at this level will not greatly affect the outcome. Second, estimates can be made by several persons, such that each individual estimate can be made by the most qualified person available. Finally, a provision exists to reiterate the analysis, changing the input at each iteration so that the effect of these changes can be determined. Such a sensitivity analysis can determine where the accuracy of estimation should be improved.

The alternative to the analytic and quantified approach is an estimate on a grosser scale which human limitations prevent from taking into account more than four or five dimensions of any problem simultaneously. While many managers are qualified by their experience, wisdom, and intuitive ability to make such decisions, the inefficiency of most large organizations attests to the fact that such men are scarce. Those who are so gifted are characterized by a high degree of organizational ability within themselves which is characteristic of the analytical quantitative techniques that are stressed here.

Lest we leave the wrong impression, quantitative techniques applied to safety are only one part of the answer. Hence we stress quantification as merely a tool in safety decision making. As the complexity of modern technology has grown, this tool has become essential to managing the safety function. That this tool must be integrated into a philosophy of systems management is stressed by the presentation of systems and control theory first.

This book is designed to present an organizational thought structure for creative use in solving problems of systems safety. The applications involve safety organizations, inspections, budgets, etc. The presentation of the theory of machine guarding, and the other provisions of the safety standards are beyond the scope of this present text. Such courses form an excellent compliment to the information contained here; however, they should not be considered as a prerequisite to this material.

I would like to express my appreciation to my many associates who contributed to the content of this book. Special thanks must go to Dr. George H. Brooks and Dr. Louis B. Trucks, who offered suggestions and constructive criticism in several phases of the manuscript development. Finally, greatest credit must go to my wife, Carolyn, who not only typed the manuscript, but also provided the emotional support necessary for this entire effort.

DAVID B. BROWN

SYSTEMS ANALYSIS AND DESIGN FOR SAFETY

Safety Systems Engineering

SYSTEMS ANALYSIS AND SYNTHESIS

1.1 Introduction

Most of the material in these first two chapters will have general application to all systems engineering* work. It is very difficult to talk about the theories of systems engineering, however, without presenting some concrete examples. It is not that such theories do not exist apart from applications. Rather, it is extremely difficult for most students who have no practical experience, and for many who do, to grasp these theories apart from some exemplification. However, this should be the goal. For if the theory can be grasped and understood apart from the application, the student's ability to apply it to new situations as they arise will be unlimited.

1.1.1 Note on Definitions

A variety of papers discussing the "systems" approach to safety have been published recently. Recht (7),† Tiger (11), and Levens (3) discussed the

---

*To be defined below.
†References are given at the end of each chapter.

1

systems approach on a general level. Wissner (12) and Biancardi (1) concentrated on the organization required for translating systems concepts into reality. Rockwell (10) discussed various approaches to measures of effectiveness that are essential if the systems approach is to be applied. Peters and Hall (6) used a mathematical approach similar to those developed for reliability evaluations, and integrated cost factors into his procedures, as did Recht (8) also. Kolodner (2), Mackenzie (4), and Recht (9) discussed the use of the fault-tree technique as both a qualitative and quantitative aid to the system designer. This cross section of available literature on safety systems technique applies to the following three areas:

1. Design of the end product for safety.
2. Design of the manufacturing process for safety.
3. Design of the safety system.

An understanding of the context of the word "system" in the above papers promotes an understanding of the common use of this word in safety applications. This, however, is insufficient; for, although it is desirable to be as close to the present usage as possible, often there is disparity, if not outright contradiction, in the use of this and related words. Nevertheless, keep in mind the common emphasis upon design, which includes both the original design and design improvement.

It must be concluded that there is no perfectly correct definition or usage for the terms we are about to discuss. There are various degrees of abuse, however, ranging from ignorance to outright deception. The definitions given below are justified because they are commonly accepted, and this usage is consistently maintained throughout the book.

1.1.2 Definitions

System. Webster defines a *system* as "an aggregation or assemblage of objects united by some form of regular interaction or interdependence; a group of diverse units so combined by nature or arts as to form an integral whole, and to function, operate, or move in unison and, often, in obedience to some form of control; an organic or organized whole."

Words such as "aggregation," "assemblage," "united," "interaction," "group," "combined," "unison," "control," and so on, emphasize organization, whereas we have seen that the focal point of the systems approach is directed toward design. Thus possibly use of the phrase "systems approach" could adequately be replaced by "organized design." This definition must be qualified, however, since the systems terminology has only recently come into vogue, whereas the design terminology has been around for a long time.

The cause of this is attributed to fairly recent advances in technology. At one time a designer of bicycles, by applying the full extent of his genius, could also design a simple aircraft. Now, however, aircraft designers are so specialized that only in very rare cases do they have the immediate capability to design a new bicycle. Although it was true that the fundamental concepts of systems engineering went into the design of the first aircraft, it was simple enough that one man could handle the problem—given enough time.

Today one man does not design the total aircraft. Most large design projects are broken into smaller design tasks, which in turn lead to the necessity for an organization of these individual projects. Such is what the "systems" concept is all about.

Analysis. The word *analysis* should be defined here because it appears so often next to the word "systems." The classical definition is: the separation of a whole into its constituent elements. It is the opposite of synthesis, which is the combination of the parts into a complex whole. Current usage, however, does not maintain this strict definition, and therefore the term *systems analysis* is not restricted to the study of the system by a strict analysis. One of the vital aspects of the systems approach is the emphasis on the necessity to see how all the parts function together as a system. Technically, this is systems synthesis.

The term "systems analysis" would seem to be at least in part self-contradictory. For the system's approach requires an interdisciplinary coordination of activities to design each component and subsystem to obtain an optimal system. Analysis in its strictest sense implies the taking apart and studying each part individually. We therefore favor the use of the term "systems engineering" to cover application of the "systems approach." This is defined further below.

Engineering. In its pure sense, the word *engineering* is restricted to the activity of applying pure science. The engineer is a necessary link between the scientist, who discovers and unveils truth, and the craftsman, who produces goods and services for the betterment of mankind. The engineer takes science and puts it into a form the craftsman can understand and can operate upon.

This, of course, is an oversimplified view of the task of engineers, for occasionally, to fulfill their functions, they may be required to dabble in purely scientific investigation or in craftsmanship. Nevertheless, this view of engineers puts their functions into the proper perspective so that these functions can be both understood and evaluated.

Systems Engineering. The terms "systems approach" and "systems engineering" were used above to convey a vague concept. This subsection attempts to clarify and define these terms. *Systems approach* is a method of attacking a problem based upon systems engineering. In turn, *systems engineering* is defined to be a discipline that has as its objective the following ideals:

1. The study and evaluation of all parts of a defined system, in combination with the science and mathematics of optimization.
2. The analysis of the system to study the effects that each part has upon the operations of every other part, and the independent effect that each part has upon the working of the system.
3. The synthesis of these parts once studied and optimized to study the effects that they have upon the total system.
4. The optimization of the total system generally at the sacrifice of the optimal design of one or more of the parts.

Although design is mentioned and stressed in the safety literature, systems engineering will include all activities that must take place to obtain a functioning system. Systems engineering is thus the application of "science" to the improved functioning or design of large-scale systems. This definition will avoid the semantical problem associated with a broadening of the definition of design.

1.1.3 Systems Engineer

A great deal of discussion has arisen recently regarding the education of a systems engineer. Some of the traits necessary to perform the systems engineering function can be learned through formal education. Other characteristics must be developed through a desire to perform well the systems engineering function. This is not because these characteristics, such as management ability, are inborn. Rather, it is because systems engineering requires a high degree of adaptability, coupled with concerted effort. Therefore, if the desire is not present, no degree of education will compensate.

The *systems engineer* is defined to be the person who practices systems engineering as described above, regardless of background or title. The fact that certain people have the technical expertise and that they have practiced systems engineering for a number of years makes them particularly capable of assuming the formal position of systems engineer. Their background could be from staff or line management and their discipline could be operations research, industrial engineering, physics, or anything else.

All persons probably engage in systems engineering at some time during their career. The objective here is not to cover all the educational and management decisions associated with choosing and creating the systems engineer. Rather, it is to show that this function can and should be performed by capable professional people in the safety application. Therefore, in the discussion that follows, some of the most important traits of the systems engineer are given. Hopefully, this will enable those seeking such a position to acquire these traits. It should also help those who are assigning the task to recognize and evaluate the proper characteristics.

As far as education is concerned, two things aid tremendously. First, a well-rounded broad cultivation of experience in many areas is required. Although systems engineers cannot possibly attain depth in all areas, they must have the willingness and the ability to do so. The necessity for this in the safety application is readily apparent. Safety engineers must function to some extent in all areas of the physical facility, and they must also cope with psychological and sociological problems. The type of problem encountered is so large that a narrow education is totally insufficient for the systems engineer. Although a very narrow education does not prevent students from broadening themselves later, quite often those who are very specialized lack appreciation for other disciplines as well as the adaptability necessary for systems engineering.

A second educational requirement calls for emphasis upon quantitative and logical problem solving. A technical speciality in some branch of engineering, mathematics, or a physical science is a great asset. Although a person may initially lack the broad background characteristics, the properly motivated, technically trained individual can usually acquire the experience that is needed for the performance of the functions of a systems engineer. However, the nontechnician, whose mathematical ability breaks down at simple algebra, is not likely to acquire the quantitative expertise without great effort. This is not to say that the task is impossible, just improbable.

The last two paragraphs present a contradiction in terms and demonstrate clearly the reason that systems engineers are not merely the product of their education. Usually the student who has a "liberal" education does not acquire the technical and mathematical abilities, and vice versa. This is another reason that the systems engineer can arise from any discipline, although, as stated before, the technically trained person has a substantial advantage.

Stressed here has been the need for systems engineers to be adaptable to the task at hand. Actually, this characteristic must be developed to the point where they can acquire new specialities quickly when required. Also necessary is their willingness to do so. While fickleness and impatience are to be discouraged the systems engineer cannot be a person who hesitates to leave the comfort of a familiar job for new experiences.

This flexibility to change, coupled with the ability to master new specialities, must be tempered with an ability to resist the temptation of getting tied down with too many details. Although systems engineers must be of above-average intelligence, they need not have the comprehension of a genius. Allowing for this "shortcoming," they adjust by not getting into a level of detail that will tap the resources needed for more critical tasks. Hence they assign the detailed work to the specialist and *allow the specialist to do it.* Although they cannot lose control of the project to subordinates, neither should they try to run that part of the show that rightfully belongs to those subordinates.

This brings up another point relating to experience as a qualification. Often it is stated that the "T man," a person who has broad but shallow experience in many areas plus a deep and thorough understanding of one specialty, makes a good systems engineer. This will be true only if such a person is willing to "part company" somewhat with that specialty. Quite often such people reach their "level of incompetence" (reference 5) merely by being unable to break old habits. This continuous dabbling in old specialties prevents them from adequately performing the job to which they are assigned. They also bother their successors and prevent them from functioning properly. So, as far as experience is concerned, the person with a fair depth in all areas is generally superior to all other types.

Of all the traits that the systems engineer must possess, none is as important as the ability to communicate. To reemphasize the need for breadth, systems engineers must be able to communicate with all disciplines. This implies the ability to speak the language, and to understand what is said by people in these disciplines. Again, this is not "educated in"; it is attained by persistent effort through experience.

The ability to communicate must extend to the written word. Clear, concise work orders and progress reports are essential to maintaining the systems engineering effort. Not only must systems engineers be able to write well and quickly, they must be able to read effectively. This does not necessarily mean a 10,000-word-per-minute rate, but it does require the ability to separate concepts from details and the pertinent from the irrelevant.

The systems engineer must have the ability to see long-range goals and strive for their attainment. This is often termed "system" or "company orientation." Whatever it is called, it is, in fact, the opposite of shortsighted, immediate-gain motivation. Although self-interest and aggressive ambition are not ruled out, political expediency and the path of least resistance must have a minimal effect upon the systems engineer's decisions. The sacrifice of short-term goals for the attainment of more important, long-term benefits must become a habit.

More than a discipline, systems engineering has been described as almost a way of life. Hopefully, it is clear that, although it has been described in rather glowing terms, everyone should not strive to be a systems engineer. This is a function to be performed by a limited number of qualified people. Although everyone can benefit from systems engineering concepts and most will use them sometime during their careers, the professional systems engineering life should only be sought by those who believe that their own personal satisfaction will be served by this type of occupation.

Remember that the systems engineer cannot succeed without a fine staff of specialists, each concentrating in his own area. For each systems engineer, therefore, there must be several non-systems-engineering types. These are people who grapple with details and have minimal interaction (i.e., interaction is not their primary occupation).

The purpose of this subsection is not to sell the systems engineering career to the individual. In general the systems engineer is no more noble, no more ethical or moral, and no happier than anyone else. However, there is an obvious need for systems engineers, especially safety systems engineers, within most organizations. The characteristics of the systems engineer were given so that those in upper management could respond to the need.

1.1.4 Management, Industrial Engineering, Operations Research, and Safety

It could be argued, quite correctly, that most of the traits given above are the same as would be required of a good manager. For completeness, a brief contrast between systems engineering and three other disciplines— management, operations research, and industrial engineering—will be described. It must be emphasized again that different people use these terms in different ways, and therefore only general descriptions can be given.

Line management is concerned with the day-to-day operations of the business. Although most of the traits of the systems engineer are required of the manager, the technical depth and the broad degree of flexibility are not required in most middle-management jobs. In addition, managers must have a very detailed knowledge of the functions that they manage. Since they are quite often dealing with the work force, they must have a good knowledge of labor relations. They must be willing to participate in the day-to-day fire-fighting operations. The list could go on and on; these characteristics are presented to show a contrast between a good manager and a good systems engineer. The difference is subtle, but there is a distinct difference.

Similar factors apply in staff management assignments such as the operations research team. Operations research is directed toward research at the operational level to provide optimally designed and functioning components and systems. Systems engineering begins before the operation, even prior to design, and continues through to the ongoing system. The distinction here is vague, although, as with management, operations research requires less breadth and more depth. This is especially true today as operations research has become more quantitatively model-oriented. The systems engineer must function quite often without the aid of models, quantitative or otherwise, while using his mathematical ability in a qualitative way.

Industrial engineering is another staff function closely related to systems engineering. Industrial engineers often assume the role of systems engineers when assigned a broad project to perform. Their experience working in a variety of areas coupled with their technical training especially equip them for systems engineering. But industrial engineering is not systems engineering, and many industrial engineers engage in projects of such limited scope that they never get into what can honestly be called a systems approach. Again, it is the concentration upon their speciality, whether it be incentives, plant layouts, work-place design, and so on, which generally makes the industrial engineer of great value.

To summarize, systems engineers are obtained from the ranks of management, operations research, industrial engineering, and many other disciplines. Their decision to become systems engineers is often motivated by dissatisfaction with their specialization. In acquiring the necessary traits, systems engineers will almost certainly lose some of the characteristics that are necessary for their specialty. They will sacrifice these for the same reason that they will sacrifice component optimality for the benefit of the system—to maximize an overall objective.

Systems engineers who are engaged in safety must interact with all the above disciplines. They must draw upon the abilities of others while providing the services of their profession (systems engineering) coupled with their specialty (safety). They must have a good working knowledge of basic safety theory and a well-rounded knowledge of the plant operations within which they function.

The remainder of this book is intended to demonstrate the discipline and some characteristic techniques employed by systems engineers as they apply themselves to safety activities. The techniques presented represent examples of the types of applications that are available through systems engineering.

It is hoped that these may be thoroughly exploited and that this may lead to further developments and applications of other improved techniques.

1.2 Safety Systems

A *safety system* will be defined as the total set of men, equipment, and procedures specifically designed to be superimposed on an industrial system* for the purpose of increasing safety. This safety system may eventually make radical changes in the industrial system. Generally these changes will be of an evolutionary nature, however, since the industrial system upon which it will act is already in existence. Although a completely new design of the industrial system for safety is not ruled out, this is only one of the alternatives in the design of the safety system.

The purpose of this section is to outline a general "systems" procedure that will be useful to those responsible for designing and implementing a safety system. It is intended to circumvent the crucial task of "getting started," which is so often the hardest step in the creative process.

The need for a safety system should be obvious. Many large companies have responded to this need by establishing safety departments, writing safety manuals, purchasing equipment, and so on. However, the question should be asked: Upon what criteria are personnel added to this department, procedures written, equipment purchased? There should be an orderly process of evaluation to ensure a maximum return on the investment. The limited budget available for safety should be allocated to obtain the maximum effectiveness.

Although the procedures presented in this chapter will not guarantee optimum decisions, they will provide useful procedures for progressing in that direction. The discussion will assume that no safety system is currently in effect for the company, although the techniques presented can serve for modification as well as initial design. A general systems approach will now be presented for the system design procedure.

1.2.1 General Systems Approach**

There are a variety of systems approaches presented in the literature, both in safety applications and in general discussions of systems engineering. All of these have three things in common:

---

*The term "industrial system" is used here in a generic sense. The name of any system requiring the superimposition of a safety system could be substituted for "industrial."

**Much of the material that follows was originally published in an article entitled "Systems Engineering in the Design of a Safety System," *ASSE Journal*, Vol. 18, No. 2, 1973. Acknowledgement and credit go to the ASSE Journal and the American Society of Safety Engineers.

1. They present a general step-by-step procedure that is subject to much interpretation, depending upon the application.
2. They emphasize decision making on a general level prior to evaluation of details.
3. The procedure given is a reiterative process.

Of these three characteristics, the second has special significance in circumventing pitfalls in the design process. All possible systems cannot be evaluated in minute detail, for this would be too costly and time-consuming. The determination of *when* to plunge into the details is crucial for the success of any system design effort.

As a basic system design philosophy, a three-level approach is suggested. Table 1.1 presents the three levels in terms of the input, the action taken at that level, and the output. The emphasis on the three-level approach is not to cause action but rather to circumvent unnecessary work. The systems engineer and the designers should not get into any more detail than their current level suggests. They should finish one level, including documentation, prior to proceeding to the next. In this way they avoid the danger of getting absorbed in detail prematurely.

Table 1.1. THREE-LEVEL APPROACH TO SYSTEMS ENGINEERING

Level	Input	Action	Output
I	Alternative system functions and goals	Consider and evaluate system alternatives	Formulated definition of system objectives
II	System objectives; alternative subsystem functions, concepts, and approaches	Consider and evaluate subsystem alternatives	Subsystem approaches
III	Subsystem approaches; alternative subsystem designs	Consider and evaluate detailed design alternatives	Detailed design

Table 1.1 might suggest the possibility of a four- or five-level approach, and indeed in many large and complicated system designs even more levels would be required. The central point is still the same: the timing of a decision is as important as the decision itself. For example, the process of designing the best guarding device for a machine would not be required if the decision is

already made to convert to a completely new automatic machine which requires no additional guarding. The design process is deferred to level III, whereas the system and subsystem decisions are made at level II.

Figures 1.1, 1.2, and 1.3 depict the analysis portion of levels I, II, and III, respectively. It is important to realize that resynthesis is also necessary after the completion of each level of analysis for proper evaluation of each subsystem and component. It is at this time that trade-offs between subsystems and/or components can be performed.

Figure 1.1 shows the emphasis at level I on system objectives and functions. The concentration here is upon *what* is to be performed or accomplished

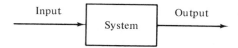

Figure 1.1. Level I Analysis of a system

as opposed to *how* it should be accomplished. Objectives worded in terms of methods should be discouraged. For example, "the reduction of recordable injuries by 10% per year" is an acceptable objective, whereas "the establishment of an effective educational system" is not, since education is a means to an end and not an end in itself. Thought at this level must be goal-oriented, not method-oriented. Figure 1.1 also emphasizes that *everything* that will transform the input into the output should be defined as part of the system.

Figure 1.2 gives a simple sequential breakdown of the system into subsystems. In actuality there might be arrows from every subsystem to every other subsystem, depicting the interaction of subsystems one toward another. For simplicity only one direction is depicted, since many sequential processes can be approximated by this model. Notice that according to the terminology described in the figure:

$$I_{i+1} = O_i \qquad \text{for all } i \tag{1.1}$$

That is, the input to any subsystem is equal to the output from the prior

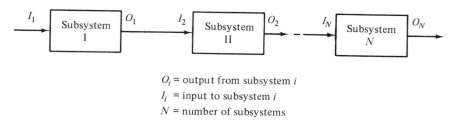

O_i = output from subsystem i
I_i = input to subsystem i
N = number of subsystems

Figure 1.2. Level II Analysis of a system

subsystem. This again emphasizes that whatever transformation takes place must occur within one of the subsystems. Hence the subsystems collectively include all of the system. It will also help organizationally if no two subsystems perform or are responsible for the accomplishment of the same function. Mathematically,

$$O_i = T_i(I_i) \tag{1.2}$$

where $T_i(x)$ represents the functional transition performed on input x by subsystem i. In other words, the output from subsystem i results from the input to subsystem i and the particular function, $T_i(\cdot)$, which subsystem i performs.

The model is fashioned in mathematical terms to increase reasoning precision. At times the abstract mathematical model can be used as part of an optimization procedure for purposes of decision making (see Chapter 8). Currently, symbols will be used merely to clarify the three-level approach to design.

Another degree of difficulty is added in level III, as depicted by Figure 1.3. Now each subsystem function is exploded into alternatives. Any one, or combinations, of these alternatives may be optimal. The systems engineer must make the decision as to which and how many alternatives are to be

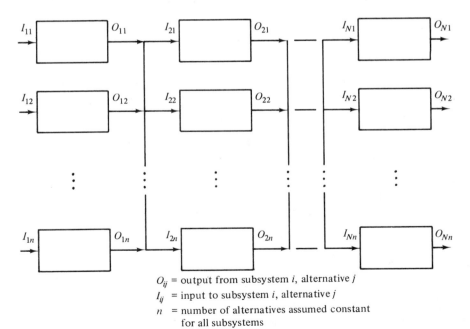

O_{ij} = output from subsystem i, alternative j
I_{ij} = input to subsystem i, alternative j
n = number of alternatives assumed constant for all subsystems

Figure 1.3. Level III Analysis of a system

exercised within each subsystem. Actually, if there are n alternatives for each of N subsystems, there will be n^N possible solutions, not counting the possibility of accepting more than one alternative in a subsystem. The problem is further complicated, as indicated in Figure 1.4, by the possibility that, although alternatives for two or more subsystems may be suboptimal in their independent contribution to the system, when linked together their contribution may be optimal. All this is not to discourage the reader from seeking optimality. It is presented to illustrate the necessity for qualitative and quantitative aids to solving these complicated problems.

The three-level approach is one qualitative method of solving the most important problems first. To illustrate, suppose that the problem is approached immediately at level III (Figure 1.4) without the benefits of levels I and II. Instead of just one "level III" problem, there would be an infinite number of them—one for each possible alternative set of objectives, and one for each alternative subsystem breakdown.

Of course, the possibility of changing objectives should not be precluded once the systems engineering process begins. On the contrary, this option is left open, as the reiteration process below will indicate. But the probability of it having to be changed is greatly reduced by proceeding in this vein.

The three-level approach of systems design will now be discussed in terms of the other two common features of systems engineering mentioned above: the step-by-step procedure and the reiterative process. Figure 1.5 presents the

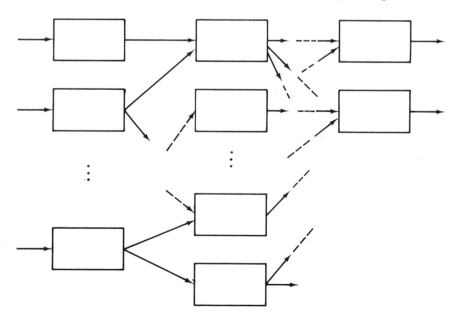

Figure 1.4. Truer representation of level III analysis of system

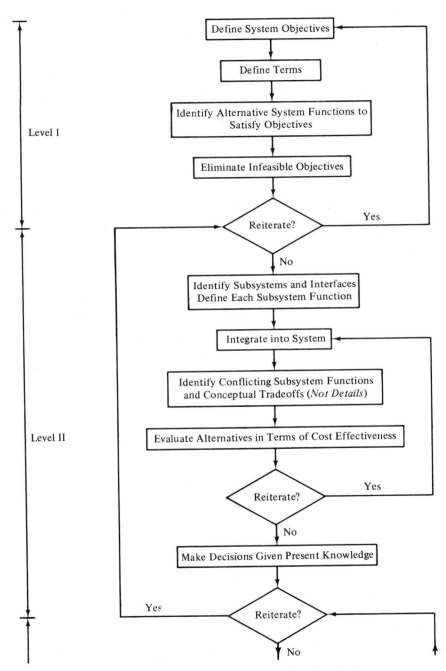

Figure 1.5. Systems engineering approach in three levels

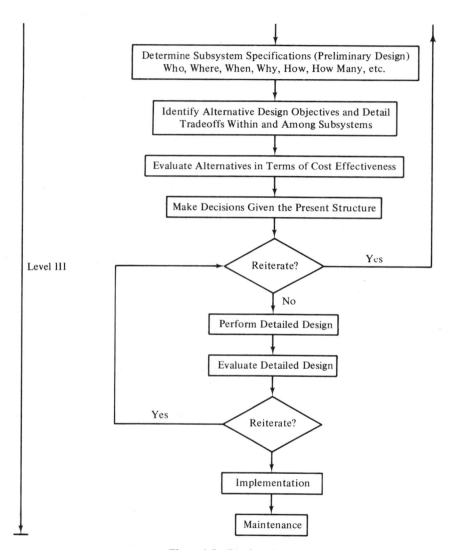

Figure 1.5. Continued

steps within each of the levels. At the end of each level, a decision is required to remain at that level for another iteration, or to proceed to the next level of design. As discussed above, this is an important decision within the process and is a direct responsibility of the systems engineer. The rationale for this decision will be discussed.

1.2.2 Level I

Within level I, precise definitions of the safety system objectives are required. It is important that these objectives be realistic and fitted to the industry to which they are to be applied. For example, zero accidents, zero injuries should be the goal of all companies. But is this goal useful for measuring the effectiveness of the safety system? If so, what should be done when this measure of effectiveness is not met? Since an infinite cost would be incurred to prevent all accidents, this goal becomes impractical as a working control component.

Practical measures of effectiveness associated with the safety system objectives (other than raw accident statistics) are required. For the measures of effectiveness to be useful, they must be translated into a correction capability—not just on the industrial process, but upon the safety system itself. Unrealistic or unmeasurable safety system objectives contribute in great measure to this problem.

The intention here is not to criticize the many fine safety systems that have been developed. However, the steps given within level I will aid if followed faithfully prior to proceeding to level II. This consists of writing down a comprehensive set of safety system objectives. These should be as specific as possible without getting into level II detail.

The next step is to clarify for all involved the exact meaning of these objectives. It may be necessary at this point to define any terms that may be ambiguous. For example, the terms "accident, injury," and so on, may have a meaning local to the company. Generalities, such as "reduction of accidents," should be made more precise and, if possible, quantitified at this step.

Once a tentative set of objectives is obtained, consideration should be given within level I to identify the *system* functions required to satisfy each objective. For example, if the objectives imply the need for safety education, then an educational function will be required within the safety system. This is the first time that thought is given to methods. The total system functions are identified, however, without going into detail as to how these functions are to be performed. Functional specification for the system is the first step in the analysis procedure.

The purpose of specifying system functions is to clarify the objectives. At times the specification of system functions brings to light decisions that must be made prior to proceeding to level II. For example, a tentative budgetary decision is required to determine monetary constraints on the system design. Since all the functions will not be funded to the maximum extent possible, trade-offs are necessary to allocate the total budget figure. Of course, specific dollar figures for each function cannot be generated yet, because the details have not been developed. However, if certain functions and objectives are seen to be infeasible at this point, they can be eliminated, thus saving the time that would go into their evaluation.

Since level I is a seemingly inconsequential step in the design process, it is often overlooked and the system objectives taken for granted. This is dangerous, especially if a large number of persons are involved in the design process. A clear definition of objectives and terms, free of contradictions and unrealistic statements, is necessary both for further design and for eventual understanding of the safety system. Since general strategic questions are resolved at level I, highest-level approval should be sought before proceeding to level II.

1.2.3 Level II

The decision to proceed to level II should be based upon a knowledge that further effort at level I will not justify the time involved. Higher management approval should be sought at level I, especially for the purposes of budget approval. As level II proceeds, decisions requiring more detailed analysis and less philosophical reasoning will be encountered, thus reducing higher-level interaction.

In level II the total system functions specified at level I will provide a guide for further analysis. The system will be divided into subsystems, and each will be identified in terms of the function that it performs. The relationships among these subsystems should be clarified so that they can be integrated to form the safety system. At this point conflicting subsystems are identified. Conflicts arise from the following:

1. Two subsystems that perform the same or nearly identical functions.
2. The functions of one subsystem adversely affect the operation of another subsystem.

The subsystems at this point have been evaluated only in terms of general feasibility and cost. Brainstorming sessions to determine alternative means for accomplishing system objectives might prove fruitful. There should be no prescreening of alternatives, since this could stifle creativity and prevent the introduction of a potentially good idea, at least upon modification. Once this is completed, however, it is necessary to evaluate, consolidate, and possibly eliminate some of the subsystems.

It is very difficult to assign a numerical cost to a possible alternative subsystem. It is even more difficult to assign a value of effectiveness in terms of dollars or even reduced accidents. Nevertheless, these factors are going to be taken into consideration (whether explicitly or not) in making decisions among alternative subsystems. The specification of values of cost and effectiveness forces the decision maker to structure his thought process to a greater degree than is otherwise possible. Even though the estimates are not

100% accurate, the process of making them serves a purpose in itself. For this reason, a cost and a "figure of effectiveness" should be estimated for each alternative. The figure of effectiveness need not have a physical meaning, such as reduced accidents, but it should be a relative measure of utility of the subsystem in terms of satisfying the objectives listed at level I. Further quantitative processing of these estimates will be discussed in Chapter 8.

Prior to making decisions and possibly eliminating subsystems from further consideration, it may be necessary to determine how the overall system would appear with such eliminations. This is necessary because of the interdependence of the subsystems. For example, in terms of cost effectiveness, an inspection subsystem may be very costly and it may not stop many accidents by itself. Thus the tendency may be to eliminate such a subsystem from further consideration. However, when looking at the total system, many other subsystems involving evaluation and correction will not function properly if the inspection subsystem is not present. Therefore, the decision maker should revise his estimate of the relative value of this subsystem. Hence the need exists for a reiterative process within level II prior to making decisions among alternative subsystems.

After the first round of subsystem decisions is made, a decision to proceed to level III is required. Possibly the subsystems were not identified properly, or additional subsystems may be required. Even at this point, flaws within the system objectives might be uncovered. The process may have to be restarted at the very top, analogous to a filtering process in which the fluid is poured back through the filter once again to increase its purity.

At the end of level II, the system designer should be satisfied that further work at this level will not be justified by the time required. Infeasible or unrealistic subsystems should be eliminated so that further analysis of these in level III will not be required.

1.2.4 Level III

At level III, the actual details of the system design are produced. This takes into account alternatives both among and within subsystems. Hence another dimension is added to the evaluation process.

The third level is divided into two parts: preliminary design and detailed design. Preliminary design may leave some areas unresolved and need not be to a final degree of detail. The purpose here, as before, is to delay the evaluation of possibly irrelevant details.

The same procedure used in level II for evaluating the cost effectiveness of alternatives can be applied here also. Now, however, a knowledge of the details enables the designer to specify the costs and benefits more accurately. Again reiteration back through the process may be required to correct improper assumptions at prior levels.

After level III is completed, the system should be ready for implementation, given that funding is adequate. Further steps of implementation and maintenance will be required for continuous operation. Once implementation takes place, the safety system design should be continuously evaluated to ensure that proper corrections are made for its ongoing improvement.

The three-level approach presented above reveals the procedure for developing an effective safety system. In the next section the necessary elements of the safety system will be detailed. This will be followed by a discussion of the management environment essential to the proper functioning of the safety system. Finally, some legal considerations will be discussed.

1.3 Safety Control Systems

1.3.1 Control Systems

Accidents occur when men, machines, or materials in some way become uncontrolled to the point where physical damage or injury results. Certainly if the entire industrial system were in complete control, no accidents would result. Thus it is reasonable to approach the design of a safety system in terms of control. By analyzing control, the necessary parts of the safety system can be determined.

The three essential elements necessary for control are:

1. A goal or standard—in a realistic form such that its attainment and realization can be progressively visualized.
2. A means of measurement—to determine if the goals and standards are being met.
3. A means of correction—to get the system* back on standard toward the goal when the measurements indicate that this goal is not being adequately met.

If any of these three elements of control are missing, control cannot be attained, and reaching the goal becomes a matter of chance. Between the elements of measurement and correction is the important function of *evaluation*: the process of translating control measurements into useful information for the correction process.

The elements of control are present in *any* system under control. To illustrate, consider the normal car and driver on the road as the system. Assume that the driver intends to keep the car in one lane. Thus the standard is the lane, or possibly 1 foot from each edge of the lane of travel. Measure-

---

*"System" here applies to both the industrial system and the safety system itself.

ments are taken frequently, if not constantly, by the driver to ensure that he is on standard. If a gust of wind or the banking of the roadway moves the car to a point where the driver feels that the standard is not being acceptably met, a correction is made by turning the wheel, modifying the speed, or whatever. Finally, if the driver is suddenly blinded, the tie rod breaks, or any other element of control is removed, control will obviously be lost and staying in the proper lane will become purely a matter of chance.

Note that there is generally some tolerance or play in any control system. The automobile was generally free to wander to within 1 foot of the edge of the lane. Slight variations of a random nature did not indicate an out-of-control situation. However, if the automobile went over the lane line against the will of the driver, this certainly would have been considered out of control. This concept of "tolerance" and "control limits" is important and it will be taken up again in Chapter 6.

The safety control system is composed of the standards, measurements, evaluations, and corrections necessary to obtain the safety objective. A large number of accidents indicates a shortcoming in the safety control system. An analysis of the system should reveal which part of the control system is deficient so that improvements can be made. The objective of the systems approach to planning is to make such improvements prior to the installation of a full-scale safety program. This eliminates the costs incurred from installation, breakdown, change, and reinstallation. In other words, it pays to design the system properly the first time. And, although changes will be required after installation, all attempts should be made to keep them to a minimum.

Figure 1.6 depicts one functional view of a safety system in terms of the elements of control. Arrows represent information flow, with feedback coming from events within the production system itself. Such a conceptual framework aids in allocating the funds available for safety use in a "good," if not optimum, way. Each of the blocks is, in a sense, competing for funds (level II). In addition, each of the blocks may have a variety of alternatives for its accomplishment (level III).

1.3.2 Example

We shall now present a brief enumeration of the design process. Limited space prohibits development of a safety system in full detail, since that would require a rigorous analysis of a particular type of system, which is not the intent of this book. Therefore, this example is taken through level II, at which point the procedures for level III are described in terms of the example.

Consider a moderate-sized production-oriented company, or a division of a large company, which allocates approximately $50,000 per year to the implementation and maintenance of a safety system. As indicated in Figure 1.5, the first step in the development of such a system is to define the objec-

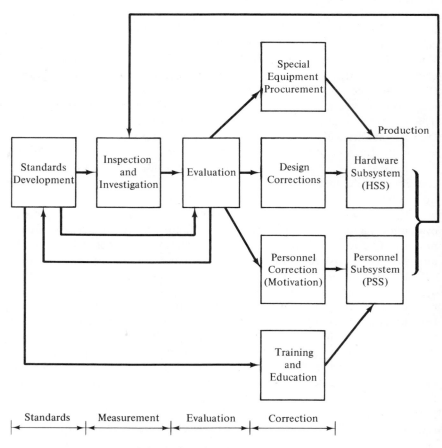

Figure 1.6. Basic safety control system

tives so that a clear understanding of purpose is apparent to all involved.

The following four objectives of the system are proposed by management:

1. To form the organization required to fully comply with the Williams–Steiger Occupational Safety and Health Act of 1970.
2. To reduce reportable injuries by a factor of 10% of the past year's rate, starting from the date of implementation.
3. To merge the new system into the ongoing safety program with a minimum of organizational changes.
4. To maintain at least the present level of production throughout the implementation of the system.

The above typical objectives will serve as a basis for further development. Prior to their evaluation, however, a thorough definition of terms should be performed. For example, such questions might arise as the following:

1. Comply: Does this mean to do nothing more than the act requires (such as in the area of recordkeeping)? How is compliance measured prior to an OSHA inspection?
2. Reportable injuries: Does this mean according to the OSHA definition or previous company practice?
3. Ten per cent of the past year's rate: What if the previous year was extremely bad?
4. Minimum of organizational changes: Does this mean that no organizational changes can be made? More precision is required.

These, again, are typical examples of questions that might require a more thorough definition of terms. They demonstrate how a seemingly good set of objectives might be inadequate. Although there are no universal answers, since each company might want to proceed differently, for the remainder of this example it will be assumed that a clear definition of terms is available.

The next step is to identify alternative system functions to satisfy the system objectives. In this regard Figure 1.6 will be useful in ensuring that all the functions for control are present. The specific departmental or functional breakdown (subsystem design) will be deferred until level II. A link should be made at level I, however, between the objectives and the system functions. These are listed for the example objectives in Table 1.2.

The identification of system functions clarify the objectives as much or more than the definition of terms. Certain flaws in the objectives might now be proposed, such as:

1. Is the plant (or company) already in conformance with OSHA? If so, possibly the proposed safety system should be more aggressive and attempt to go beyond the current standards. This is especially true in industries that lack the necessary standards (e.g., new processes or highly sophisticated techniques). Also, if the company's accident rate is significantly below the industrial rate, this might bring about other gains (less chance of inspection, public relations, etc.).
2. Is the reduction of *recordable* injuries indicative of safety (and health) progress? Might this lead to the recording of fewer minor injuries, which in turn could lead to a loss of valuable data? Is there a more accurate measure, such as lost workdays? Possibly noninjury accident recording could also lead to better prevention.
3. The minimum-disruption guideline might itself be a disruptive influence in the design of an aggressive safety system. It should be understood that the implementation of the system should be per-

Table 1.2. EXAMPLE RELATIONSHIP BETWEEN OBJECTIVES AND SYSTEM FUNCTIONS

Objective	Function	Relationship
1. Compliance with OSHA	1. Standards development	Standards will come from OSHA; additional interpretation and development might be required.
	2. Inspection	Necessary to ensure that OSHA standards are being met.
	3. Corrections: equipment and personnel	If OSHA standards are not met, corrections should be made.
	4. Training and education	To inform all workers of OSHA standards.
2. Reduce injuries by 10%	1. Evaluation	Determine which are the most frequent and critical accidents.
	2. Corrections: equipment and personnel	Action taken on most critical areas first.
3. Minimum disruption	1. Training and education	Make everyone knowledgeable of changes.
4. Maintain present production	1. Inspection and evaluation	To determine effect on production.
	2. Equipment procurement and other corrections	To maintain production.

formed without *unnecessary* disruption. However, there might be a clear conflict between this and the other objectives.

4. The "maintenance-of-present-production" objective neglects to mention cost per unit. Changes in the process might slow production, requiring overtime to maintain the present production rate. Under the current objective, this would be acceptable. Is it?

The four seemingly straightforward objectives become rather weak when subjected to analysis. At this point the objectives will be redesigned to eliminate inconsistencies and other shortcomings. Suppose that the reiteration process has continued a sufficient number of times, and management has produced the following three objectives, which they feel are sufficient:

1. To form the organization of the safety function necessary to achieve a superior degree of safety, equal to or surpassing the basic OSHA requirements.
2. To reduce lost workdays by at least 10% of the lowest of the last five years. In addition, to report and record all injuries (lost workday or not), and to record as many noninjury accidents as possible.
3. To maintain and evaluate both accident and production records with the goal of maintaining present production and cost per unit as well as the safety and health goals mentioned above.

The decision is now made to proceed to level II and to structure the organization necessary to carry out these objectives. A full description would require many pages of documentation and would be specific to the particular industry. For this example one general organization of subsystems will be summarized. First, the functions that are to be performed must be specified and, based upon the judgment of higher management, tentative figures allocating the budget should be associated with each function. Table 1.3 presents this schedule based upon the functions given in Figure 1.6 and a budget of $50,000.

Table 1.3. ALLOCATION OF BUDGET TO FUNCTIONS

Function	Investment
Standards	$ 2,000
Inspection	10,000
Evaluation	6,000
Special equipment procurement	8,000
Design corrections	20,000
Motivational corrections	2,000
Training and education	2,000
	$50,000

The next, and possibly the most difficult, step is to integrate the functional subsystems into the industrial system. This will depend heavily upon the current management structure, and no general solution can be given. However, consider two alternative structures: (1) an autonomous safety department performing all the functions in Table 1.3, and (2) a committee structure coordinated by a safety director drawing upon existing departments (such as quality control, computer processing, personnel, etc.) to supply the required services. These are extreme examples, but they are useful in demonstrating the design procedure. The final design will, in most cases, be between the two extremes.

Assume for the present that the first alternative will be adopted. Certain conflicts may arise between the current production control functions (e.g., quality control, incentive policies, etc.) and the safety control system. Conflicts may also arise within the safety control system itself, such as between special equipment procurement and design corrections or between motivational corrections and training and education. At this level these conflicts should be resolved and jurisdictional areas clearly defined. For example, a compromise might be made to allow (or require) the quality control department to perform the statistical analyses and enter into many of the judgments required in evaluating inspection results and accident reports.

This would result in an economical arrangement that would be free of conflicts.

Elimination of redundancy within the safety organization will also help. For example, the consolidations of the more "psychological" functions, motivational corrections and training and education, might be useful, depending on the company.

These examples demonstrate the types of decisions that are made at level II. Although alternatives should be evaluated on a cost-effectiveness basis, often no measures of costs or effectiveness are available. Therefore, judgments will be made realizing that the level II decision process may be reiterated if subsequent details indicate that this is necessary.

In level III the process of assigning tasks to individuals is initiated. At this time the exact organization will be clarified. For example, the limited budget of $50,000 is going to prohibit staffing an autonomous "safety department" unless two very versatile and experienced persons could be acquired at relatively low salaries. This would allow very little for the purchase of other equipment and services required. A search of the job market should be conducted to determine specific alternatives.

Assume that such a search reveals the impossibility of securing the type of staff necessary to establish an autonomous department. Depending on the structure formulated at level II of the system development, it might be required to reiterate level II given new information such as this.

Complete job specifications will be written in level III to ensure that each of the functions discussed in level II will be carried out. Consider the following two examples:

1. Safety Director: will be assigned to the coordination of the total safety system. He will report directly to the Vice President of Manufacturing. He will have primary responsibility for the execution of the purposes of the safety system. This will include the preparation of the budget and the authorization of all expenditures in the area of safety as well as the securing of necessary company services and the coordination of these services.

2. Quality Control Manager*: will receive weekly reports on accidents and inspections from the safety department. His staff will work together with the safety director to ensure that the forms are designed properly to aid processing and to obtain the maximum utility from the information available. The quality control department will design statistical evaluation techniques applicable to the information supplied and will be involved with the safety department in the monthly evaluation of the information. They will also determine the relation-

---

*Note that these are only the responsibilities of the quality control manager with respect to safety. Quality control and safety are viewed as separate departments.

ship, if possible, between this safety information and production standards of quality and quantity.

In the completed specification of the design there will be job descriptions for all departments or individuals involved as well as a specification of any equipment that will be purchased initially. The details will be specified in varying degrees, depending upon the flexibility to be given to the individuals involved. For example, in the two job descriptions given above, the first was more general than the second, in order to give the safety director the flexibility that he needs in order to function properly. The responsibilities of the quality control manager with respect to safety had to be specified precisely to make the relationships between the departments clear. As time goes on, more detailed information can be added upon a first or second revision. For example, the specific data-collection forms should be included, as well as the specific procedures for processing and evaluating the data.

Once the initial level III preliminary design work is completed, a total evaluation of the design should be performed in terms of cost–effectiveness, where effectiveness is measured in terms of the stated objectives. If necessary, all or part of the design process should be repeated. The result will be as complete a set of specifications for the safety system as current information will allow. This will not complete the design process, since modifications should continue through the implementation of the system. However, when the detailed design is completed, management can proceed with the assurance that they have given due consideration to designing an adequate and efficient safety system.

1.4 Organization and Management of Safety

Since the primary objective of this book is the application of quantitative techniques to safety, no effort will be made to duplicate the many fine works dedicated to safety management. However, the previous sections of this chapter have opened the door to many management concepts. This section will formalize these concepts so that the appropriate management environment can be established for an effective safety system.

1.4.1 Psychology of Management

The purpose of management is to organize men, machines, and materials in such a way as to accomplish specific goals. If these goals are acceptably attained, the manager is appropriately rewarded. If not, he is either removed from his position, or else *de facto* position is removed from him. The higher up in management, the greater the responsibilities of policy decisions.

Responsibility implies risk. Hence two things are required for an individual to progress up the ladder of management: (1) the individual's own willingness to accept the risks involved, and (2) the determination of higher levels of management that the individual has the capability to perform the new function and will be successful.

The concept of risk is introduced early in this discussion because of its inherent relationship to safety. Risks to the manager are generally not physical, although through the manager's psychological makeup, risks can produce definite mental and physical health problems. Nevertheless, these management stresses are not ordinarily considered within the realm of safety problems.

The risks that a manager takes are generally of a financial nature. These involve both himself personally and the financial status of the company. Quite often in management these two are so closely linked that they become the same. This is especially true at the very top level of management, since here the individuals are investing their money with a gain or loss depending entirely upon the success of the company.

This has been a rough and oversimplified model, and many exceptions could probably be discussed in monopolistic companies and governmental organizations. It does provide a basis from which a principle can be drawn, however.

RULE 1: *Management personnel are rewarded for their success roughly in proportion to the amount of financial risk (both personal and company) involved in achieving that success.*

A second concept must be introduced before the essence of this first principle can be appreciated. On the surface, when financial risks are discussed, the concept of the hard, cold money man wheeling and dealing may come to mind. But money is much more than the ability of its possessor to demand and receive goods and services. To the majority of individuals, who have worked hard for their income, their money represents their own stored-up life. It is payment for a part of their life which they have sold to their employer.

Again the exceptions to the above concept should be duly noted. Many persons obtain their wealth through inheritance, and some by illicit or immoral means, whether legal or not. Other very fortunate individuals enjoy their occupations to the point where they would perform it whether or not they received remuneration. But these are exceptions, and quite often diluted exceptions, of what we will generalize as Rule 2.

RULE 2: *Generally man perceives of money, whether consciously or not, to be a stored-up quantity of that portion of his life for which it was given.*

Thus most men freely trade a good portion of their lives, usually eight hours per weekday, for money. This is done because the individual views the overall return (i.e., life with money minus the eight hours of "freedom") to be superior to life without money and the things it can buy. To prove this rule for a given individual, all that is required is to remove his payment and see how long he remains on the job. Obviously he considers the money that he makes *at least* as valuable as the time he worked for it.

Now, when Rules 1 and 2 are combined, added insight into management actions with regard to safety can be obtained. The man who takes financial risks for the company is at the same time subjecting himself to financial risks in roughly the same proportion. But his personal financial risks can be viewed as a risk of a stored-up quantity of his own life. Hence it is only reasonable to assume that he will try to minimize the probability of the loss involved in this risk.

It has been recognized for some time that the successes enjoyed by organizations under capitalism are largely the result of each individual performing according to what he perceives to be his own self-interest. As long as the self-interest of the individual is channeled into the best interest for all of society, progress will result. The free market is a good example. An excess of a particular product forces the prices down, resulting in losses to investors. These investors seek other, more profitable investments. A shortage of another product forces prices up, making the investment in that area appear lucrative. Thus self-interest draws investors toward the production of those goods most demanded by society.

But we must face facts; we do not live in a dream world. Quite often the self-interest of the individual goes too far. The "free" market is never totally free. Other considerations, such as the preservation of resources for future generations, must be considered. Similarly in the area of safety: individuals may act in their own self-interest to the detriment of others. They may not know or they may not care about the effect upon others. At this point it is necessary for society itself, in the form of government, to impose regulations to prevent the situation from getting out of hand.

In safety the problem involves a transference of risk. A manager may believe that certain safety investments are a threat to his success. To improve his probability of success, he transfers the risk to the workers. The moral and legal problems that come about are quite complex. The manager may free himself from moral guilt by reasoning that the labor market is "free." The workers, after all, do not have to stay on the job. They can get other jobs if they think the risk is too high. (Of course, this is not always true.) Second, he may reason that to invest a great deal in safety will put his company out of business. Then the workers will not even have a choice. Reasoning like this, he transfers the risk from himself to the worker.

The reason that this is such an intriguing problem is that we are not dealing with absolutes. Some degree of danger is inherent in any human activity; and it is society in general, not the manager, who insists upon taking risks. Neither is the worker totally guiltless in accepting and even needlessly taking risks for financial or social gain. A strategic approach to this problem will be presented in Chapter 10 once the other concepts of quantitative decision making are discussed. This chapter continues by discussing the management and legal controls necessary for preventing unbridled transference of risk from taking place.

1.4.2 Management Control

Section 1.4.1 started by stating that the purpose of management is to organize men, machines, and materials in such a way as to accomplish specific goals. Since goals are one of the essential elements of control, it is reasonable to expect the other two elements to be present in effective management. A new term, "management by objectives," has stressed the need for developing long-range, intermediate, and operational goals to manage successfully. Old-line managers, however, recognize that not only goals, but measurements and corrections, have been essential elements in management control since before the days of Noah.

With larger organizations comes the necessity for the delegation of authority. This was brought about because an organization grew to the point where one person could not perform all of the measurement and correction required. The earliest recorded delegation of authority and responsibility is recorded in the Old Testament (Exodus 18: 13–27), indicating once again that this concept is not new; and the abuses and misuses of this concept are as old as the concept itself.

When management systems break down it is because one or more of the elements of control are not being properly performed. Authority and responsibility should only be delegated when it strengthens the measurement and correction functions of management control. A good example is the old complaint of the manager who is given the responsibility to accomplish an objective but who is not given the appropriate authority. Here the measurement and correction elements are weakened, since a barrier is placed in the path of translating control measurements into corrective actions. Corrections flow either unjustifiably to middle management, or directly past middle management, causing a psychologically demoralizing situation to all involved.

Of course, the delegation of authority must have its limits defined by the responsibilities assigned and the latitude necessary to discharge those responsibilities. Managers must be controlled by higher levels of management to ensure that they are not (1) going beyond the degree of authority given, (2) failing to exercise the degree of latitude provided, and/or (3) exercising authority within the right bounds but improperly. The first two can be judged directly by methods described below. The third, however, is highly subjective and requires a direct measurement against intermediate objectives. Since individual methods vary, criticism of methods defeats the purpose of delegation. Hence patience and trust must be maintained until success can be measured directly.

The first two items can be subjected to control from the outset given that a general management policy is available. The technique for accomplishing this task is called "management by exception." Simply stated, it consists of definition of areas in which the middle manager has flexibility to make decisions and exercise judgment. Beyond those points, he is to obtain advice from upper levels. This provides a control on the manager to ensure that he goes neither too far nor falls short in discharging his responsibility. It should also be recognized as a flexible tool for management development, since, once the manager proves to be competent in a given area of decision making, his role can be increased.

Before concluding this section on management control, some discussion of the motivating behavior of workers is in order, since they often contribute to the occurrence of accidents. It is impossible to force individuals to do something against their will. Rather, the will must be changed such that the alternative to the desired action is thought to be less desirable than the action itself. This is a generalization of the "principle of rewards" and covers both positive and negative incentives. The "principle of rewards" simply states that generally actions that are rewarded will be repeated and actions that are punished will not be repeated. This is again an oversimplification, but it provides a basis upon which we may build.

The above being generally true, authority must proceed from the lower to the upper levels of management. That is, authority only exists to the point where subordinates allow it to exist. This is universally true, although not generally accepted; and for this reason many management control systems fail.

In terms of control, when this failure occurs it is the correction system that is breaking down. A simple statement or order is assumed to correct a situation. However, if those given the order do not respect the authority of the issuer, no corrective action will take place.

This comes about for a variety of reasons. The first cause could be management itself. Unless upper management gives total and consistent support to middle-management functions, authority will be eroded. For example, if the safety function is performed purely by a staff group, the directives of this group must carry equal weight with those prescribed by line management. Methods for accomplishing this will be given in the next section.

A second factor that could erode authority is the lack of proper incentives, both positive and negative. Many cannot understand how this could be a problem in safety, since obviously no one wants to get hurt. A visit to many independent logging operations where sawyers are paid strictly by the thousand board feet of timber felled would provide great insight into this problem. Here there is a conflict in objectives. To increase productivity the sawyer might accept a given risk not otherwise taken. Most large companies recognize that in the long run the highest of safety standards will increase productivity. The *collective* risk of poor safety practice *to the company* is intolerable. However, to the individual, the risks that he takes from day to day seem small when compared to the increased paycheck at the end of each week. A similar phenomenom occurs when auto drivers take risks by speeding, passing on hills, and so on.

A third factor that often negates the manager's authority is often described as "informal lines of authority." This comes about because of the social nature of man. The incentive for social acceptance often transcends financial and even physical well-being. In certain groups social acceptance is attained through a display of "daring." The term "groups" is used here to designate an informal set of individuals who choose of their own volition to associate together. The group will generally have a leader and supporting secondary leaders, all completely informal. Rarely is this informal leader the same as the formal manager to whom the group reports.

Common methods for dealing with this informal organization vary from frequent transfers to promotion of the group leader. Usually they have little effect. The main thing to realize is that the group must be motivated together simultaneously. For example, if a group formerly revered an act as "daring" but now has been motivated to regard the same act as "stupid," that act will not be repeated. It is usually advisable to identify and win over the group leader. However, manipulation of the leader may cause him to lose his status. Therefore, the most advisable approach is the simultaneous motivation of the group as a whole.

This section has discussed some of the basic principles of management control. In the next section these principles will be applied to specific problems involving the management of the safety function.

1.4.3 Management Strategies

In this section we shall discuss the management environment necessary for the effective functioning of a safety system. Obviously, one organizational chart will not serve every organization. Therefore, a discussion of the most general management strategies concerning the safety organization will be given.

Regardless of the particular organization, the integration of the safety system throughout the organization is essential if the safety function is to be effective. To view safety as a separate function organized under a vice-presidential office may be impressive to government inspectors and public relations interests. However, if safety is truly organized as a *separate* function, its effectiveness is doomed.

Participation within the safety function must be obtained across organizational lines. Responsibility and authority must be delegated to those individuals who are most suitable for accomplishing this task. Very rarely does a staff group have the "clout" required to influence all levels of workers.

With this thought in mind, however, the necessity for a safety staff function must be admitted. Safety concepts are getting increasingly complicated. Safety and health standards are difficult to read and interpret as well as being voluminous by nature. The equipment and measurements required for many health studies require specialists in these areas. No company can afford to integrate into their line management the rare safety and health expertise that is available. Hence a staff group must be formed in the large company, or a staff position for a specialist in the small company, to provide the expertise necessary to serve the entire company.

The apparent conflict between staff organization and integration can be resolved if the proper emphasis is given to safety by all levels of management. This brings us to the first strategy of safety organization: *Although specialized functions must be performed by staff personnel, the ultimate authority and responsibility for the safety of the work force must rest with their immediate supervisors.* Of course, it is understood that the individual who delegates responsibility is still accountable for the results. Hence responsibility is distributed up and down the line chain of command.

Where does this leave the staff people? Certainly they, too, must bear the responsibility for a plant's safety record. And, without authority and responsibility, the safety staff will in all probability become a waste of money. Thus certain guidelines must be established to ensure that (1) responsibility and authority are placed with staff personnel, but also that (2) the responsibility and authority delegated to line managers is not "shared" with the staff. The reason for this should be obvious. Shared responsibility ultimately results in "buck passing," which in essence means no responsibility at all.

To resolve this problem, recognize that to initiate action for safety requires the implementation of what we will call a safety countermeasure. The primary staff function consistent with safety control is to (1) identify safety and health hazards (i.e., measurement), and (2) recommend countermeasures (i.e., corrections). (Note that a knowledge of standards and a formulation of goals and objectives is assumed.) Now the identification of hazards with respect to prescribed standards is the responsibility of the *staff* through plant inspections. However, the identification of hazards unique to the operation of the system must be the responsibility of the line manager. This is obvious since (1) the staff personnel cannot be on hand under every circumstance, and (2) the staff personnel cannot be as intimately familiar with the various operations as is the line manager.

Now consider the second function of the staff (i.e., to recommend countermeasures). Here the countermeasures logically fall into two categories: (1) operational and (2) special. Operational countermeasures fall into the realm of those common sense items that should be observed during the normal course of operations. On the other hand, special countermeasures may require procedural development, special equipment, or even a termination of work. These fall squarely upon the shoulders of the staff personnel to implement. So, once again there is a logical division of authority and responsibility between the line and staff management.

Let us summarize the above as a second management strategy. *Staff safety personnel are responsible for identifying hazards with respect to standards, and for implementing special countermeasures beyond the normal operational countermeasures required by day-to-day operations.*

Before leaving this particular strategy, some emphasis on the method by which the staff discharges its responsibility is in order. Effective communications are essential and will be discussed below. Also, the responsibility described in the strategy above implies the authority required. A reasonable exception tool here would be a budget under the control of the safety staff. Additional expenditures beyond the budget, and possibly large expenditures within the budget, could be approved on an exception basis.

Now the two strategies given above do not solve the problems of separating responsibilities without other operational tools. Formal communication lines are essential to prevent implied transference of responsibility. Documentation of actions tailored to the organizational requirements is essential. This may utilize various approaches, both formal and informal. However, to describe the interrelationships in terms of authority and responsibility, two example forms will be discussed.

Figure 1.7 provides a mechanism through which the safety staff can discharge its responsibility. Systematically, and at random intervals, the safety staff personnel should inspect all operations. When a discrepency is found between company, OSHA, or other applicable standards, this should be noted on the form presented. The actual standard violated should be indicated along with an understandable description of the hazard. The corrective action should be listed. If there are several alternatives for correction, this should be handled by the methodology to be discussed in Chapter 2 or 8. Here it is assumed that the best alternatives have been resolved such that a recommendation to higher levels of management can be made. The last three columns involve the cost of the improvement and the source of funds. If the improvement will be funded out of the safety budget, then generally no additional approval is required and the countermeasure will be implemented. On the other hand, if other funds (labeled operational) are to be used, higher-level management approval is required. Notice that this relieves the safety staff of their responsibility in this matter, since they have at this point done everything in their power to secure the improvement.

Of course, if there is a no-cost improvement, such as a change in method, communication will go in the opposite direction. That is, the formal documented recommendation will be sent to the manager in charge of the operation. Again, this will relieve the staff personnel of their responsibility and place it upon the shoulders of line managers.

XYZ COMPANY

SAFETY INSPECTION FORM

Date _____ Department _____ Manager _____

Inspected by _____

Standard	Description of hazard	Corrective action required	Cost	Source of funds	
				Safety	Operational

Safety director _____

Approval _____

Figure 1.7. Example safety inspection form

It may seem that the safety staff is "passing the buck" according to this procedure. This is not the case, however, when the two strategies above are considered. Note that the staff still bears *all* the responsibility for an accident where: (1) they fail to discover a violation of standards and hence an accident results, or (2) they make a wrong recommendation, which is implemented but does not bring about the desired effect. In the final analysis, of course, higher levels of management will have to judge the value of the safety staff.

Figure 1.8 gives an example of a form to initiate action from the line manager. This form is sent to the safety staff group for action. The safety staff then proceeds to complete the more detailed information of Figure 1.7 and take appropriate action, as described above. The purpose of this form is to relieve the line manager of his responsibility when unreasonably hazardous conditions exist. Once this form is sent to the safety staff, they are responsible to evaluate the situation and recommend a solution.

The purpose of the responsibility structure discussed above is to bring about an effective working relationship between line and staff management. This is done by taking advantage of the expertise that each possesses: the line

XYZ COMPANY

SAFETY ACTION REQUEST FORM

Date _____   Department _____   Manager _____

Description of Hazard	Action requested

Figure 1.8. Example safety action request form

manager's intimate knowledge of operations coupled with the staff's expertise in safety.

Of course, a scene could be pictured wherein the two groups did nothing but send forms back and forth—each recommending actions on the other's part. When this becomes a problem, both should seek immediate higher-management intervention. At that point, the problem would be resolved by (1) accepting one or both of the suggestions proposed, or (2) reprimanding one or both of the conflicting parties. Personalities and politics should never be allowed to influence decisions affecting the safety and health of men.

In concluding this section, the shortcomings of the above strategies and procedures should be emphasized. No documentation or procedures, whether it be a book or a set of forms, can substitute for good "on the job" management. Hence the above has been given to stimulate the thoughts of management and possibly provide some new insights into old problems. If these principles stated are considered by management, however, it cannot but increase their ability to reduce accidents. Finally, appropriate innovation will always be required to tailor the procedures and controls to the particular industry involved.

1.4.4 Legal Considerations

The purpose for presenting the management strategies above was to ensure that adequate control was within the organization so that the safety system could function properly. Quite often mention was made of "higher levels" of management in resolving difficulties and obtaining funding. In situations where society in general believes that these "higher levels" are either (1) not responding or (2) responding inconsistently, government intervention is often requested. One of the most recent and far-reaching examples of government action in the United States is the Occupational Safety and Health Act of 1970.

The discussion of legal questions will be deferred until Chapter 10. No attempt is made in this book to interpret or in any way modify the safety and health laws. By applying the techniques presented here, however, it is hoped that the spirit of the law can be better kept.

Legal considerations are mentioned here, however, so that they can be recognized as but another mechanism to bring about safety control. Laws provide standards, measurements, and corrective action on the most general scale. For the most part they do not affect a direct solution to safety problems. Rather, they bring about a social climate in which top management is motivated to provide a safe and healthful work environment.

To demonstrate the secondary nature of the legal mechanism, consider the three elements of control as they materialize by law. First, the law itself is a standard which in turn references other more detailed standards for particular industries. Generally, the keeping of the letter of the standards will improve safety conditions. But Congress recognized in 1970 that the burden of innovation was not to be placed on the standards but on the employer. The first duty of the employer was not to keep the letter but rather the spirit of the law. Section 5 of the Occupational Safety and Health Act of 1970 states that "Each employer shall furnish to each of his employees employment and a place of employment which are free from recognized hazards that are causing or are likely to cause death or serious physical harm to his employees." The second duty was to comply with the standards promulgated under the Act.

Similarly, in the area of measurement, the legal aspects provide rather indirect means for accident reduction. Note that the measurements under the 1970 act fall into three categories:

1. Inspection of catastrophic accidents (one or more fatalities or five or more hospital admissions in any one accident).
2. Inspections without warning on any industrial establishment.
3. Mandatory recordkeeping and reporting.

Although category 1 gets to the heart of a specific problem, by the time it takes place it is usually too late. It serves to prevent a repeat accident and possibly fix the blame. Inspections without warning are also secondary in their effect in that they serve more as a deterrent than as a direct accident-prevention mechanism. One reason for the "without warning" requirement is that relatively few establishments can be inspected. Thus to motivate compliance there is neither advance warning nor trial inspections. Finally, the mandatory recordkeeping and reporting performed under the direction of the Bureau of Labor Statistics serves a very useful purpose only if they are used on an ongoing basis by management. The statistics accumulated nationally are good for overall direction, but they do not provide the individual manager with the details required to plan a policy of hazard abatement. Methods to augment and use the records required by law will be presented in Chapter 7.

The correction segment of the law consists of fines, jail sentences (or the threat thereof), and mandatory shutdown in cases of extreme hazards. These are only as effective as the inspection mechanism that leads to their implementation. In turn, the inspection with respect to the standards are no better than the standards themselves. It may be possible to have a perfect system of control here but still not progress toward the desired goal. Using the automobile for an analogy, it may be in perfect control but on the wrong road. Such would be the case if the standards, to which measurement and corrections are being made, do not reflect the safety needs of the industry upon which they are imposed.

At this point a distinction should be made between the two types of legal mechanisms: (1) fault systems that require restoration of damages, and (2) direct regulations, as discussed above. Although some interaction of these two mechanisms exists, they are separated by law into the civil and criminal categories, respectively. The OSH Act of 1970 specifically states that it is not to be construed to change workmen's compensation, common law, and statutory requirements. Thus consideration should be given to the other aspect of legal motivation: the lawsuit.

In civil actions certain elements are quite clear. Usually a definite loss has taken place and a claim is being made. The judge or jury weighs the "reasonableness" of the actions of the plaintiff and the defendant, along with economic and social costs of hazards and their various abatements in order to draw conclusions on a case-by-case basis. The results of one case, in providing a precedent, would motivate actions of individuals who might be faced with similar problems in the future. Clearly this is a different type of control mechanism than the direct regulation proposed by OSHA. The merits of the different mechanisms are beyond the scope of this text, but the interested reader should see reference 13.

The concept of civil action has been introduced mainly because its consequences are feared much more than the consequences of most criminal actions in the safety area. Insurance, whether private or government sponsored, is usually employed to overcome this risk. By statute, workmen's compensation* has taken a heavy burden from the shoulders of the employer. However, there is not the same statutory protection in the area of product liability or personal negligence. In either case, those who fail to employ all safety techniques at their disposal, including systems analysis techniques, place themselves in an unfavorable light. Similarly, a record of noncompliance with safety standards cannot help but place a defendant in a compromising position, whether it be in the occupational, traffic, or any other safety area.

One other significant role of law should be emphasized before closing. The OSH Act of 1970 established a mechanism for promulgating industrial safety and health standards to companies throughout the country. These standards form an essential element in the development of an effective safety system. Their promulgation will be useless, however, if there is only token compliance on the part of industry. In those cases where standards are self-defeating, or where the economic and social costs are unjustified, appeal mechanisms exist so that changes can be made. OSHA also looks to the industries themselves for review and guidance.

In summary, the civil and criminal law mechanism applied to safety only works effectively if it provides a deterrent to future accidents. Those who view the law as an end in itself obscure the spirit and tend to misuse the letter of the law. Only about half of occupational accidents, and probably even a smaller proportion of traffic and other public accidents, can be affected by standards. The point is not that the laws, law makers, or law enforcers are deficient or incompetent. Rather, the shortcoming is in the expectation that the legal control system can do the whole job. It can only establish the environment whereby each individual company (or other functional entity) can set up its own safety control system. When properly executed the law will ensure that no unfair advantage is given to those who are negligent in this regard.

SELECTED REFERENCES

1. BIANCARDI, M. F., "Design for Product Safety," *ASSE Journal*, Apr. 1971.

2. KOLODNER, H. J., "The Fault Tree Technique of System Safety Analysis as Applied to the Occupational Safety Situation," *ASSE Monograph 1*, June 1971.

---

*Note that generally workmen's compensation laws vary from state to state. State statutory and case law should be consulted for specifics.

3. LEVENS, E., "Hazard Recognition," *ASSE Monograph 1*, pp. 26–31, June 1971.

4. MACKENZIE, E. D., "On Stage—For System Safety," *ASSE Journal*, Oct. 1968.

5. PETER, L. J. and R. HULL, *The Peter Principle*, William Morrow & Company, Inc., New York, 1969.

6. PETERS, G. A. and F. S. HALL, "Design for Safety," *Product Engineering*, Sept. 13, 1965.

7. RECHT, J. L., "Systems Safety Analysis: An Introduction," *National Safety News*, Dec. 1965.

8. RECHT, J. L., "Systems Safety Analysis: Failure Mode and Effect," *National Safety News*, Feb. 1966.

9. RECHT, J. L., "Systems Safety Analysis: The Fault Tree," *National Safety News*, Apr. 1966.

10. ROCKWELL, T. H., "A Systems Approach to Minimizing Safety Effectiveness," *ASSE Journal*, Dec. 1961.

11. TIGER, B., "Weapons System Safety Assurance," *ASSE Journal*, Feb. 1969.

12. WISSNER, I. E., "How System Safety Relates to Industrial Safety," *National Safety News*, May 1966.

13. *Law and Contemporary Problems*, Vol. 38, No. 4, 1974.

QUESTIONS AND PROBLEMS

1. Define system, engineering, synthesis, analysis, and systems engineering.

2. State the purposes of systems engineering and describe why it is needed in safety today.

3. Why is systems engineering a loosely defined discipline?

4. Contrast systems analysis and systems engineering in the technical definitions as well as common usage.

5. Contrast the work of a manager, an industrial engineer, an operations research specialist, a systems engineer assigned to computer system development, and a systems engineer assigned to safety system development.

6. State the primary goal and secondary objectives of the systems approach.

7. Draw a diagram illustrating the iterative nature of the three-level approach as it might be applied to the redesign of an automobile instrument control panel for safety.

8. List the activities in each level of the three-level approach to system development.

9. What criteria are used to determine when to proceed to the next level?

10. In applying the three-level approach, who makes the following decisions, at what level, and upon what criteria are they based?
   (a) Definition of system boundaries.
   (b) Definition of subsystem boundaries.

(c) Decision to proceed to the next level.

(d) Specification of who, what, where, and so on.

(e) Evaluation of subsystems to resolve conflicts.

(f) Determination of a final operating budget.

11. In what way does the three-level approach given resolve shortcomings at previous levels caused by lack of detailed information?

12. State the three elements of control and give an example of each if the system to be controlled is the demolition of an old building.

13. How is the function of evaluation within the context of occupational safety performed?
    (a) At the plant level.
    (b) By the government.

14. Give examples in a large building fire-control system of subsystems that adequately perform each of the three functions of system control.

15. Draw an information flow chart for the subsystems obtained in problem 14.

16. List two factors that are generally required for a manager's promotion.

17. State the differences between risks, with respect to personal finances and safety, taken by upper management, middle management, first-line management, and the union worker.

18. Using Rules 1 and 2 as premises, state the logical conclusion that is implied.

19. Compare the freedom of the labor market and the acceptance of risk in the occupational environment with the freedom of the automobile driver and his acceptance of risks.

20. What dangers are incurred in management when responsibility is delegated? State the controls essential to prevent each.

21. What causes the informal lines of authority to be established and how can this problem be controlled?

22. Why is a safety staff necessary? What dangers lie in too much staff responsibility?

23. Define the point at which line-management responsibility ends and staff responsibility begins in the area of safety.

24. Draw a diagram that details the ways in which safety countermeasure proposals progress to management implementation.

25. In what way and at what levels do safety laws bring about control?

CASE STUDIES

1. Apply the three-level approach to safety system development to a system with which you are familiar, such as a set of traffic intersections, or an industrial plant in which you have worked.

2. Set up the organization for the safety function in a plant or service organization with which you have a working knowledge.

HAZARD ANALYSIS AND COST
EFFECTIVENESS

2.1 Introduction

Many textbooks, articles, standards, and manuals have been written that detail specific system design requirements and/or modifications that should be made for safety. A knowledge of these principles as they apply to the safety engineer's company is essential. For example, the *Accident Prevention Manual for Industrial Operations** contains a machine safety checklist with 15 design points. One of these is: "Design the machine so that corners and edges are rounded." Even though such checklist specifications become almost second nature to the system safety engineer, they should be reviewed periodically in the design and improvement of man–hardware interfaces.

Hazard analysis consists of three parts: (1) hazard identification, (2) logical procedures for formulating countermeasures, and (3) selection of the best countermeasures to implement. Concentration here will be upon one method of proceeding from raw data inputs through policy formulation. For a review of other methods as well as an excellent hazard-analysis checklist, the reader is referred to Hammer (6). Also, for the formulation of countermeasure

---

*Sixth edition, National Safety Council, Chicago, 1972.

2

strategies and also the identification of hazard points, such papers as that written by Haddon (5) will be useful. This chapter attempts to provide a logical approach to the application of the principles presented in the literature. No attempt will be made to repeat these principles here, since the volumes written provide adequate presentation. It should be recognized, however, that more than a knowledge of principles is necessary to bring about a safe environment. A *procedure* must be followed whereby the principles on paper become reality. This is never as easy as it might seem.

The objectives of this chapter go beyond the mere presentation of a technique, however. This chapter should be viewed as one of the intermediate stepping stones from the purely qualitative approach to safety to the almost entirely quantified approaches presented in subsequent chapters. Any quantitative technique must start with a qualitative foundation. For example, the measurements in the inspection function must be preceded by the identification of a hazard. This can only be done by qualitative analysis.

In addition to hazard identification, qualitative evaluation is essential for the establishment of alternative means of safety improvement. In the new design of a system, this would take the form of alternative designs to accomplish the goals of the system. In any case, something other than quantitative techniques are required. A rich background in safety countermeasures is also a necessity in this endeavor.

Thus a qualitative systems safety analysis is required to identify hazards and to propose countermeasures. This analysis is hardly an end in itself, however. The data obtained by such an analysis can be further processed to enable trade-offs to be evaluated. The first step is a type of qualitative or semiquantified cost-effectiveness evaluation. This, of course, leads the way to further quantification and toward optimization.

This chapter continues by describing various approaches to hazard analysis. These are discussed from the occupational and traffic-safety points of view. The latter sections of the chapter describe a cost-effectiveness technique that is of particular value in comparing alternatives for a given location or function. When several locations or functions are to be considered simultaneously, the technique is still useful, but it will not lead to a unique solution as will the methods described in Chapter 7.

2.2 Basic Concepts of Hazard Analysis

The first step in eliminating a hazard is its recognition. Historical data may indicate common accidents at particular locations. Certain accident causes may surface in the processing of accident records. Nevertheless, the judgment of the experienced safety analyst is the most important ingredient in hazard identification. It is his responsibility to go beyond the historical data to (1) determine locations with high accident potential which may not necessarily have glaring accident histories (e.g., in new operations); (2) identify severe hazards which, although they have a small probability of occurrence (and thus little or no history), could lead to major accidents; and (3) eliminate from undue considerations hazards that are relatively unimportant.

This brings up a second important aspect in hazard analysis. Total hazard elimination is a very noble goal. Yet, as long as the human element is present in a system, perfection will be impossible to attain. The recognition that hazards exist at different degrees of severity and accident causation leads to the concept of eliminating the most "important" hazards first. Thus an evaluation to determine the importance of hazards is essential if there is to be any boundary whatsoever on the problem.

The evaluation of a specific hazard involves the estimation of its probability of occurrence as well as the severity of injury and/or property damage that will result if the hazard leads to an accident. This will be formulated further in this chapter and quantified in Chapter 5. Once this is done, it is possible to compare one hazard with another, a ranking that will be useful in cost-effectiveness evaluations.

One more step, which overlaps evaluation, is necessary if hazard analysis is to be successful. That is, countermeasures for each hazard must be proposed and further evaluated. This adds another dimension to the problems since a decision is required which again requires trade-offs among alternatives.

The pragmatic may view this as a long process of evaluation or "red tape," which stands in the way of progress. And indeed, countermeasures are required not next year or next week, but right now. However, without the systems view of the safety objective, resources may be exhausted ineffectively on but a small part of the problem. On the other hand, common remedial action, especially if it is inexpensive and of proved value, should proceed *without* the formal hazard analysis or the quantitative analyses presented in this book. Where the outcome is indeed obvious, and it is known beforehand that the figures will only prove the obvious, immediate unencumbered action should take place.

In many situations the obvious and inexpensive solutions have been implemented and there is still a safety problem. It is to these situations that the analytical techniques are to be applied. The usefulness of these techniques will be clarified during the journey upon which we are about to embark.

2.2.1 General Hazard Analysis

General hazard analysis is provided by an overall look at the industrial system under consideration, to identify and isolate those safety problems that require more detailed analysis. The industrial plant lends itself through departmentalization and formal organization to analysis. However, hazards quite often tend to be interdepartmental in nature. Quite often they surface at the interface of departments. Thus the analyst should not necessarily analyze by departments. Rather, the analysis should focus on the identification and evaluation of the hazard itself.

General hazard analysis is also required in other areas of safety, such as highway or aircraft safety. Here the broad analysis might identify locations that exhibit high potential hazards, or parts of the aircraft that are prone to failure. The eventual purpose of general hazard analysis—to identify highly critical areas in need of further analysis—should always be kept in mind.

Generally systems as a whole possess hazards by their very nature. Systems that involve the use of temperature extremes, power sources, movement with high momentum, poisonous gas, and so on, have easily identifiable hazard points. In general hazard analysis, these points should be defined and documented.

Other than the nature of the system itself, two other avenues are available to identify critical points for general hazard analysis: (1) historical records and (2) general investigations. Accident records are required by law in most industrial environments. These records should not be an end in themselves.

Rather, they should serve to point out problem areas for further analysis. If sufficient detail is not provided by government regulations, it should be added. For example, OSHA regulations require the type of accident, location, and severity in lost workdays (among other things) to be recorded. This provides adequate detail on those accidents which are recordable. However, noninjury accidents, as well as first-aid cases, are not recordable. Yet these are indicators of causes that may later manifest themselves in severe injury or even fatal accidents.

Unless all indicators of out-of-control situations are adequately recorded and translated into corrective actions, the safety control system will be lacking. Other indicators that might be produced from the same causes as accidents include (1) unusual delays, (2) damaged goods, (3) cost overruns, (4) customer complaints, and (5) employee complaints. Figure 2.1 demonstrates that the same cause might manifest itself by various intermediate indicators prior to causing accidents and injuries. If these indicators are detected and analyzed quickly, the cause may be eliminated or at least abated to the point of preventing the accident. Figure 2.1 should be adapted to the particular industry; the list of indicators are simply examples that can be expanded to fit the application.

This leads to the second avenue to identify critical points in the industrial system: general investigations. For, although many of the above intermediate

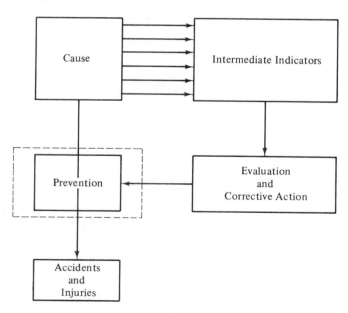

Figure 2.1. Methodology by which intermediate indicators lead to accident prevention

indicators, such as customer complaints, can be monitored on an ongoing basis by historical records, many indicators must be directly observed by the trained eye. "Trained" here is meant to imply two possibly diverse backgrounds. The trained eye of the safety professional can often detect that which line personnel do not recognize to be an intermediate indicator. However, the experienced craftsman or lowest-level employee can often detect an out-of-control situation that may be unrecognizable to the safety specialist. Thus the wise safety director will multiply his capability possibly a hundred fold by having eyes and ears all over the plant.

General hazard analysis should include the full range of activities for which the safety staff is responsible. To summarize, its input sources are threefold: (1) obvious hazards due to the nature of the process, (2) hard recordkeeping data required by law and augmented to reveal intermediate indicators, and (3) investigations that stress the cooperation of line and staff personnel. The output of general hazard analysis will be a priority list of hazards that will be used to guide the sequencing of future detailed hazard analyses.

Since this procedure is at such a general level, the rules for generating priority order must be qualitative and possibly subjective. However, as mentioned above, this is an essential step in the direction of quantification. It is not recommended that a numerical analysis be performed at this point, although costs and benefits should bear heavily within the decision maker's mind. Rather, the details of each hazard should be placed on a standard form or General Hazard Analysis Card, as shown in Figure 2.2.

GENERAL HAZARD ANALYSIS CARD   No. _____

Prepared by _____   Date _____

Hazard Description _____

_____

Departments:

Severity	Probability	Cost	Action
O Nuisance	O Unlikely	O Prohibitive	O Defer
O Marginal	O Probable	O Extreme	O Analysis
O Critical	O Considerable	O Significant	O Immediate
O Catastrophic	O Imminent	O Nominal	Date _____

Figure 2.2. Example of a "hazard card"

Before explaining the General Hazard Analysis Card, the purpose of general hazard analysis should be reviewed. The primary purpose is to establish a plan of action to lead to further analysis and quantification. However, there may be a secondary purpose at this point. It was emphasized at the end of the last section that obvious decisions can be made without the use of formal analysis. That is, when the outcome of a detailed quantitative analysis is known beforehand, there is no need to waste time doing it. General hazard analysis may bring to the surface several very low cost countermeasures that should be implemented immediately. This point will be given further consideration as the procedures are developed.

The General Hazard Analysis Card represents a very cursory view of the hazard. The data are placed on a card so that all the hazards can be arranged easily by priority. Room is provided at the top for a description of the hazard, the name of the person who prepared the card, and the date the hazard was identified. Also there is room for the names of departments affected. In filling out the remainder of the form, representation should be obtained from all affected departments. Thus the decision of which hazard to rectify immediately, which to analyze further, and which to defer until the future will be, at least in part, a group decision.

The remaining four items on the card are (1) severity, (2) probability, (3) cost, and (4) action. Each of these requires a check in one of the categories given. The categories are arranged so that the highest priorities are lowest on the list. For example, if the severity was *catastrophic*, the probability of an accident by the hazard was *imminent*, and the cost was *nominal*, action should be immediate.

Thus each hazard should be ranked according to its anticipated severity and to its probability of causing an accident. Probability here is relative and must be viewed in terms of a similar time period for all hazards.

The matter of cost requires particular attention. Since alternatives have not yet been recommended, the cost is unknown. There may be a variety of ways to abate the hazard, ranging from very inexpensive to very costly, each with a different effectiveness. Here again, however, the current state of knowledge of the line and staff personnel should be pooled to roughly estimate the cost in comparison with other hazards. All estimates at this point are relative to one another and need not be accurate in the absolute sense.

The Action column should not be completed until all general hazard cards are accumulated and sorted. Those at the bottom of the priority list might be deferred until more important hazards are analyzed further. As mentioned, those hazards that can be abated without further analysis and at small cost should be marked for immediate action regardless of where they may appear on the hazard priority list. A more detailed form would then be completed, such as that given in Figure 1.7, and forwarded to the appropriate individual for action.

The majority of hazards identified will fall into a category that requires further analysis prior to decision making. Analysis takes time, and therefore top-priority hazards should be analyzed first in the ongoing system. Mention was made above of sorting and arranging the cards in priority order. This is not a difficult process at this point, since the hazards have been studied and also placed in the severity–probability–cost categories.

One point is worth mentioning before we elaborate further on the technique of sorting. It is of no value to assign numerical quantities to the categories and multiply, add, or otherwise arrive at a quantitative priority. The data have been obtained in a qualitative way; and only by keeping them in detailed form can they be useful in making a qualitative judgment.

For a preliminary sorting, a useful technique is to place all the catastrophic severities together, followed by the critical, marginal, and nuisance cards, respectively. Then, within *each* of these groups, place the imminent probability followed by the considerable, probable, and unlikely, respectively. Finally, do the same thing for cost (least cost first). Thus the cards will be sorted by cost within probability within severity class. This makes severity the largest factor and cost the least important.

Remember that this is only a preliminary sorting. Once this is accomplished, the analysts should go through the cards ordered as above. Check, for example, the first against the second and ask the question: Should the first hazard really be considered prior to the second? Here a subjective opinion based upon all past knowledge of the two hazards is required. Switches should be made if it is felt necessary, regardless of the card ratings. After the first two cards are arranged, the second card should be compared with the third. Whenever a switch takes place, check the card before the one that was switched to ensure that the hazard should not be increased in priority. In this way, one pass through the deck should be sufficient to order the general hazards.

When the sorting is completed, the analyst should be able to go all the way through the cards without making any switches. At this point a priority list can be drawn up based upon the order of the cards. Detailed analyses will begin at the top of the list. Several hazards should be considered concurrently. Budget limitations may prevent all hazards from being totally eliminated, and lower-priority hazards may demonstrate better cost effectiveness. Therefore, the priority set up for detailed hazard analysis will not necessarily be the priority for funding countermeasures. This will be further recognized when cost effectiveness is discussed. Prior to that, however, it is necessary to complete the discussion of hazard analysis.

2.2.2 Detailed Hazard Analysis

At this point general hazard analysis should be completed, and a number of general hazards are ready to be analyzed further. Those hazards with obvious inexpensive remedies and those that are not significant have been eliminated. Further, those remaining have been ranked according to a qualitative estimate of priority. The procedure now is to analyze these hazards further to evaluate the effects of countermeasures. This will be used eventually in a cost-effectiveness procedure to produce a policy for implementing countermeasures.

There are two ways in which hazardous activity can be analyzed: within the hazard or within the activity. The procedure explained below performs these simultaneously. An example of an analysis of a general hazard may be the operation of pouring molten metal. The hazard, as identified in the general hazard analysis, is the molten metal itself and the possibility of its contact with man. This hazard can be analyzed into such elements as spilling, overflowing, malfunction of handling mechanism, and so on. On the other hand, the activity itself could be analyzed into elements of lifting, transporting, pouring, and so on. Within each of these activity elements there may be one or more hazard elements present.

This leads to a worksheet matrix model such as the one given in Figure 2.3. Each General Hazard Analysis Card will have several associated detailed hazard analysis forms, one for the original operation and one for each proposed countermeasure or combinations of countermeasures. The two forms should be cross-referenced in the spaces provided.

The activity elements should be determined first, because these will often provide an aid to determining the hazard elements. Here standard task analysis used in motion-and-time study can be used. The following points should be observed in dividing the job into elements:

1. The task must be observed in its entirety several times before analysis takes place. Observe workers of average skill.
2. Pick a logical starting point in the sequence of events.
3. Beginning with the first action at this point, define the element that is being performed.
   (a) Do not select activity elements so large as to defeat the purpose of the analysis—to isolate each part of the activity that contains each hazard element.
   (b) Do not select activity elements so small that one hazard element may overlap many activity elements, unnecessarily complicating the analysis.
4. Repeat step 3 for subsequent activity elements, ensuring that no necessary activity is either excluded (gaps) or included twice (overlaps).
5. Where options or simultaneous activities are required, analyze each separately and then combine when listing (see examples below).
6. List the elemental tasks on the Detailed Hazard Analysis form in the order in which they occur. Ensure that the starting and ending points of each element are clearly defined.

DETAILED HAZARD ANALYSIS WORKSHEET

General Hazard Analysis No. _____ Page _____ of ____

Department(s) _____ Prepared by _____

Task or Activity _____ Date _____

Countermeasure _____ Cost _____

Activity Elements	1	2	Hazard Elements 3 4 5 6 ...
1			
2			
3			
⋮			
(May also include combinations)			

Hazard Summary			

Figure 2.3. Detailed hazard analysis worksheet

Although this same type of analysis is used in time study, the elemental breakdown may not be exactly the same. Nevertheless, past time studies of an operation can be very useful in establishing the job elements. The reader is referred to the many books in the area of motion-and-time study for additional information on this subject (e.g., see references 3 and 7). Before discussing the hazard analysis portion of the detailed hazard analysis, an example of the methodology of determining elemental tasks will be given.

Example 2.1 Determination of Elemental Tasks

Because of the many occupational examples of elemental activity analysis in the literature, the above methodology will be applied to a detailed hazard analysis in the area of traffic safety. Consider a typical undivided intersection of two four-lane urban roads. There are currently no traffic signals at the intersection, only stop signs on one of the roads and caution signs on the other. No other signing is present.

There are a number of possible "tasks" that could be analyzed. Suppose that a general hazard analysis has shown this to be a hazardous intersection, because vehicles make left turns onto the through street. Accidents consist of cut-offs, rear-end crashes, and a few head-on collisions. The particular task to be considered here consists of the operator's actions in making a left turn assuming that he is approaching the intersection in the right lane from one block away.

The following is a breakdown by activity elements for a hazard analysis:

1. Driver checks left lane to see if he can merge left.
2. If clear, driver merges left. If not clear, repeat elements 1 and 2. Simultaneously go to 3.
3. Driver notices stop sign; decelerates for stop.
4. If the driver could not merge left, he would not make a turn. Assuming that merge is completed, driver applies the brakes and brings the vehicle to a stop in the left lane.
5. Driver looks both ways for oncoming traffic.
6. Driver looks ahead for oncoming traffic.
7. When all three ways are clear, driver executes a left turn.

This breakdown is not unique. An element could have been added for the driver's decision to accelerate or decelerate for merging. The preference here was to refrain from unnecessary detail. With the definitions given above, Figure 2.3 can be partially completed. The elements under Activity Elements might appear as follows:

Activity Elements . . .
1
1, 3
2, 3
3
4
4, 5
5
6
7

Elements that are performed simultaneously should be so listed that the special hazards which occur can be isolated and evaluated. Notice that this is not overlapping, which was warned against above, since no one activity element definition repeats the activity of another.

To this point general hazard analysis has identified a hazard for further analysis and the activities surrounding the general hazard have been analyzed. The final analysis step is to break down the hazard itself. While dividing the activities into elements, some detailed hazard analysis had to be considered. The activity elements were chosen according to this end.

A list should be made at this point of all the hazard elements associated with each activity or task element. Each of these will generally be a part of the hazard, or one manifestation of the hazard, described on the corresponding General Hazard Analysis Card (Figure 2.2). Together they should constitute all the ways that the hazard could cause accidents in the operation being studied.

Once the hazard is broken into elements, the elements should be ranked according to priority. Hazards that are judged to be the most critical will be listed first. Severity, exposure time, and probability of a resulting accident should be considered in ranking the hazard elements. Here the experience of both line and staff personnel will be useful, both in identifying and in ranking hazard elements.

When the hazard elements are ranked, the numbers in Figure 2.3 under Hazard Elements can each be assigned a hazard, the most critical first. In the body of the matrix a symbol will be placed if the hazard occurs in that element. A suggested set of symbols is given in Table 2.1. The Hazard Summary at the bottom of the form will contain the appropriate summary symbol for the hazard element.

Table 2.1 HAZARD ANALYSIS SYMBOLOGY

Symbol	Meaning
X	Hazard is present in activity element.
R	Hazard has been reduced by the countermeasure.
$R_1, R_2, R_3, \ldots$	Hazard has been reduced to the degree indicated by the subscript.
E	Hazard has been eliminated for activity element.
(blank space)	Hazard was never present for the activity element.
I	Hazard has been increased by the countermeasure.

This will result in a pictorial view of the hazard elements present during the original operation and for each countermeasure. Generally marks farther to the right indicate less critical hazards. No attempt should be made to quantify the number of marks and so compare various general hazards at this point. Rather, the total picture should be used to gain insight into the problem prior to proceeding to further evaluations.

This completes the analysis portion of the evaluation procedure. The following steps summarize the hazard analysis to this point:

1. Compile a list of hazards from historical records, general investigations, the nature of the operations, and intermediate indicators. Take immediate action with respect to obvious countermeasures.
2. Based upon a cursory investigation and collective knowledge of operations, complete the General Hazard Analysis Card for each hazard identified.
3. Sort the hazard cards, placing the most critical first, to form a priority list for detailed investigations.
4. Perform detailed hazard analyses on all hazards that require it. Those which can be eliminated include (1) hazards for which the countermeasures are obvious and inexpensive, and (2) hazards that are of significantly less importance than the average hazard, based upon the judgment of the analyst.
5. Perform a detailed hazard analysis, which will consist of the following:
   (a) Classification of the activity into activity elements.
   (b) Classification of the general hazard into hazard elements.
6. Complete the Detailed Hazard Analysis Worksheet for the original system and each countermeasure, using a set of symbols such as those given in Table 2.1.
7. Perform the synthesis steps as outlined in the remainder of this chapter.

Example 2.2 Continuation of Example 2.1

Example 2.1 presented a task analysis for a driver negotiating a left turn from a stop street onto a through street. The details will not be repeated here. Assume that accident records, on-site investigations, and analysis with respect to the elemental tasks lead to the following hazards, in order of importance:

1. Right-angle crash
2. Head-on crash
3. Cut-off crash
4. Rear-end crash

The original system is first summarized with respect to the elemental activities of Example 2.1 and the four elemental hazards listed (Figure 2.4). Two countermeasures will also be analyzed for this example: (1) a four-way stop, and (2) a traffic signal with a leading left-turn light. Figures 2.5 and 2.6 present these analyses. Note the use of *R* for a reduced hazard when, in fact, the total hazard element has not been eliminated. If degrees of difference in the various reductions are important, subscript notation can be used.

DETAILED HAZARD ANALYSIS FORM

General Hazard Analysis No. _____10_____ Page _____1_____ of ____3____

Location or Department _____Intersection 8th & Rt. 22_____

Task or Activity ____Left turn____

Prepared by ____D. B. Brown____ Date ____8/12/74____ Cost ____0____

Countermeasure ____Original____

Activity Elements	Hazard Elements				
	1	2	3	4	
1					**Hazard Elements:**
1,3				X	1 = right angle
					2 = head on
					3 = cut off
2,3			X	X	4 = rear end
3				X	
4				X	**Activity Elements:**
					1. Driver checks left lane
4,5					2. Merges left
					3. Decelerates for stop
5					4. Brakes and stops
					5. Checks both ways
6					6. Checks ahead
					7. Makes left turn
7	X	X			
Hazard Summary	X	X	X	X	

Figure 2.4. Example of detailed hazard analysis

In the first countermeasure activity elements 5, 6, and 7 change slightly in interpretation. It is no longer necessary to ensure that the oncoming traffic is perfectly clear, since they are now required to stop. Of course, the good driver will not take it for granted that everyone is always going to stop. However, it is assumed here, in order to have a basis for evaluating countermeasures, that the law is going to be obeyed. This being so, the chance of a right-angle crash or a head-on crash is virtually eliminated. The other elemental hazards are judged to be not affected by the countermeasure.

DETAILED HAZARD ANALYSIS FORM

General Hazard Analysis No. ___10___ Page ___2___ of ___3___

Location or Department _____

Task or Activity _____

Prepared by ___David B. Brown___ Date ___8/12/74___ Cost ___$800___

Countermeasure ___4-Way stop___

Activity Elements	Hazard Elements				Hazard Elements:
	1	2	3	4	
1					1 = right angle
1,3				X	2 = head on
					3 = cut off
2,3			X	X	4 = rear end
3				X	
4				X	
4,5				X	
5					
6					
7					
Hazard Summary			X	X	

Figure 2.5. Example of hazard analysis for 4-way stop

For the second countermeasure, the activity elements must be redefined. Here the countermeasure changes the tasks completely and the original set of actions will not suffice. The following new elements are defined:

1. Driver checks left lane to see if he can merge left.
2. If clear, driver merges left. If not clear, repeat elements 1 and 2. Simultaneously go to 3.
3. Driver notices light. If yellow or red, decelerate for stop. If green, decelerate for turn.

DETAILED HAZARD ANALYSIS FORM

General Hazard Analysis No. _____10_____ Page _____3_____ of _____3_____

Location or Department _____

Task or Activity _____

Prepared by _____D. B. Brown_____ Date _____8/12/74_____ Cost _____$10,000

Countermeasure _____Traffic signal; leading left_____

Activity Elements	Hazard Elements				
	1	2	3	4	
1					Hazard Elements:
1,3				R	1 = right angle
2,3			X	R	2 = head on
3				R	3 = cut off
4					4 = rear end
5					
6		R		X	
Hazard Summary		R	X	R	

Figure 2.6. Example of hazard analysis for signalization

4. If the driver could not merge left, he would not make the turn. Assuming that merge is completed, driver brings vehicle to turning position in left lane.
5. Assuming that light is yellow or red, driver stops. On a leading left-turn signal, driver makes left turn.
6. Assuming that light is green, driver checks ahead for oncoming traffic and when clear makes left turn.

In Figure 2.6 an analysis is given for the new activity elements. The probability of a head-on collision has been reduced but not eliminated since on a green light there is still free traffic flow through the intersection. Also, the presence of a light warns all motorists to slow to a stop when the light changes, and thus it is estimated that the rear-end-crash potential has been reduced for task elements 1, 2, and 3. However, for task 6 the probability of a rear-end crash is significant. The analyst estimates a net reduction, however, as indicated by the R in the Hazard Summary. Hazard element 3 has not changed for either of the countermeasures.

The example was of necessity oversimplified for illustrative purposes. Other drivers' actions should also be considered simultaneously at this intersection. However, this example does point out certain uses of detailed hazard analysis not previously mentioned. The following side benefits of detailed hazard analysis are significant:

1. It enables the analyst to classify hazards according to when and where they take place in an operation so that the severity and exposure duration can be better estimated.
2. It provides an effective means for getting to the primary cause of accidents as opposed to intermediate causes and symptoms.
3. It brings to the surface countermeasures that may not be considered in a more general analysis. For example, cut-off accidents were not affected by the two countermeasures discussed above. This might bring to light the necessity for additional signs or painted stripes to eliminate or reduce the hazard element. This may not have been discovered without the detailed analysis.
4. Detailed hazard analysis by task elements demonstrates how each countermeasure affects the task. If changes increase hazards in other areas, this is noted.

Probably the most important use of detailed hazard analysis is in performing cost-effectiveness evaluations among and within hazards. This will be demonstrated in the next section. In the final section a comparison is made between detailed hazard analysis and fault-tree analysis, and an example is given of the application of detailed hazard analysis to occupational safety.

2.3 Cost Effectiveness in Hazard Elimination

The general and detailed hazard analyses discussed above lead to a qualitative picture of safety problems and their resolution. One problem remains, the selection of alternatives that will provide the greatest degree of safety for the given budget constraint. This problem could be restated as being one of selecting alternatives that will provide the desired degree of safety at the lowest cost. However, the solutions to these problems will not always be the same; and the desired degree of safety is absolutely no probability of accidents ever, a noble goal for which the cost would be infinite. Therefore, the only cost-effectiveness problem discussed will be the one where the budget is assumed to be fixed. This does not prevent concurrent solutions for several budget constraints to be produced simultaneously, thus providing the decision maker additional latitude in specifying the budget.

The inputs into the cost-effectiveness procedures presented here consist of the detailed hazard analyses discussed above. These were obtained through qualitative judgments, hopefully made by a broad representation of staff and line management. Thus no attempt is made in this chapter to produce a quantitative result that may be interpreted as optimal under a given set of assumptions. Rather, the procedure given is meant to channel the thinking of the decision maker such that the best qualitative decisions can be made and a consensus of opinion obtained.

The main purpose of cost-effectiveness evaluations is not to produce results that can be shown to be optimal. It is a tool for rational thinking. And, just as there are dozens of procedures for producing hazard analyses, there are probably hundreds of procedures for performing cost-effectiveness evaluations. The one presented here should be tailored to the specific application.

Two things are essential in the evaluation of cost effectiveness:

1. Alternative means must exist for increasing the safety of a system.
2. Reasonable constraints must be imposed upon the problem.

Generally there are at least two alternatives at every level of safety improvement: improve the system with some countermeasure, or allow the system to continue operating as is. If a countermeasure is not available, the problem is: to operate or not to operate. Thus closing the system down altogether becomes one alternative. Quite often several countermeasures can be employed, and this is where cost effectiveness becomes useful.

The reasonable constraint criteria were discussed in connection with the specification of a budget constraint. In this context, the fact remains that systems (other than pure safety systems) are generally not in business primarily for safety. If it were true that "safety is our foremost concern," most systems would close down altogether. The relationship between financial risks and safety risks was discussed in Chapter 1 and will be taken up again in Chapter 10. The point here, however, is that through the working of the system of concern, a level of risk for safety will be defined. For a given budget constraint, the more intelligently decisions are made, the lower the safety risk will be. The absence of utopian goals, and the integration of reasonable safety goals into the other systems goals, will help to increase the intelligence used in arriving at safety decisions.

Thus the system itself quite often determines what these reasonable constraints are. What may be totally intolerable on a production line could be considered routine procedure in a field operation. Risks in military operations are not to be compared with risks in industrial operations.

To account for these differences, the cost-effectiveness procedures outlined below should be integrated into the overall cost-effectiveness evaluations for the system. This is beyond the scope of the present discussion. However, most of the general literature on cost effectiveness utilizes safety as one of the criteria for total system evaluation.

2.3.1 Matrix Display—Evaluations within General Hazards

The purpose of a matrix or tabular presentation is to provide the decision maker with cost-effectiveness information in a concise and logical format. The information is not processed in the sense that a solution is spelled out by mathematical manipulation. Rather, the matrix format seeks to arrange the information so that the obvious can be eliminated from consideration and more concentration can be given to important problems.

Two uses of the matrix format will be considered. In this section we shall compare various countermeasures within a given general hazard. The next section deals with comparisons of alternative countermeasures among various general hazards. At this point it is important to recognize that these two analyses must be performed concurrently, since decisions on one level will affect decisions on the other. This will become obvious as the methodology is revealed.

The cost-effectiveness matrix for comparisons within general hazards is set up as in Figure 2.7. The specific hazards as determined by analysis in the

COST-EFFECTIVENESS EVALUATION FORM

		Hazardous Conditions Exposure						
Countermeasure	Cost	1	2	3	4	5	6	7
1								
2								
3								
4								
.								
.								
.								
.								
.								
N								

Figure 2.7. Format for the cost-effectiveness evaluation

detailed hazard analysis are listed at the top of the page in order of importance. The same ordering and numbering can be used as in the detailed hazard analysis. The considerations for numbering and arranging hazard elements were given in Section 2.2.2.

The alternative countermeasures are listed vertically in order; the one most effective in eliminating the first (or highest-ordered) hazard element(s) is listed first. This is termed the *northwest-corner rule*, because the upper left-hand corner of the matrix will generally be completed first. The purpose of this arrangement is to provide an easy way to determine those countermeasures which are not cost-effective. A countermeasure at the top of the matrix with a low cost would be a good bet. One at the bottom of the matrix with a relatively high cost would obviously be inferior.

Consider the results of Example 2.2, which are recorded in Figure 2.8. Note that the results are for the assumptions of Example 2.2 and are not implied to be in any way generally applicable. Here the four-way stop was assumed to eliminate the two most important hazard elements. A traffic signal eliminates right-angle crashes but leaves the possibility of head-on

Countermeasure	Cost	Hazardous Conditions Exposure			
		Right Angle	Head On	Cut Off	Rear End
4-way stop	$800/6 yr = $133/yr	—	—	X	X
Traffic Signal	$10,000/15 yr = $667/yr	—	R	X	R
Original System	—	X	X	X	X

— = elimination of the hazard element probability
R = reduction of the hazard element probability
X = no change in the hazard element probability
I = increase in the hazard element probability

Figure 2.8. Cost-effectiveness evaluation for Example 2.2

exposure. Now, even though this cost-effectiveness evaluation seems to give the four-way stop the obvious edge, other long-range goals must be considered. One of the most crucial of these factors is life expectancy or service life.

The service life of a safety countermeasure is determined by the lesser of the following: (1) the time period that the countermeasure can physically serve its intended purpose without replacement, or (2) the time period after which changing conditions will cause the countermeasure to be obsolete. Costs in the cost column should be presented both as an immediate budget requirement and as a cost-per-year over the service life. This enables the decision maker the flexibility to perform a quick sensitivity analysis on a service-life basis. For example, in Figure 2.8, if the traffic volume were known to be increasing such that the life of the four-way stop was two years instead of six, the cost per year would increase to $400/year, thus reducing the attractiveness of this countermeasure.

Cost effectiveness adds a dimension beyond detailed hazard analysis. Now several countermeasures can be viewed simultaneously to help the reasoning process of the decision maker. Quite often two or more countermeasures will be so close in effectiveness that a clear-cut decision cannot be resolved. The question should be asked: Does the increased effectiveness of the higher-priced alternative warrant the additional cost to be incurred from it? More investigation, or the simultaneous trials of two countermeasures, might sometimes be called for.

The approach discussed above is recommended despite the fact that it will not guarantee optimality. However, it is a straightforward and rational basis to arrive at a conclusion. Those making the final decision can see and understand this basis. It should be clear that every X, R, or – in Figure 2.8 does not have equal weight. If so, the results could be quantified and optimal results obtained. However, the degree of importance of each hazard element as well as the degree of accident reduction is assumed to be unknown, and thus it must be estimated. This technique allows the decision maker to keep these things estimated qualitatively (by ordering) and still proceed to a logical conclusion.

2.3.2 Matrix Display—Evaluations among General Hazards

Hopefully conspicuous by its absence in the above discussion was the statement that the decision maker should rush to implement the alternative that the cost effectiveness revealed to be best. Such action would violate the primary goal of the systems approach. It is necessary to synthesize the total system to fully evaluate the effect of all parts. Here the implementation of a good countermeasure with respect to one general hazard could have detrimental effects upon the system in that the resources (i.e., the finite budget) would be depleted to the point where other more important countermeasures could not be implemented.

At this point it is necessary to complete the cost-effectiveness synthesis by simultaneously considering all general hazards of the system. The format for performing this evaluation is the same as that given in Figures 2.7 and 2.8. Now, however, that general hazards will be listed under Hazardous Conditions Exposure rather than the hazard elements. In Section 2.2.1 the methodology was given for arranging the most important general hazards first. This arrangement should be continued here within the cost-effectiveness evaluation.

Under Countermeasures the specific countermeasures will be listed as they were listed in the former cost-effectiveness comparisons. At this point, however, obvious losers should be eliminated from consideration. The vertical ordering of countermeasures should be according to the northwest-corner rule, as before. Now, however, there is only one column per general hazard, whereas there were several, one for each hazard element, in the previous evaluations. This should pose no problem except for apparent ties within general hazards. However, since the countermeasures were arranged within general hazards according to effectiveness, this order should be preserved when posting these countermeasures to the new list.

It may seem at this point that the listing according to the northwest-corner rule would be quite trivial. And this would be the case if all countermeasures were local to a particular general hazard. However, some countermeasures will affect two or more general hazards simultaneously. For example, first-aid courses to key personnel would affect the ultimate severity of injuries caused by any hazard. Very often the identical countermeasure will appear within the detailed hazard analysis of several general hazards. When this occurs, that countermeasure should be listed for this final evaluation. Although it may not appear to be cost-effective in the individual evaluations, it may be a "winner" when its cumulative effect is seen.

Because a countermeasure is listed first in the vertical listing of the matrix should not imply that its effectiveness is superior to those below. This is true in all cost-effectiveness evaluations of this sort. The northwest-corner rule only gives a starting point for the evaluation. It pinpoints obvious winners and losers, but it does not determine the best policy.

This can probably be best seen by an example. Figure 2.9 presents a

Countermeasure		Cost		Hazardous Conditions Exposure						
	Budget	Life	Cost/Year	1	2	3	4	5	6	7
* 1	$1,000	1	1,000	E	X	X	X	X	X	X
* 2	5,000	8	625	E	X	X	X	X	X	X
3	3,000	1	3,000	X	E	E	X	X	X	X
4	15,000	10	1,500	X	E	X	X	E	X	X
5	8,000	10	800	X	X	E	R	X	X	X
6	30,000	5	6,000	X	X	E	X	X	E	E
7	200	1	200	X	X	R	X	X	X	X
8	4,800	3	1,600	X	X	X	E	E	X	X
9	2,500	5	500	X	X	X	E	R	R	R
10	1,800	7	257	X	X	X	X	E	E	X
11	24,000	5	4,800	X	X	X	X	X	E	E
12	15,800	10	1,580	X	X	X	X	X	X	E

*Cannot be implemented simultaneously

Figure 2.9. Example cost-effective matrix for evaluations among general hazards

possible matrix output wherein there are 12 proposed countermeasures and 7 general hazards. Countermeasure 1 is listed before 2 because it was so ordered in the evaluation within general hazards. Countermeasure 3 comes afterward because it eliminates a lesser hazard, hazard 2. Countermeasure 3 comes before 4 because it also affects hazard 3, whereas 4 affects hazard 5. Countermeasures 5 and 6 might present a problem. However, the analyst decided that the reduction of hazard 4 was more important than the elimination of hazards 5 and 6. And so the ordering continues on down the list.

Consider countermeasures 1 and 2 as they compare to 3. The northwest-corner rule dictated that 1 and 2 should be first. However, the elimination of both hazards 2 and 3 by countermeasure 3 might render it superior. In the absence of quantitative data, this must be determined by the judgment of the decision maker. It is essential that the decision maker have a major role in assembling the cost-effectiveness matrix, since its full impact cannot be obtained without a knowledge of the underlying details not presented by the matrix.

The final product of cost-effectiveness evaluation is the specification of policy: that is, a list of countermeasures to be implemented during the upcoming time period. The first step in this regard is to mark those counter-measures which are mutually exclusive (i.e., they cannot be implemented simultaneously). For example, two countermeasures at one hazard location, such as: (1) install four-way stop signs, and (2) install traffic signals, may be presented in the final matrix. Countermeasures within different general hazards will not usually be mutually exclusive.

The decision of which countermeasure to select of those which are mutually exclusive should be made simultaneously with the selection of the other alternatives. This is because cost will play a heavy part in determining which alternative to implement. Consider Figure 2.9 again. Here counter-measures 1 and 2 are considered to be mutually exclusive so that the alternatives are to select 1 or 2 or neither for implementation. As far as effectiveness is concerned, 1 has the advantage, since it is listed first. However, both virtually eliminate general hazard 1. Therefore, the big question is: How much money is available, or can be obtained, in this year's budget? Counter-measure 2 is a better long-range investment because of its service life. If financing can be arranged, it will be superior to countermeasure 1 from the cost standpoint. On the other hand, the increased effectiveness coupled with the lower current outlay required for countermeasure 1 may make it quite attractive.

The decison of 1 versus 2 cannot be made independently of the total budget allocation, as discussed above. Suppose that the total budget constraint is \$35,000 in Figure 2.9. One approach would be to isolate the most costly countermeasures—4, 6, 11, and 12—and then perform individual trade-offs. For example, countermeasure 4 could be replaced by 3 and 8 at half the cost. Another approach, at least for getting started, is to go down the list selecting the minimum cost alternative to eliminate each of the hazards. For this example, countermeasures 1, 3, 5, 9, 10, 11, and 12 can do the job for $56,100, which is clearly over the budget. However, countermeasure 3 covers hazard 3 and thus can eliminate countermeasure 5. Judgment must be used here to ensure that the elimination is with comparable effectiveness. Similarly, countermeasure 11 can be eliminated by 10. This brings the total cost down to $24,100, which is well within the budget.

At this point all general hazards are virtually eliminated, with the following alternatives: 1, 3, 9, 10, and 12, with a slack of about $11,000. The budget allocation can now be fine-tuned to improve effectiveness or reduce long-range cost. For example, the substitution of countermeasure 2 for 1, assuming no loss in effectiveness, will save the present worth of $375/year. This also assumes that either 1 or 2 would be in effect for the next 8 years. Here countermeasure 1, which must be repeated yearly, has the advantage of flexibility.

Assume that the decision maker favors countermeasure 2. The new policy is 2, 3, 9, 10, and 12, at a cost of $28,100. There are various other options within the $35,000 budget that could be exercised. For example, alternative 8 could be substituted for 9 if this were thought to be advantageous. Alternative 7 could be added at very low cost. The process of evaluation between the various alternatives should continue until the decision maker is satisfied that no better policy can be produced.

During this process the decision maker should constantly refer to the more detailed and specific cost-effectiveness evaluations discussed in Section 2.3.1. A reiteration of these detailed analyses should be considered constantly. The possibility of pulling countermeasures formerly discarded into the final analysis should also be considered. This may become particularly necessary when there is a tight budget constraint.

The example above was of necessity brief, to illustrate the theory involved in cost-effectiveness applications to safety. In a large company, or in a government organization, the proposed countermeasures, as well as the general hazards, may run into the hundreds. A qualitative tool is essential when quantitative techniques, such as those discussed in the remainder of this book, are not possible to implement. The next section presents an example from industry.

In concluding this section, the reader's attention is drawn to an overview of the system presented above. It began with an analysis of the system under consideration to resolve the general hazards. Then each general hazard was further analyzed, using job or task analysis, to the point where hazard elements were defined. This analysis was used directly to provide an input into a first synthesis step of cost effectiveness within general hazards. In turn, a subsequent synthesis of the total system of countermeasures was performed using the first synthesis step as input. The final output was a policy statement specifying the countermeasures to be implemented. This process would be repeated periodically and updated to give additional consideration to nonimplemented countermeasures as well as those not yet proposed.

The approach given has intentionally been restricted to qualitative evaluation throughout, with the possible exception of the synthesis steps, where costs were introduced. The budget calculations can be performed easily with nothing but a desk calculator. The main goal, however, was not computational simplicity. It was to provide the intermediate link between subjective judgment and quantitative techniques. Hopefully, this tool will be more than an end in itself.

2.4 Comparative Example

Rarely does the subject of hazard analysis arise in the literature that it is not accompanied by a discussion of fault-tree analysis. We have chosen to hold off the discussion of fault-tree analysis until Chapter 5 in order to introduce the concepts necessary for its use as a quantitative technique. As part of the discussion in Chapter 5, an example is presented that involves the fault-tree analysis of a grinding operation. In order to compare the techniques of this chapter with fault-tree analysis, that example will be introduced here.

For simplicity, assume that a general hazard analysis has isolated three general hazards for additional analysis. It is the second of these, the one of intermediate importance, that will be discussed in detail here. It involves eye injuries caused by chips generated by a grinding operation. Records show that operators have not been injured, since they generally wear safety glasses. Other workers, without safety glasses, frequently come into the proximity of the grinding operation to obtain or return tools to a nearby rack. Occasionally someone will come in for other purposes. The following elemental task analysis was obtained by observing the operation for approximately 1 hour:

1. Machine operating tool disengaged.
2. Machine operating tool engaged.
3. Nonoperator in area of grinding.
4. Operator shuts down machine because of interruption.
5. Machine not operating because of other reasons.

Note that for this analysis it is the mutual operations of the operator and the other worker that are of importance. The elemental task breakdown is not like that of a time-study analysis, because the accidents are not happening during the normal occurrence of the operation. If there were other hazards identified in the records for this operation, the analysis might be more complete—going into the detailed operation of grinding (i.e., starting the machine, engaging the tool, adjusting the feed, and so on).

The general hazard, as identified by the accident records, can now be divided into its elements. Four elements are identified, in order of their estimated importance:

1. Chips from grinding operation enters eye.
2. Dust from grinding operation enters eye.
3. Dust or chips cause other injuries.
4. Distraction of operator causes accident.

Note that elements 3 and 4 were not obtained from the records. That is, the cause of the accidents on record has lead the analyst to attribute potential hazards to the operation which were not recorded. This is one of the primary benefits of the detailed hazard analysis—the identification of hazards before

DETAILED HAZARD ANALYSIS FORM

General Hazard Analysis No. _____2_____ Page _____1_____ of ____4____

Location or Department _____Grinding_____

Task or Activity ____Nonoperator using tool rack_____

Prepared by _____David B. Brown_____ Date _____9/15/74_____ Cost 0

Countermeasure _____Original operation_____

Activity Elements	Hazard Elements				Hazard Elements:
	1	2	3	4	
1					1. Eye injury, chips
1,3		X	X	X	2. Eye injury, dust
					3. Other injury, chips or dust
2					4. Operator distraction
2,3	X	X	X	X	
3,4					Task Elements:
					1. Machine operating, tool disengaged
3,5				X	2. Machine operating, tool engaged
					3. Nonoperator in area
4					4. Operator shuts machine down
					5. Machine not operating
5					
Hazard Summary	X	X	X	X	

Figure 2.10. Hazard analysis: original operation

they cause accidents. Note that the fourth hazard element could potentially cause a far greater accident than any that were recorded.

The results of the detailed hazard analysis of the current operation are given in Figure 2.10. Here an X indicates that the hazard is present in its normal degree. The hazard elements are absent altogether in those job elements in which the machine is not operating or when the nonoperator is absent. One exception is that the operator could be distracted during a setup or adjusting operation.

Figure 2.11 analyzes the first proposed countermeasure: Operator shuts

DETAILED HAZARD ANALYSIS FORM

General Hazard Analysis No. ___2___ Page ___2___ of ___4___

Location or Department _____Grinding_____

Task or Activity _____Nonoperator using tool rack_____

Prepared by _____D B Brown_____ Date ___9/15/74___ Cost $500/year

Countermeasure _____Stop operation when interrupted_____

Activity Elements	Hazard Elements 1	2	3	4	
					Hazard Elements:
1					1. Eye injury, chips
1,3		E	E	X	2. Eye injury, dust 3. Other injury, chips or dust
2					4. Operator distraction
2,3	E	E	E	X	
3,4					
3,5				X	Task Elements:
4					1. Machine operating, tool disengaged 2. Machine operating, tool engaged
5					3. Nonoperator in area 4. Operator shuts machine down 5. Machine not operating
Hazard Summary	E	E	E	X	

Figure 2.11. Hazard analysis: stop operation when interrupted

down machine whenever nonoperator is in area. The cost of $500/year is the estimated cost of delays caused by such shutdowns. This would not come from the safety budget, and it will be handled separately during the cost-effectiveness analysis.

Figure 2.12 analyzes the second proposed alternative to the original system: relocate the tool rack. This will greatly reduce the hazard, but it will not eliminate the problem entirely since there will still be other "visitors" to the area who could be subjected to the hazards. However, a cost will be incurred for additional time since the grinder operators were the primary

DETAILED HAZARD ANALYSIS FORM

General Hazard Analysis No.   2   Page   3   of   4

Location or Department   Grinding

Task or Activity   Nonoperator using tool rack

Prepared by   D.B. Brown   Date   9/15/74   Cost $1,500 + $1,000/year

Countermeasure   Move storage area

Activity Elements	Hazard Elements				
	1	2	3	4	
1					**Hazard Elements:**
					1. Eye injury, chips
					2. Eye injury, dust
1,3		R	R	R	3. Other injury, chips or dust
					4. Operator distraction
2					
2,3	R	R	R	R	
3,4					
3,5				R	**Task Elements:**
					1. Machine operating, tool disengaged
4					2. Machine operating, tool engaged
					3. Nonoperator in area
5					4. Operator shuts machine down
					5. Machine not operating
Hazard Summary	R	R	R	R	

Figure 2.12. Hazard analysis: move tool storage area

users of the tool rack. The cost of moving the rack is $1,500, and the cost per year of extra time for the grinder operators to go out of their area to the rack is $1,000.

Since the two alternatives are not mutually exclusive, a third alternative can be formulated which includes both. Figure 2.13 is used in this analysis. A separate analysis is required here because neither the effects nor the costs are additive. Note especially that the cost includes the $1,500 for moving the rack. But the yearly cost is not the sum of $1,000 plus $500 that would be true if the

DETAILED HAZARD ANALYSIS FORM

General Hazard Analysis No. ___2___ Page ___4___ of ___4___

Location or Department _____Grinding_____

Task or Activity _____Nonoperator using tool rack_____

Prepared by _____D.B. Brown_____ Date ___9/15/74___ Cost $1,500 + $1,100/year

Countermeasure _____Combination of countermeasures_____ (stopping operation and going outside)

Activity Elements	Hazard Elements				
	1	2	3	4	
1					**Hazard Elements:**
1,3		—	—	R	1. Eye injury, chips
2					2. Eye injury, dust 3. Other injury, chips or dust 4. Operator distraction
2,3	—	—	—	R	
3,4					
3,5				R	**Task Elements:**
4					1. Machine operating, tool disengaged 2. Machine operating, tool engaged
5					3. Nonoperator in area 4. Operator shuts machine down 5. Machine not operating
Hazard Summary	—	—	—	R	

Figure 2.13. Hazard analysis: combination of alternatives

first two countermeasures were cumulated. Rather, since there will be far fewer interruptions and less call to stop the grinding operation, the yearly cost is estimated at $1100.

Having completed the general hazard analysis and the detailed hazard analysis, the next step is a cost-effectiveness analysis within each general hazard. For general hazard 2, the one analyzed above, Figure 2.14 presents the summary. Decisions to reduce the number of alternatives should be made at this point if possible. There is no "right" answer. The procedure is set up to crystallize the thought process—not to force one answer.

Suppose that the decision is made at this point to consider counter-measures 1 and 3 in the final synthesis step. Also, suppose that similar analyses had been performed in the other two general hazard areas. The results of the final synthesis step might look like Figure 2.15. Note that countermeasures 1 and 3 from Figure 2.14 form countermeasures 3 and 4 of the evaluation among hazards. When placed in the context of the total evaluation, some sacrifice of information is necessary. The four marks, one for each hazard element, must be reduced to one mark for the hazard. Various rules could be proposed, such as majority rule, worst of the hazard elements, and so on, which would have advantages and disadvantages. Since no rule will work all the time, it is preferable to leave this to the judgment of the analyst. The final decision should not be based upon the final matrix above. Rather, the final matrix should be considered simultaneously with all supplementary analyses.

The selection of the final policy for implementation would proceed as described in the preceding section. It is interesting to note in Figure 2.15 that the differences between countermeasures 3 and 4, which were analyzed in detail above, seem to be relatively insignificant when they are placed in the final matrix. Actually, the decision not to accept countermeasure 2 in Figure

COST-EFFECTIVENESS EVALUATION (WITHIN GENERAL HAZARDS)

Countermeasure	Budget	Life	Cost/Year	Hazardous Conditions Exposure			
				1	2	3	4
1. Combination 2 and 3	$1,500	15	$1,100	–	–	–	R
2. Stop Operation	–	–	500	–	–	–	X
3. Move Tool Rack	1,500	15	1,000	R	R	R	R
4. Original Operation	–	–		X	X	X	X

Figure 2.14. Example of cost-effectiveness for general hazard 2

COST-EFFECTIVENESS EVALUATION (AMONG GENERAL HAZARDS)

Countermeasure				General Hazard		
	Budget	Life	Cost/Year	1	2	3
1	500	3	167	—	X	X
2	9.000	5	1,800	—	X	—
3	1,500	15	1,100	X	—	X
4	1,500	15	1,000	X	R	X
* 5	2,500	3	833	X	X	—
* 6	1,500	2	750	X	X	—
* 7	3,000	4	750	X	X	—

* Mutually exclusive alternatives.

Figure 2.15. Example of cost-effectiveness for comparisons among general hazards

2.14 essentially determined the outcome. The final policy would, of course, depend upon the budget constraint. It seems highly unlikely at this point, however, that the $100/year could be very significant.

The above example was given for two reasons: (1) to provide a complete review of the hazard analysis and cost-effectiveness procedure, and (2) to provide a comparison of this qualitative technique with the quantitative technique of fault-tree analysis to be presented in Chapter 5. Care should be taken that the comparison be of the methods used and not the specific costs. The units of cost for fault-tree analysis are not as arbitrarily set as in cost-effectiveness evaluations. More precision and specific probabilistic effects are used in estimating a numerical value of effectiveness or benefit in fault-tree analyses.

SELECTED REFERENCES

1. *Air Force Systems Command Design Handbook*, DH 1–6, 4950/TZH, Wright-Patterson Air Force Base, Ohio, Jan. 20, 1972.

2. ALFANDARY-ALEXANDER, MARK (ed.), *Analysis for Planning Programming Budgeting*, Washington Operations Research Council, Potomac, Wash., 1967.

3. BARNES, RALPH M., *Motion and Time Study*, John Wiley & Sons, Inc., New York, 1968.

4. ENGLISH, J. MORLEY (ed.), *Cost Effectiveness*, John Wiley & Sons, Inc., New York, 1968.

5. HADDON, WILLIAM JR., "Energy Damage and the Ten Countermeasures Strategies," *Human Factors*, Vol. 15, No. 4, Aug. 1973.

6. HAMMER, WILLIE, *Handbook of System and Product Safety*, Prentice-Hall, Inc., Englewood Cliffs, N.J., 1972.

7. NIEBEL, BENJAMIN W., *Motion and Time Study*, Richard D. Irwin, Inc., Homewood, Ill., 1962.

QUESTIONS AND PROBLEMS

1. Why is qualitative analysis necessary in the systems approach to safety?

2. Why is it necessary to go beyond historical records in hazard analysis?

3. What general considerations are given to the ranking of hazards? Of countermeasures?

4. Name some intermediate indicators that indicate potential hazards in the car-on-the-road system.

5. What sources provide the input into general hazard analysis? What is the output?

6. Why does the hazard card contain qualitative information instead of actual probabilities, costs, and expected loss in dollars?

7. Chapter 1 mentioned the necessity for integration. What steps toward integration are performed in using the general hazard cards?

8. Why is it necessary to analyze hazardous activities in two ways?

9. Why should the activity analysis precede the hazard analysis?

10. In what ways would unrealistic goals prevent a cost-effectiveness evaluation from attaining its desired goal?

11. How does the service life of a countermeasure affect the cost-effectiveness evaluation?

12. What prevents identical symbols in the cost-effectiveness evaluation from having identical meanings?

13. How can a countermeasure that appears not to be cost-effective in the comparison within hazards become cost-effective in the comparison between hazards?

CASE STUDY

Perform a hazard analysis on a system with which you are familiar—your company, school, or some other recreational facility with which you are involved. Make assumptions regarding the costs of various countermeasures. Assume a number of yearly budget allocations and plan an overall 5-year strategy.

LOGICAL ANALYSIS*

3.1 Introduction

In the common nonwar environment, no one wants to cause pain either to himself or to his fellow man. The occurrence of an accident despite this obvious lack of intent indicates that there is a logical flaw in some part of the reasoning process. In Chapter 2 this was described as out of control. In either case, a breakdown in the cause–effect chain of reasoning is apparent. In this chapter the goal is to establish a procedure for evaluating the logical reasoning process. Although this procedure brings a great benefit in itself, the groundwork is laid for Chapters 4 and 5, in which the logical methods developed here are given further, and possibly more significant, application.

The methodology of this chapter is based upon the concepts of Boolean algebra. Currently the most prominent use of Boolean algebra is in the design of the switching circuits used in electronic digital computers (reference 7). Its original application, however, was in the context of logical reasoning (reference 3). It is commonly used to solve the logical portion of problems involving

---

*Some of the material in this chapter was first published in the *Journal of Safety Research;* see reference 5.

3

probability calculations (reference 2). All of these applications can be of great benefit to the safety engineer in his goal of reducing accidents. Of particular current interest is the use of Boolean techniques in fault-tree analysis (reference 6).

3.2 Technical Overview

A Boolean algebra variable, usually represented by a capital letter, represents a distinct event or fact. For example, we may let A represent the event that the chain on a certain machine breaks. If this occurs, we say that $A = T$ or A is true. If the event fails to occur, we say $A = F$ or A is false. Of course, there must be some finite time during which the system is under consideration; and there is a probability associated with event A, although it is often unknown.

A Boolean algebra variable has two modes, true or false (occurrence or nonoccurrence). Similarly, a function, also having but two modes, can be formed as a combination of Boolean algebra variables. To do this, the AND, OR, and NOT operators are used. These follow the rules as given in Tables

3.1, 3.2, and 3.3, respectively. A function will have a truth value, and therefore another variable can be assigned to represent the function.

For example, consider A determined by the following function:

$$A = BC + D\bar{E}$$

To determine the truth value of A, the truth value of the other four variables must be known. The order of priority for evaluating the function is: (1) NOT, (2) AND, and (3) OR. Therefore, if $B =$ false, $C =$ true, $D =$ true, and $E =$ false, A could be determined by using the truth tables (1, 2, and 3) as follows:

$$A = FT + T\bar{F}$$
$$= FT + TT$$
$$= F + T$$
$$= T$$

Table 3.1* AND OPERATOR

X	Y	XY (X AND Y)
T	T	T
T	F	F
F	T	F
F	F	F

* $T =$ true (occurrence), $F =$ false (nonoccurrence).

Table 3.2 OR OPERATOR

X	Y	$X + Y$ (X OR Y)
T	T	T
T	F	T
F	T	T
F	F	F

Table 3.3 NOT OPERATOR

X	$\bar{X}$ (NOT X)
T	F
F	T

The order of precedence for operators can be modified by using parentheses, in which case the inner parentheses are evaluated first. For example:

$$A = B\overline{((C + D)E)}$$
$$= F\overline{((T + T)F)}$$
$$= F\overline{(\overline{TF})}$$
$$= F\bar{F}$$
$$= FT$$
$$= F$$

With experience the writing down of every step becomes unnecessary. Also certain rules become apparent and logical so that the experienced analyst can make many immediate simplifications. For example, let X, Y, and Z represent any three Boolean variables; the simplification rules given in Table 3.4 can be easily proved by substituting all possible values for each variable. Since there are only two possible values for each variable, it is easy to cover all possibilities. By applying the operator rules given in Tables 3.1, 3.2, and 3.3, the validity of each rule can be demonstrated.

Table 3.4 SIMPLIFICATION RULES

I.	$XX = X$	$X + X = X$
II.	$X\bar{X} = F$	$X + \bar{X} = T$
III.	$XY + XZ = X(Y + Z)$	$(X + Y)(X + Z) = X + YZ$
IV.	$XY + X\bar{Y} = X$	$(X + Y)(X + \bar{Y}) = X$
V.	$X + XY = X$	$X(X + Y) = X$
VI.	$X + \bar{X}Y = X + Y$	$X(\bar{X} + Y) = XY$

The simplification rules are given for completeness. They should be understood, but not necessarily memorized, in order to gain a working knowledge of Boolean algebra. Concentration here will be on the map and on tabular methods used to deal with Boolean expressions. For a more comprehensive coverage of Boolean expressions, an elementary text on the subject should be consulted.

Example 3.1 Machine-Safety-Guard Application of Boolean Algebra

To illustrate the use of Boolean algebra as a logical tool, consider a meeting called by a division manager to determine whether to purchase a certain production machine. Attending the meeting are the plant manager, a foreman, the plant safety engineer, and others. When the discussion turns to safety, the following points are brought up by various people around the table:

1. The machine is chain-driven and, although a guard is present, it could be removed.
2. The chain is under a great deal of tension under normal operating conditions. It could be expected to wear out and break periodically.
3. It is possible that chips from a nearby process could get into the chain mechanism, causing it to break prematurely.
4. If the chain breaks, the present guard may not be able to protect surrounding workers, depending on a variety of circumstances.

Of course, in such a meeting the discussion could go on and on, considering a variety of other factors as well. For this example this subset of comments will be considered. The question is: Should the manufacturer be requested to provide more protection, or should the machine be accepted as is? To better understand the problem, we define the following Boolean algebra variables:

A: chain guard is removed.
B: chain wears out and breaks.
C: chips from nearby process break the chain.
D: guard is sufficient to protect the operator in any eventuality.

Further, let X represent the presence of a hazardous situation (not necessarily an accident). The following Boolean expression can be written:

$$X = A + \bar{D}B + \bar{D}C$$

which can be factored to

$$X = A + (B + C)\bar{D}$$

To prevent the hazardous situation, X must be prevented from becoming true. This means that A must be false; at the same time D must be true, or both B and C must be false concurrently. Table 3.5 presents the alternatives in terms of accepting or rejecting the machine. The final decision will rely heavily upon economic considerations.

Although Boolean algebra does not determine whether to accept the machine, it does provide a means for formulating the problem in such a way that the decision can probably be readily made at this point with little further analysis.

Table 3.5 CLARIFICATION OF BOOLEAN ALTERNATIVES

Event	*If Machine Is Accepted As Is* *Action Required*	*If Machine Is Not Accepted* *Action Required*
A	Guarantee through some means that guard will not be removed unnecessarily	Get manufacturer to provide locking device on guard
B	Periodic inspection, preventive maintenance, and replacement	Obtain heavier chain or engineer to relieve tension
C	Add additional shield to protect chain from chips	Obtain guard that will keep chips off
D	Reinforce guard to protect against chain breaks	Obtain guard that will protect against chain breaks

Note: Some combination of manufacturer and in-house modifications might also be sufficient. In any event, action is required with respect to *A*, and action is required either with respect to *D* or else with respect to *both B* and *C*.

3.3 Map Method of Simplification

3.3.1 Mapping Techniques

When Boolean expressions become quite complicated, the use of the basic rules also becomes tedious. For expressions with few (six or less) variables, the map method works extremely well. It can also be used for greater than six variables, but at that point a computerized approach is recommended. Since the map method of simplification demonstrates some of the basic theories of Boolean algebra, this will be covered first. In the next chapter tabular and computerized methods will be demonstrated as they affect reliability computations.

According to Marcus (7), the use of charts for simplification was introduced by E. W. Veitch and later modified by M. Karnaugh to the form presented here. A Boolean algebra map is defined to be a complete spatial representation of all possible events. Karnaugh is credited with the positioning of the events within this model such that advantage could be taken of adjacent events. To illustrate, two events, *A* and *B*, can be mapped concurrently as given in Figure 3.1. Note that, for convenience, 1 and 0 are used to represent true (occurrence) and false (nonoccurrence), respectively. An alternative map for *A* and *B* is given in Figure 3.2 without the descriptions within the spaces.

Let us consider the mapping of the function $T = A + B$. Figure 3.3 illustrates the procedure. The function is taken term by term. First consider *A*. Since, by rule IV of Table 3.4, $A = AB + A\bar{B}$, in order to represent *A* in

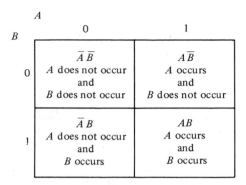

Figure 3.1. Map for *A* and *B*

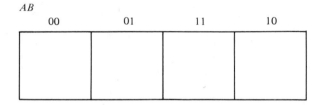

Figure 3.2. Alternative mapping for *A* and *B*

its entirety, a covering must be made of *both* the space represented by *AB* and that represented by $A\bar{B}$. In other words, *A* is in its true mode and *B* is in *all* its modes. Similarly, when *B* is covered, both *AB* and $\bar{A}B$ are covered. Hence $T = A + B$ is represented by the bottom map. Note that the space *AB* has actually been covered twice. This does not matter, however, although its appearance once is sufficient.

The two-variable map can easily be extended to three or four variables. Each time a new variable is added, the number of total spaces must be doubled. Hence a three-variable map, as given in Figure 3.4, requires 8 spaces and the four-variable map of Figure 3.5 requires 16 (i.e., 2^n, where *n* is the number of variables).

Rules for putting a Boolean function onto the map can now be presented. These are as follows:

Step 1. Multiply the function out so that it is in terms of sums of products. The procedure for doing this will be described below. All the examples given are in terms of sums of products (i.e., AND groups separated by OR functions).

Step 2. Take one term at a time and "cover" the spaces on the map which completely describe that term.

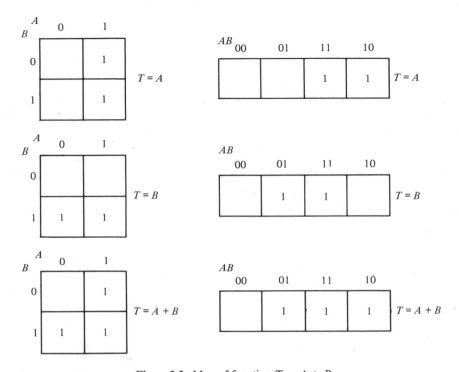

Figure 3.3. Map of function $T = A + B$

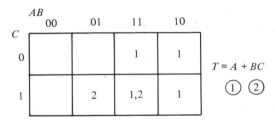

Figure 3.4. Three variable example

For each term, consider first the modes of the variables which appear. If a variable appears in its affirmative (unbarred) mode, all cells associated with the affirmative of that variable are candidates for covering. They may not all be covered, however, if other variables in that term further restrict the covering.

For example, consider $T = A + BC$, which is the sum of two products A and BC. The first term A is in the affirmative, so all spaces labled by A in the affirmative are candidates for covering. Since this term is not conditioned by

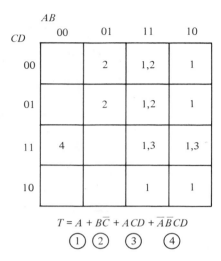

$$T = A + B\bar{C} + ACD + \bar{A}\bar{B}CD$$

① ②   ③   ④

Figure 3.5. Four variable example

any other variables, they *all* must be entered in *all* their modes. This is indicated by the 1 placed in the appropriate spaces of Figure 3.4. In the second term, *BC*, *B* is first considered (middle two columns of map), and all modes of *B* become candidates for covering. However, in this case *B* is further modified or conditioned by *C*. Hence *only* those spaces in affirmative *B* that are also in affirmative *C* will be covered. This is indicated by the 2 placed in the appropriate spaces of Figure 3.4.

This procedure may seem trivial with two variables, but when mapping four or five variables simultaneously, a systematic procedure is helpful. Consider $T = A + B\bar{C} + ACD + \bar{A}\bar{B}CD$ given by Figure 3.5. Since *A* is alone, all combinations of all other variables must be covered within *A*. In the second term, $B\bar{C}$ is covered by first restricting attention to the eight spaces of the affirmative *B*, and then considering only the four of these eight, which are also within $\bar{C}$. This is depicted on the map by 2. In the third term, *ACD*, attention is again restricted to affirmative *A*. It is further restricted to that part of *A* within *C*. And finally, that part of *AC* within *D* is covered. The fourth term, $\bar{A}\bar{B}CD$, can easily be mapped, since it will occupy only one space. This can be read directly from the coding scheme, that is, 00 on top and 11 on the side, since $\bar{A}\bar{B}CD = 0011$.

With a little practice and a minimal amount of knowledge of Boolean algebra, the procedure for entering functions on the map becomes quite simple. Before proceeding, however, it would be well to explain in more detail step 1 of the mapping procedure. The examples above illustrate the ease of mapping a simple function that is a sum of products. Suppose, however, that the Boolean function is a product of sums, as will be the case later in many fault-tree applications. Consider, for example, the product of sums $T = (A + B)(C + D)$.

There are two approaches. The first is to map the function directly, realizing that $(A + B)$ means all of A *or* all of B, and that this is restricted by $(C + D)$, all of C *or* all of D. This is illustrated in Figure 3.6. Alternatively,

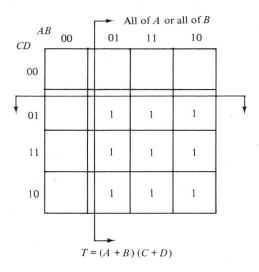

$$T = (A + B)(C + D)$$

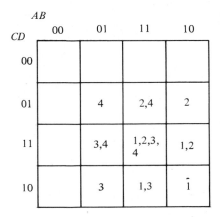

$$T = AC + AD + BC + BD$$

Figure 3.6. Mapping of $T = (A + B)(C + D)$

the function could be multiplied out variable by variable; that is, $T = (A + B)(C + D) = AC + AD + BC + BD$. This would be entered on the map in the manner illustrated in Figure 3.6. Of the two methods, the second is recommended for consistency and simplicity. Step 1 was given to obtain the sum of products form prior to mapping.

Example 3.2 *Mapping of $T = (A + \bar{B})(C + \bar{D})(A + C)$*

Map the following expression:

$$T = (A + \bar{B})(C + \bar{D})(A + C)$$

Step 1.   Obtain the expression as a sum of products.

$$T = (AC + A\bar{D} + \bar{B}C + \bar{B}\bar{D})(A + C)$$
$$= ACA + A\bar{D}A + \bar{B}CA + \bar{B}\bar{D}A + ACC + A\bar{D}C + \bar{B}CC + \bar{B}\bar{D}C$$

Applying rule I of Table 3.4 and arranging variables alphabetically yields

$$T = \underset{\text{①}}{AC} + \underset{\text{②}}{A\bar{D}} + \underset{\text{③}}{A\bar{B}C} + \underset{\text{④}}{A\bar{B}\bar{D}} + \underset{\text{⑤}}{AC\bar{D}} + \underset{\text{⑥}}{\bar{B}C} + \underset{\text{⑦}}{\bar{B}C\bar{D}}$$

Step 2.   Each term is now entered on a four-variable map as given in Figure 3.7. Numbers have been used to illustrate the location of each of the seven terms on the map.

Notice that certain terms, notably 4 and 7, occupy outside corners. These opposite outside edges should be considered as adjacent since only one variable changes. A final caution should be heeded when setting up the map.

CD	AB 00	01	11	10
00			2	2,4
01				
11	6		1	1,3,6
10	6,7		1,2,5	1,2,3,4, 5,6,7

Figure 3.7. Map of $T = (A + \bar{B})(C + \bar{D})(A + C)$

Be sure, as in all the examples above, that only one variable changes for each adjacent block. The sequence 00, 01, 11, 10 is used exclusively here. The importance of maintaining this sequence will become apparent as simplification is discussed.

3.3.2 Map-Simplification Techniques

Once the methods of mapping a Boolean function to a chart are understood, the method of simplification becomes very straightforward. In a sense, simplification involves "taking off" the function from the map using the same rules (rule IV, Table 3.4). However, the particular method of consolidation is of importance here. The general rule is to form groups as large as possible from those spaces covered. This leads, in turn, to the simplest Boolean expression possible which still maintains the necessary truth value.

Before investigating this method consider the objective of simplification a bit further. Using Figure 3.7 in Example 3.2, a large number of legitimate Boolean functions could be constructed equivalent* to the original $T = (A + \bar{B})(C + \bar{D})(A + C)$. For example, each individual space could be considered as one term, from left to right, top to bottom, to obtain the following function:

$$T = \bar{A}\bar{B}CD + \bar{A}\bar{B}C\bar{D} + AB\bar{C}\bar{D} + ABCD + ABC\bar{D} + A\bar{B}C\bar{D}$$
$$+ A\bar{B}CD + A\bar{B}C\bar{D} \qquad (3.1)$$

[Note that those spaces covered more than once (for example, $A\bar{B}C\bar{D}$, which was covered seven times) need only be removed from the map *once*; see rule I, Table 3.4.] Now, the above lengthy function is logically valid despite its redundancy. However, a simplified form of this expression would aid the understanding of the logic that it contains. Therefore, the map is used to observe where simplifications are possible. Here the use of the "one-variable-change" coding system is exploited. Any two adjacent spaces may be combined; the result is a cancellation of the letter that appears in both of its modes. For example, in simplifying the expression given in Example 3.2 and mapped in Figure 3.7, the combinations given in Figure 3.8 could be made. The resulting Boolean function is

$$T = A\bar{C}\bar{D} + \bar{A}\bar{B}C + ACD + AC\bar{D} \qquad (3.2)$$

Note that the specific numbers have been omitted from Figure 3.8 since the frequency and specific source of the covering is inconsequential. Note further

---

*Equivalent here means that regardless of the truth values of A, B, C, or D, the function will *always* have the same truth value as the original.

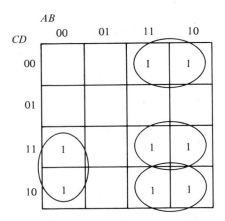

Figure 3.8. Combinations for Figure 3.7 in sets of two

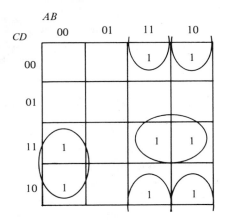

Figure 3.9. Alternative combinations for Figure 3.7

that this particular combination is not unique. Figure 3.9 also presents combinations in sets of two. The Boolean function for this choice is

$$T = AB\bar{D} + A\bar{B}\bar{D} + \bar{A}\bar{B}C + ACD \tag{3.3}$$

These two, as well as several other legitimate functions, are perfectly equivalent; they have the exact same truth value. The choice is at the discretion of the analyst.

Combinations in sets of two are a direct application of rule IV in Table 3.4. A comparison of Equations (3.1), (3.2), and (3.3) demonstrates the simplification procedure. As long as there are no uncovered spaces included and all the covered spaces are included, the new function remains equivalent.

Combinations in sets of two greatly reduce the number of terms and the length of each term in the expression. But further simplifications can be made. Combinations can be made in sets of four provided that the proper adjacency is present. Later it will be demonstrated that any adjacent set in an even power of 2 can be combined. Figure 3.10 shows three ways that a combination of four may lead to the further simplification of Equations (3.1)–(3.3). These three equations are equivalent in truth value to each other as well as to Equations (3.1)–(3.3). Again, however, the size of the expression has been reduced.

If it were required to make combinations in mutually exclusive sets, the procedure would stop right here. Indeed, in subsequent uses of Boolean maps, this will be the goal. Now, however, it is desirable to obtain the briefest possible expression. Therefore, since the truth value of the expression is maintained, it does not hurt to overlap combinations. The simplest expression will be formed of the largest possible combinations, since by selecting the largest possible combinations, the analyst will guarantee the fewest number and shortest terms.

Continuing with the above example, Figure 3.11 illustrates the full simplification of any or all of the expressions in Example 3.2. Usually there are a variety of "simplest expressions," all equivalent in truth value and simplicity. In this particular example, however, this is the only one, with three terms of two variables each.

Although the example presented above demonstrates the use of maps for simplification, the original Boolean function presented in Example 3.2, $T = (A + \bar{B})(C + \bar{D})(A + C)$, was not simplified. It is still three terms of two variables per term. However, one is a product of sums, and the other is a sum of products. Using the map method it is possible to transform from one form to another as well as generating any one of the intermediate forms.

In summary, the following step-by-step procedure is recommended for simplifying a Boolean expression:

1. Write the expression as a sum of products.
2. Denote the spaces on the map that represent each term in the expression.
3. Combine the spaces covered on the map with the largest possible (fewest) combinations.
4. Read the new simplified expression from the map.

Example 3.3   Illustration of the Simplification Procedure

Consider the following Boolean expression:

$$T = AB + B(C + \bar{D}) + A\bar{B}D + AB\bar{C}D$$

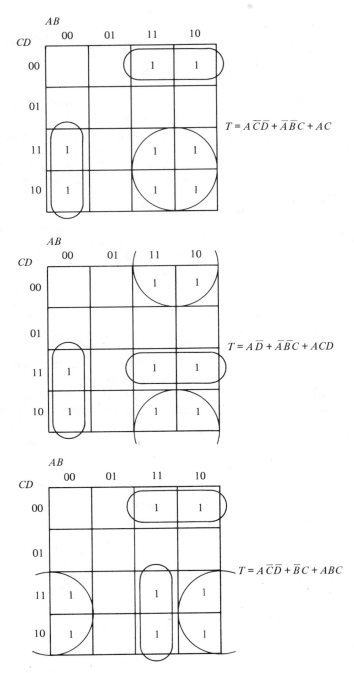

Figure 3.10. Alternative combinations in sets of two and four

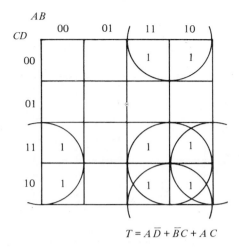

$$T = A\bar{D} + \bar{B}C + AC$$

Figure 3.11. Complete simplification of Equations (3.1)—(3.3)

Using the four steps above results in the following:

Step 1. $T = AB + BC + B\bar{D} + A\bar{B}D + AB\bar{C}D$

      ①   ②   ③   ④   ⑤

(*Note:* Numbering of terms is for reader convenience only and is not required for the procedure.)

Step 2.

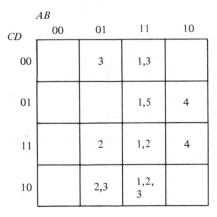

Step 2.

Step 3.

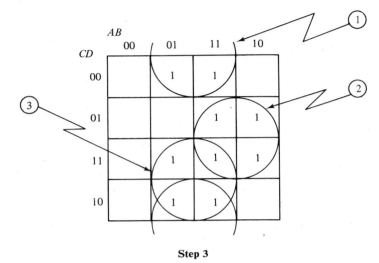

Step 3

Step 4. $\quad T = B\bar{D} + AD + BC$
$\qquad\qquad$ ① $\qquad$ ② $\qquad$ ③

3.3.3 Five- and Six-Variable Maps

The map method of simplification is extremely useful for up to four variables. As the number of variables increase above four, the ease of usage drops, owing to the lack of adjacency. For completeness, five- and six-variable maps are presented here. In a subsequent section, a more general tabular method is presented.

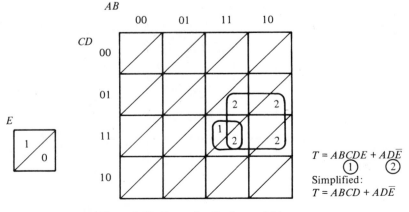

$T = ABCDE + AD\bar{E}$
$\qquad$ ① $\qquad$ ②

Simplified:
$T = ABCD + AD\bar{E}$

Figure 3.12. Example of a five variable map

A five-variable Boolean map can be depicted by Figure 3.12. Here each of the 16 spaces of the four-variable map is subdivided to accommodate the fifth variable. The expression $T = ABCDE + AD\bar{E}$ is charted on this map and simplified. Notice that combinations cannot be made whenever adjacent spaces appear. This is because for some adjacent spaces more than one variable changes. The exception is within a given square, such as within $ABCD$ in the example. Figure 3.13 gives examples of proper and improper combinations.

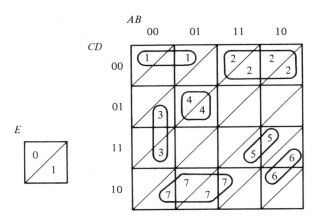

Acceptable Combinations: 1, 2, 3, 4
Incorrect Combinations: 5, 6, 7

Figure 3.13. Proper and improper combinations on the five variable map

The six-variable map is further subdivided, and the number of spaces are again doubled, to $2^6 = 64$. Figure 3.14 gives an example of the use of a six-variable map. As before, care must be taken to ensure that proper combinations are made. Figure 3.15 gives examples of proper and improper combinations.

There are other techniques for handling more than four variables (see reference 7). Rather than presenting them, a tabular technique which can be programmed on a computer is presented at the end of the chapter. This technique may also be easier for some people to use than the five- and six-variable maps. It can be used for any number of variables, given enough time and/or enough computer storage.

Although most safety problems will involve a large number of variables, the portion requiring simplification can usually be isolated from many of the other variables, thus reducing the problem to one that can be solved without a computer.

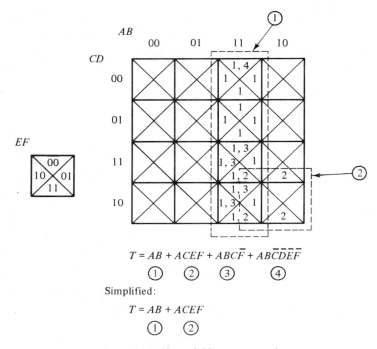

$$T = AB + ACEF + ABC\overline{F} + AB\overline{C}\overline{D}\overline{E}\overline{F}$$
①   ②   ③   ④

Simplified:

$$T = AB + ACEF$$
①   ②

Figure 3.14. Six variable map example

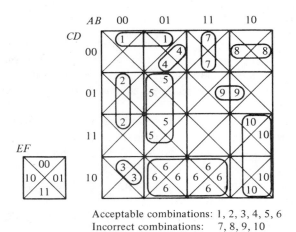

Acceptable combinations: 1, 2, 3, 4, 5, 6
Incorrect combinations: 7, 8, 9, 10

Figure 3.15. Proper and improper combinations on the six variable map

For example, consider the expression

$$X = A + BC + B\bar{D} + EF + EFGH + E\bar{G}H$$
$$= A' + B'$$

where A' and B' are subexpressions of X, such as

$$A' = A + BC + B\bar{D}$$
$$B' = EF + EFGH + E\bar{G}H$$

Now A' and B' can be reduced by hand and then recombined. Since no variable in A' appears in B', no further simplification is required. The procedure would still be valid if some of the variables appear in more than one subexpression. However, upon recombination a final simplification step would be required to determine if the interaction between the subexpressions could lead to further simplification.

3.3.4 Examples of Boolean Algebra Application

Consider, for example, a car–truck head-on collision on a partially divided four-lane highway. The car was clearly on the wrong side of the divider, and the four men in the car were killed. The following are reports from investigators and witnesses:

Medical Report: The blood tests made for drunkenness on the driver of the auto are borderline and inconclusive. The driver may or may not have been drinking. Past medical history shows that the driver did have a condition such that, when he got excited, especially when he laughed hard, he became dizzy, to the point where he might be unable to tell the right side of the road from the left.

Witness No. 1, Bartender: The driver and three other fellows came into the bar and stayed about 15 minutes. The driver did not order a drink, and I did not personally see him take a drink. However, I believe he had been drinking because when he left the bar he was in a jovial mood, laughing and joking with the other three.

Witness No. 2, Patron: I was the only one in the bar when the others came in. I completely concur in every detail with what the bartender had to say. Not only that, but I noticed that one of the other fellows in the car also had a bottle. So I believe that he probably had a drink shortly after he left the bar, if not before.

Investigator No. 1: In the absence of proof to the contrary we must assume that the driver was sober. However, because he was joking with his friends, chances are he got dizzy and could not discern the right side of the road.

Investigator No. 2: It has not been shown that the driver was laughing to the point that would cause dizziness. However, a Keep Right sign, which could have prevented the accident, was dimly lit. A person who had consumed any alcohol at all would see the sign just a second too late. This is what I believe happened.

Witness No. 3, in Car Following: The accident car was in front of me. Just prior to the accident another car pulled out from the side in front of the accident car. It seemed to me that that is what caused the accident. They were quite far ahead of me, so I couldn't be sure. I believe that if the driver of the accident car had been drunk or dizzy, he would have hit the car that pulled out. Instead, he took a calculated risk and drove on the other side of the divider. The risk just didn't pay off.

The following Boolean variables can be used to describe the possible causes of the accident:

A: had been drinking.
B: medical problem; laughing caused dizziness.
C: had been drinking in car as well as at bar.
D: was joking and laughing very much with friends.
E: Keep Right sign too dimly lit for slightly intoxicated man to see.
F: forced onto wrong side by another car.

Beginning with Witness No. 1, a Boolean expression can be written for each of the opinions. Since any one of these opinions would describe events sufficient to cause the accident, the opinions can be linked together by an OR function as follows:

$$T = AB + ABC + \bar{A}BD + A\bar{B}E + \bar{A}\bar{B}F \tag{3.4}$$

A brief term-by-term explanation is as follows:

AB: had been drinking, and laughing caused dizziness.
ABC: had been drinking in car as well as at bar, and laughing caused dizziness.
ĀBD: had not been drinking, was joking, and laughing caused dizziness.
A B̄ E: had been drinking, had no medical problem, and Keep Right sign could not be read.
Ā B̄ F: had not been drinking, had no medical problem, and took calculated risk to avoid car.

This Boolean expression can now be charted in a six-variable map, as given in Figure 3.16. Note that Equation (3.4) reduces to

$$T = AB + BD + AE + \bar{A}\bar{B}F$$

The first simplification, which is the dropping of *ABC* in the first expres-

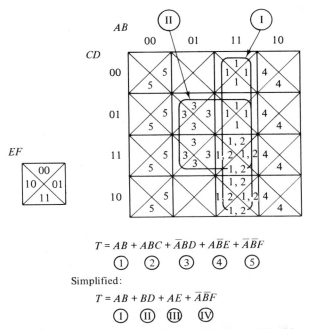

$$T = AB + ABC + \overline{A}BD + A\overline{B}E + \overline{A}BF$$

     ①   ②   ③   ④   ⑤

Simplified:

$$T = AB + BD + AE + \overline{A}\overline{B}F$$

     (I)   (II)   (III)   (IV)

where III comes from combining 4 with the lower left of 1 and
IV comes directly from 5

Figure 3.16. Simplification for the six variable example

sion, should be obvious since it is irrelevant where the driver was drinking, if in fact he was drinking. This also follows from simplification rule V in Table 3.4. The other simplifications are not so obvious and would require some expansion and rearrangement to apply the rules of Table 3.4. However, tabular and graphical methods make the task of simplification much easier.

Using the simplified expression, the irrelevant facts can be overlooked in favor of a concentrated effort on the pertinent facts. The following is an explanation of the simplifications from a logical standpoint:

ABC: Reduced to *AB*, since it is irrelevant where the drinking took place.

ĀBD: reduced to *BD*, since it is irrelevant whether the driver had been drinking or not in the situation where he was laughing very hard, thus bringing on dizziness in either event.

A B̄E: reduced to *AE*, since obviously the presence of the medical problem would contribute to not being able to read the sign, and hence its absence was not a condition of the accident.

ĀB̄F: this cannot be reduced since, if the driver took the risk discussed, he would have to be in full command of his faculties.

Note again that Boolean algebra does not tell which of the opinions is correct. It assumes that each opinion has a rational basis and then provides a means for simplification to eliminate redundancies between opinions. For example, the "calculated risk" opinion, which includes the fact that the driver must have been sober, could be false. As a matter of fact, the contradictory nature of several of the opinions with each other indicates that someone is in error. Although Boolean algebra does not determine who is in error, it does provide a tool for simplification so that a concentrated effort can be placed on continued determination to resolve the pertinent facts. Without such a logical approach, quite a bit of time could be spent in running down irrelevant leads.

3.4 Synthesis of Boolean Expressions

In Chapter 5 the use of Boolean algebra in the quantitative evaluation of fault trees will be discussed. Closely related to this is the use of Boolean algebra to synthesize the events that surround an accident. This differs from the usage in the example above in that the purpose is no longer concerned with isolating the *one* cause. Rather, it is recognized that several causes working simultaneously resulted in a particular accident of a particular severity. The objective is to put together the various causes in terms of a Boolean expression that can be further evaluated according to probabilistic methods discussed in Chapter 5.

Accidents with multiple causes can usually be analyzed such that the causes are organized according to the source of the cause. For example, certain causes are characteristic of the environment in which the accident takes place. Others are brought about by the human element. Still others may be chance occurrences of the outside environment. If a Boolean expression can be written for each, these expressions can be synthesized to establish a total expression for the accident itself. To demonstrate this procedure, consider Example 3.4.

*Example 3.4 Example of Boolean Synthesis**

The following facts are to be considered in this example:

1. The time was 8:40 A.M. on a Sunday.
2. A single car, occupied by the driver only, struck the concrete uprights of a bridge pier.

---

*This example was taken from "Multidisciplinary Accident Investigation Summaries," U.S. Department of Transportation, *DOT HS-600 622*, Vol. 2, No. 4, page 143. It was an actual case summary of a car–fixed object accident. Most of the other accidents given in the DOT summaries are amenable to similar analysis.

3. The driver was pronounced dead on arrival at the hospital 15 minutes after the crash. He suffered a compound comminuted fracture of the skull after being thrown from the vehicle and his head struck on a concrete pillar.

4. The roadway was wet, and it was possibly raining hard during the accident; it was daylight.

5. The roadway was a divided, limited-access highway with three lanes each way.

6. The accident took place 200 feet before an exit ramp to the right. The uprights that were struck were off the road to the right.

7. Traffic controls consisted of speed-limit signs (45 mph), broken white-lane-divider lines, solid white line dividing roadway from 10-foot shoulder, and exit signs.

8. The vehicle was a 1964 Pontiac Catalina, recently inspected, that had traveled 72,843 miles. Mismatched tire sizes and tread types may have contributed to the accident.

9. At impact the left front door detached as a result of broken hinges. The driver was thrown from the vehicle. Lap belts were not being used at the time of the accident. The car spun around 180 degrees clockwise.

10. The driver, 19 years of age, was alleged to suffer blackouts and nystagmus. He was familiar with the area. Alcohol tests proved negative.

11. According to a witness, the vehicle was traveling at 90 to 100 mph when the vehicle veered to the right beneath the overpass and struck one of the concrete uprights.

12. After impact the left rear tire was seen to be deflated.

The "pertinent" facts for Boolean analysis are as follows (with the applicable item number in parenthesis):

A: wet roadway contributed to loss of control (4).
B: raining hard, limited visibility (4).
C: confusion between bridge uprights and exit ramp (6).
D: improper traffic controls (7).
E: mismatched tires caused loss of control (8).
F: weak hinges (9).
G: lap belts not secured (9).
H: shoulder belts not installed.
I: driver illness caused accident (10).
J: speed contributed to loss of control (11).
K: left rear tire blew out and caused accident (12).
L: there were no guardrails to prevent direct impact with the bridge piers.

Although the Boolean operators are stated as facts, there is no implication that they are all true. In fact, the following "contradictions" are significant:

$\bar{C}$: driver was familiar with the area, no confusion.

$\bar{D}$: traffic controls were standard for this type of highway; speed limit (45 mph) was adequate.

$\bar{E}$: mismatched tires would only cause loss of control in sudden braking; no evidence of this was available.

$\bar{K}$: deflation of left rear tire would have cause a pull to the left, which was not observed. Also, if the tire blew out, this would make the effect of mismatched tires insignificant, further strengthening the probability of $\bar{E}$. The deflation of the left rear tire could have been an effect of the car spinning around after impact.

Now the objective is to synthesize a Boolean expression for the accident in terms of each contributing factor. First consider the hazard caused by the weather expressed as W, where

$$W = A + B$$

A second hazard could have been caused by the roadway, given by R, where

$$R = C + D$$

A third hazard could have been caused by the vehicle, given by V, where

$$V = E + K$$

A fourth factor will be charged to shortcomings (not necessarily intentional) of the driver. These could be in terms of errors in judgment, illness, or intentional violation of the law. Calling this factor P,

$$P = I + J$$

Finally, there are factors that did not cause the accident but did add to the severity. Calling these severity factors S,

$$S = F + G + H + L$$

In the absence of other evidence, it will be assumed that the concurrence of events W, R, V, P, and S lead to the particular accident. Thus, letting X be the occurrence of an accident of this type and severity,

$$X = WRVPS$$
$$= (A + B)(C + D)(E + K)(I + J)(F + G + H + L)$$

This Boolean expression is in its simplest form and would be best left as a product of sums. Methods by which further quantification can be obtained using probabilities will be discussed in Chapter 4.

Before leaving this example, consider the reduction that would result if only the presence of an accident was considered, and not the total event occurrence X. Letting Y be the accident only,

$$Y = WRVP$$

Similarly, those who are to perform modification on the roadway, vehicle, or driver might restrict their considerations to those pertinent subexpressions.

3.5 Tabular Method

As can be seen from the example, accident investigations often require more than six or seven variables. In these cases the problem of simplification increases in complexity, and a tabular approach might be easier. The tabular method has the advantage of providing a step-by-step procedure that can be applied regardless of the number of variables.

The basic principle applied is still exemplified by the expression $X = XY + X\bar{Y}$, where X may be one or more variables functionally ANDed together. In general, the approach is to take each term and first expand it, as is done in the mapping procedure. Then these terms resulting from the expansion are combined and the simplified expression results. To facilitate the procedure, we employ a coding scheme that uses binary numbers.

Binary numbers can be used to produce all possible combinations of an expression. For example, in Table 3.6, all the binary numbers from 0 to 31 are presented. Suppose that the positions of the five digits are numbered from right to left, and that A is assigned to position 5, B to 4, C to 3, D to 2, and E to 1. Now, letting the value in the variable position indicate its truth value, all possible combinations of the truth value of $ABCDE$ are generated. For example 18, or 10010 in binary form, would represent $A\bar{B}\bar{C}D\bar{E}$.

Another useful property is that the *unit* digits of any binary representation can never be matched, digit for digit, by any lesser number. For example, the number 11 has ones in positions (from the right) 1, 2, and 4. No other number less than 11 can have ones in position 1, 2, and 4. The numbers 15, 27, and 31 also have ones in position 1, 2, and 4, but they are greater than 11.

It is this principle which enables binary numbers to aid in simplifying Boolean expressions. Prior to giving the step-by-step procedure, a simple example will be given. Consider the Boolean expression

$$T = AB + CD + \bar{A}BC \qquad (3.5)$$

$$\quad \text{①} \qquad \text{②} \qquad \text{③}$$

Table 3.6 BINARY NUMBERS
0–31

Base 10	Binary Numbers
0	00000
1	00001
2	00010
3	00011
4	00100
5	00101
6	00110
7	00111
8	01000
9	01001
10	01010
11	01011
12	01100
13	01101
14	01110
15	01111
16	10000
17	10001
18	10010
19	10011
20	10100
21	10101
22	10110
23	10111
24	11000
25	11001
26	11010
27	11011
28	11100
29	11101
30	11110
31	11111

This is easily simplified on a map, as in Figure 3.17, where the appearance of A is eliminated from the last term. Consider now a tabular representation of all the spaces represented in the map, allowing the numeric representation to dictate the binary representation, as indicated in Figure 3.18. In other words, the modes of A, B, C, and D dictate the values of the four digits, respectively. The third column of Table 3.7 shows how the expression can be represented tabularly. All that remains is to combine the individual spaces in a tabular way, and the process will be complete.

This may seem complicated, but the nature of binary numbers enables the following simple step-by-step procedure to be used:

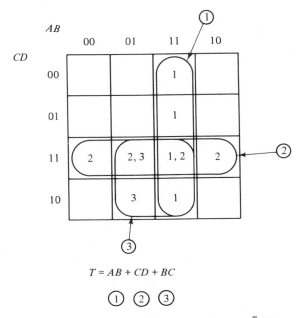

$$T = AB + CD + BC$$

① ② ③

Figure 3.17. Simplification of $AB + CD + \bar{A}BC$

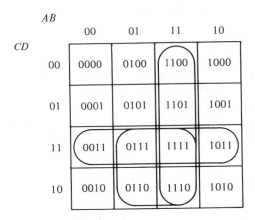

Figure 3.18. Binary representation of each element on map

Step 1.   Transform the Boolean expression into a set of binary numbers, term by term, according to the following rules:

(a) If the variable appears in its unbarred mode (e.g., X), assign its position a 1.

Table 3.7 TABULAR REPRESENTATION OF EQUATION (3.5)

Base 10	Binary Numbers	Equation (3.5)	Term
0	0000		
1	0001		
2	0010		
3	0011	*	2
4	0100		
5	0101		
6	0110	*	3
7	0111		
8	1000		
9	1001		
10	1010		
11	1011		
12	1100	*	1
13	1101		
14	1110		
15	1111		

(b) If the variable appears in its barred mode (e.g., $\bar{X}$), assign its position a 0.

(c) If the variable does not appear, assign its position an asterisk (*).

(*Note:* See examples in Table 3.8.)

Step 2.   Post each term to an ordered list of binary numbers from 0 to $2^n - 1$, where n is the number of variables in the expression. This will consist of merely denoting those binary numbers which are in the expression. To do this, mark all binary numbers that have identical digits in each of the positions without an asterisk. For example, if *1*0 is a term in the expression, then the following would be marked: 0Ĭ0Ŏ, 0Ĭ1Ŏ, 1Ĭ0Ŏ, 1Ĭ1Ŏ. These are easy

Table 3.8 EXAMPLES OF STEP 1 CODING

Total Number of Variables	Term	Code
4	ABC	111*
4	$ABC\bar{D}$	1110
4	$BC\bar{D}$	*110
5	ABD	11*1*
5	$A\bar{B}E$	10**1

to pick out by tracing down a binary-number list, starting with the number obtained by replacing the asterisks with zeros and proceeding downward.

Step 3. Combine terms by comparing successive numbers. Two numbers can be combined if they differ in *one position only*. Wherever this occurs, take the two numbers, combine them, and post the new representation to a new list. In combining numbers, put an asterisk (*) in the position that had the uncommon digit. Combine each term with as many other terms as possible. If a term cannot be combined, post it to the new list as is.

Step 4. Using the new list established in step 3, once again go through all combinations, seeking those representations which vary in only one digit. Make the combination, placing an asterisk (*) in the position of the uncommon digit, and post the new term to another list. Continue this process until all possible combinations have been made.

Step 5. The reduced expression can be formed from the last list of step 4. Each remaining binary–asterisk representation is one term. If the position of a variable is occupied by a 1, it will be in its affirmative, unbarred mode. If it is 0, it will be in its barred mode. An asterisk in the position indicates that the variable should not appear.

Example 3.5 Tabular Reduction

Consider the Boolean expression simplified in Example 3.3:

$$T = AB + BC + B\bar{D} + A\bar{B}D + AB\bar{C}D$$

Before proceeding to step 1, note that this function is expressed as a sum of products. Although not specifically emphasized, this is also a requirement of the tabular method.

Step 1. Since there are four variables, the transformation is performed as follows:

$$T = AB + BC + B\bar{D} + A\bar{B}D + AB\bar{C}D$$
$$= 11** + *11* + *1*0 + 10*1 + 1101$$

Step 2. The expression is expanded as in Table 3.9. Note that the last column of this table was added to inform the reader why the term is in the expression. This serves no other function and can be eliminated as soon as proficiency is obtained in generating all terms from the binary-number listing.

Table 3.9 STEP 2 POSTING FOR EXAMPLE 3.5

Base 10	Binary Numbers	Numbers in Term	Reason for Being in Term
0	0000		
1	0001		
2	0010		
3	0011		
4	0100	0100	*1*0
5	0101		
6	0110	0110	*11*, *1*0
7	0111	0111	*11*
8	1000		
9	1001	1001	10*1
10	1010		
11	1011	1011	10*1
12	1100	1100	11**, *1*0
13	1101	1101	11**, 1101
14	1110	1110	11**, *11*, *1*0
15	1111	1111	11**, *11*

Step 3. Attention is now directed to the expanded representation of the expression. Combinations are made by comparing each number with each succeeding number. The combinations made are given in Table 3.10.

Comparisons need only be made between each term and all others of higher value.

Table 3.10 STEP 3 COMBINATIONS

Base 10	Binary Number in Term	Combinations	
4	0100	4–6	01*0
6	0110	4–12	*100
7	0111	6–7	011*
9	1001	6–14	*110
11	1011	7–15	*111
12	1100	9–11	10*1
13	1101	9–13	1*01
14	1110	11–15	1*11
15	1111	12–13	110*
		12–14	11*0
		13–15	11*1
		14–15	111*

Step 4. The new list of combinations (above) is now used for further combining. To illustrate this, the combinations above have been assigned a number in Table 3.11. These numbers are merely for demonstration purposes and need not be included as part of the procedure. Comparisons of this secondary type are even easier to make, since the asterisks must be in the same

Table 3.11 STEP 4 COMBINATIONS

Number	Combination	New Combination	
1	01*0	1–10	*1*0
2	*100	2–4	*1*0
3	011*	3–12	*11*
4	*110	4–5	*11*
5	*111	6–11	1**1
6	10*1	7–8	1**1
7	1*01	9–12	11**
8	1*11	10–11	11**
9	110*		
10	11*0		
11	11*1		
12	111*		

column for a match to take place. Note that only one digit may change for a combination to be made. In the new combinations of Table 3.11 there are duplicate combinations. These can be eliminated since they merely indicate a double coverage of the appropriate block. Hence the terms reduce to *1*0, *11*, 1**1, and 11**. Since no more combinations can be made, step 4 is complete. If two of the terms varied by only one digit, further simplifications could be made.

Step 5. The retransformation is performed as follows:

$$\overset{*1*0 \qquad *11* \qquad 1**1 \qquad 11**}{T = \quad B\bar{D} \; + \; BC \; + \; AD \; + \; AB}$$

This answer is consistent with Example 3.3, by the map method, although here there is no guarantee of minimum number of terms.

Example 3.6 Five-Variable Tabular Reduction

A five-variable example will now be given to further illustrate the procedure. Since five variables are involved, it will be necessary to use a table of binary numbers from 0 to 31. Consider the following expression:

$$T = AB + ABC + \bar{A}BD + A\bar{B}E$$

Each step will now be given without explanation.

Step 1. $T = AB + ABC + \bar{A}BD + A\bar{B}E$

Binary representation: $11*** + 111** + 01*1* + 10**1$

Step 2.   See Table 3.12.

Table 3.12 STEP 2 POSTING FOR EXAMPLE 3.6

Base 10	Binary Numbers	Numbers in Term
0	00000	
1	00001	
2	00010	
3	00011	
4	00100	
5	00101	
6	00110	
7	00111	
8	01000	
9	01001	
10	01010	10
11	01011	11
12	01100	
13	01101	
14	01110	14
15	01111	15
16	10000	
17	10001	17
18	10010	
19	10011	19
20	10100	
21	10101	21
22	10110	
23	10111	23
24	11000	24
25	11001	25
26	11010	26
27	11011	27
28	11100	28
29	11101	29
30	11110	30
31	11111	31

Step 3.   See Table 3.13.

Table 3.13 STEP 3 COMBINATIONS FOR EXAMPLE 3.6

New Number	Combination	
1	10–11	0101*
2	10–14	01*10
3	10–26	*1010
4	11–15	01*11
5	11–27	*1011
6	14–15	0111*
7	14–30	*1110
8	15–31	*1111
9	17–19	100*1
10	17–21	10*01
11	17–25	1*001
12	19–23	10*11
13	19–27	1*011
14	21–23	101*1
15	21–29	1*101
16	23–31	1*111
17	24–25	1100*
18	24–26	110*0
19	24–28	11*00
20	25–27	110*1
21	25–29	11*01
22	26–27	1101*
23	26–30	11*10
24	27–31	11*11
25	28–29	1110*
26	28–30	111*0
27	29–31	111*1
28	30–31	1111*

Step 4.   See Table 3.14.

Step 5.   The reduced term is

$$*1*1* + 1***1 + 11***$$

or

$$T = BD + AE + AB$$

Table 3.14 STEP 4 COMBINATIONS FOR
EXAMPLE 3.6

New Number	Combinations from Step 3 (Number Assigned in Step 3)	
1	1–6	01*1*
2	1–22	*101*
	(2–4	01*1*)
3	2–23	*1*10
	(3–5	*101*)
	(3–7	*1*10)
4	4–24	*1*11
	(5–8	*1*11)
5	6–28	*111*
	(7–8	*111*)
6	9–14	10**1
7	9–20	1*0*1
	(10–12	10**1)
8	10–21	1**01
	(11–13	1*0*1)
	(11–15	1**01)
9	12–24	1**11
	(13–16	1**11)
10	14–27	1*1*1
	(15–16	1*1*1)
11	17–22	110**
12	17–25	11*0*
	(18–20	110**)
13	18–26	11**0
	(19–21	11*0*)
	(19–23	11**0)
14	20–27	11**1
	(21–24	11**1)
15	22–28	11*1*
	(23–24	11*1*)
16	25–28	111**
	(26–27	111**)

Note: Those combinations above which produce the same results need only be considered once. Hence, only the first result is numbered; subsequent common results are denoted by parentheses. Further combinations are as follows:

New Number	Combinations (Numbers Assigned Above)	
1	1–15	*1*1*
	(2–5	*1*1*)
	(3–4	*1*1*)
2	6–14	1***1
	(7–10	1***1)
	(8–9	1***1)
3	11–16	11***
	(12–15	11***)
	(13–14	11***)

3.6 Closure

Considerable time has been spent in this chapter in a rigorous explanation of Boolean simplification. The tabular technique, which is time-consuming and tedious, was presented also. This was done to lay the foundation for Chapter 4, since most of the methods used in reliability analyses and in calculating probabilities are a direct application of the theories presented in this chapter. And, although the tabular method is time consuming to do by hand, in the long run it is more powerful than the simple mapping techniques. This results from its adaptability to computerization.

SELECTED REFERENCES

1. ARNOLD, BRADFORD H., *Logic and Boolean Algebra*, Prentice-Hall, Inc., Englewood Cliffs, N.J., 1962.

2. BOOLE, GEORGE, *Studies in Logic and Probability*, Open Court Publishing Co., La Salle, Ill., 1952.

3. BOOLE, GEORGE, The *Mathematical Analysis of Logic*, Basil Blackwell, Oxford, 1951.

4. BROWN, DAVID B., "A Computerized Algorithm for Determining the Reliability of Redundant Configurations," *IEEE Trans. Reliability*, Vol. R-20, No. 3, Aug. 1971.

5. BROWN, DAVID B., "The Use of Boolean Algebra in Safety," *J. Safety Res.*, Vol. 4, No. 4, pp. 155–159, 1972.

6. KOLODNER, HERBERT H., "The Fault Tree Technique of System Safety Analysis as Applied to the Occupational Safety Situation," *ASSE Monograph 1*, June 1971.

7. MARCUS, MITCHELL P., *Switching Circuits for Engineers*, Prentice-Hall, Inc., Englewood Cliffs, N.J., 1967.

QUESTIONS AND PROBLEMS

Simplify the following using map and tabular methods.

1. $AB + A\bar{C} + \bar{B}\bar{A} + \bar{D}\bar{C}\bar{B} + \bar{B}CD$

2. $ABC + BCD + CDE + ACE + \bar{A}\bar{B}CD\bar{E}$

3. $(A + B)(\bar{A} + C)(AC + \bar{D})$

4. $(A + B + C)(\bar{A} + B + D)$

5. $ACF + DCF + ADC + ABCD + AE\bar{F} + \bar{A}CD\bar{F}$

6. $AB(C + D) + CD(E + \bar{F}) + ABF + CD\bar{E}F$

7. $ABD + A\bar{C}D + AB\bar{C} + E\bar{F} + EGH + EF\bar{H}$

8. $(A + B)(\bar{C})(A + D)$

9. $(A + D + E)BCD(\bar{A} + \bar{C})$

10. $(A + B)(C + D)(E + F)AB\bar{C}$

11. Consider the following situation. A worker received an injury while operating a machine. A guard that should have prevented the accident was not on the machine. However, whether it came off at the time of the accident or was taken off before the accident is a matter of speculation. Three people were interviewed, with the following results:

   (a) Worker: I was operating the machine according to specifications when the guard must have vibrated off or something. The next thing I knew the accident had occurred.

   (b) Manager: I have told these men thousands of times to observe safety practice. Yet so often they take the guards off to clean the machine and just don't put them back. Now you can use the machine without the guard if you are real careful, but the minute you become the least bit fatigued, one slip and that is it. Either that or else he was distracted by some of the other workers. I believe if he had been operating the machine at the right speed, none of this would have happened.

   (c) Foreman: I am sure that everyone was attentive and not fatigued since the accident occurred only a few minutes after the break. However, the guard must have been taken off, since I checked it and it was securely fastened earlier that morning.

   Define Boolean events of the pertinent causes. Assuming that one or more of those interviewed is completely correct, establish the Boolean function for the accident cause. Simplify the expression. Logically explain the simplifications.

12. Synthesize a Boolean expression for the following accident facts. Obtain sub-expressions for combinations of the following causes: weather, roadway, vehicle, driver, and severity.

   (a) Accident occurred at a T intersection in daylight early on a Sunday morning. Two cars were involved.

   (b) Only the drivers were in the cars as they collided head-on. One was injured fatally; the other received minor injuries.

   (c) The weather was dry, the temperature below freezing. The roadway edges were wet from melting snowbanks.

   (d) The roadway was an undivided highway, 33 feet wide but decreased to 23 feet because of the snowbanks. The speed limit was not posted, but was 30 mph because the road was in a residential area.

   (e) The only traffic control was a faded white center line. There were no warning signs or signals.

(f) Driver 1, fatally injured, was driving a compact car. Driver 1 attempted a left turn into the path of vehicle 2, an average-sized American car.

(g) Evidently driver 1 did not see the oncoming car, owing to inattention, poor eyesight, or environmental conditions.

(h) Neither driver was using available restraints. Vehicle 1 had both shoulder and lap belts. Vehicle 2 had lap belts only.

(i) The light and fragile construction of vehicle 1 added to the severity.

PROBABILISTIC RELIABILITY CONSIDERATIONS

4.1 Intoduction

In Chapter 3 a systematic method was presented for crystallizing the logic of a given accident situation or circumstance. Although this tool aids the analyst by concentrating his attention on the pertinent facts, it does not tell him what action to take. This is especially important when limited resources prevent actions from being taken to prevent the occurrence of every possible causal event.

For example, consider the car–truck head-on-collision example of Chapter 3. Through simplification, the possible causes were clarified, but the exact cause was not specifically determined. Table 4.1 contains the possible causes and proposes some alternative solutions (i.e., actions by which future accidents of this type might be prevented). Notice that, in addition to being an incomplete list of remedies, each alternative could exist in varying degrees. Further, some of the remedies will eliminate the problem altogether, whereas others will reduce only by a small degree the probability of reoccurrence. This presents a serious problem to the decision maker, who must determine which remedies to apply as well as the extent of their application. Compound this with the large variety of accidents that can occur, either from the highway or

4

the occupational safety viewpoint, and it becomes obvious why safety problems still exist.

In this chapter the groundwork is laid for a more rigorous quantitative analysis to resolve this problem. Since accidents obviously do not occur with

Table 4.1 HEAD-ON-COLLISION CAUSES AND SOLUTIONS

Causal Event	*Proposed Remedies*
1. Drunken driving	1. Increased number of enforcement officers
	2. Driver-testing device to prevent driving under intoxication
2. Medical problem of driver	1. Medical check-ups for drivers
3. Inability to read sign	1. Improve lighting of roadway
	2. Redesign roadway to make head-on crash impossible
4. Forced onto wrong side by another car	1. Improved driver-training techniques
	2. Improved testing program

certainty, quantification under uncertainty must be applied using probabilistic concepts. Because of the close dependency in many systems between safety and system reliability, an immediate application to time-independent reliability problems is presented in this chapter. In Chapter 5 direct cost/benefit calculations are made, applying the concepts of probability through the technique of fault-tree analysis.

4.2 Basic Definitions

As discussed in Chapter 3, any Boolean event A has two modes: occurrence and nonoccurrence. In manipulating and simplifying Boolean expressions, there was a determination of the effect that the mode of any given variable would have upon the validity of the expression. For example, if A is true (occurs), what effect will that have on the expression $X = AB$? By using certain rules, seemingly complicated expressions could be reduced to help answer questions such as these.

In this chapter the objective is not to assume a given mode and examine the consequences. Instead, each Boolean variable is assigned a relative frequency with which its event is expected to occur. The problem is to determine the relative frequency with which the entire expression will "occur." Note that each Boolean variable still has only two modes. However, the relative frequency of each of these modes will now be specified.

By definition, *probability* is estimated by *relative frequency*, and hence an event's probability of occurrence is the relative frequency with which that event is expected to occur. Calculation of probabilities is often greatly simplified by keeping this definition in mind.

To better understand the concept of relative frequency, first consider absolute frequency. Suppose that a bag contains equal-sized marbles of three colors: black, white, and red. Suppose that any marble has been drawn out of the bag on a number of past occasions, and that after each draw the one drawn was *replaced* in the bag. Further, suppose that a record is kept of the outcomes, which indicates that a white marble was drawn 1,000 times. It may be useful to predict future outcomes. This would be possible if the number of each color of marble in the bag could be determined. From the information above this is not possible, since the total number of draws, or the number of other outcomes, is unknown.

On the other hand, if 3,000 draws or "trials" were made, of which 1,000 were white, it might be inferred that $\frac{1}{3}$ of the marbles in the bag were white. This would be an *empirical* measure of relative frequency with which the occurrence "white marble drawn" has been observed. The best estimate of the probability of drawing a white marble would be $\frac{1}{3}$. Notice that probability (future relative occurrence) is being estimated by using past measures of relative frequency. This method of estimating is valid only if all conditions remain the same. Even at that, it is still only an estimate, not a certainty. This is not a criticism of the *empirical* technique (in most cases it is the only one that is available). Rather, it is a warning to those who apply it to use judgment as well as mathematical prowess.

Before continuing, consider another method by which the probability of this particular event could be determined. Suppose that the bag of marbles was emptied and each color was counted. For example, suppose that two white, three black, and one red marble were found. It would be possible to obtain an a priori estimate of the probability by again using the concept of relative frequency. Here the relative frequency is the frequency with which the event is expected to occur, divided by the total number of possible events. Two are successes (white balls), and there are a total of six possibilities. Therefore, the a priori calculation of probability for this event is $\frac{2}{6} = \frac{1}{3}$. As before, it is assumed that each ball is equally likely to be drawn and that it will remain so as experimentation continues.*

Thus there are two methods for predicting the probability of an event: (1) *a priori*, from the nature of the system itself, or (2) *empirically*, from observation of past outcomes. Both involve measurements of the relative frequencies with which an event is expected to occur. Both assume randomness, the tendency of the process to continue acting in the same uninfluenced fluctuating manner as when the measures of relative frequency were taken.

Probability when applied to predicting the expectation of future accidents takes on a slightly different aspect. This is because accident situations are not limited to a fixed number of outcomes, but rather there are an infinite number of possible variations. In subsequent sections it will be demonstrated that this infinity of possible accidents can be made discrete through categorization. Prior to this, however, the general type of events that occur *in time*, rather than in a bag, will be considered.

4.3 Estimating Probability in Time

Suppose that an accident has been recorded to occur 1,000 times in the past. Could it be expected to occur again in the near future? This question cannot be answered unless two terms are further quantified, the past and the near future. Once again the measure of the absolute frequency with which an event occurred is not as important as the relative frequency. In this case, however, frequency is with respect to a given time span.

---

*This is the general concept of randomness. The laws of probability do not hold for nonrandom processes.

To relate this to the discussion above, the concept of "trials" becomes important. In the experiment with the marbles in the bag, there were 3,000 trials; that is, marbles were drawn out of the bag 3,000 times, after which they were replaced and mixed. In the case of predicting the probability of a given accident, it is also necessary to have a concept of trials. But accidents do not lend themselves conveniently to an obviously defined trial, as in the case of the marbles-in-the-bag experiment. Hence an artificial "trial" must be created to compensate for this shortcoming.

One way to do this is to let intervals of time become the trials. Care must be taken to ensure that the event under consideration will only occur once in a given trial, or that the probability of multiple occurrences is so small that they can be considered negligible as far as the calculation of probabilities is concerned. The purpose of this requirement is to maintain the concept of Boolean events, which have no provision for multiple occurrences in a trial.

Consider, for example, a particular type of accident that has occurred in the past 10 times in 100 million man-hours. Letting 1 million man hours (mmh) be considered as a trial, the probability of occurrence in any given trial is estimated by definition as the relative frequency (e.g., $\frac{10}{100} = .1$ per mmh). On the other hand, suppose that 5 mmh were chosen for one trial. In this case the probability of occurrence in one trial could be estimated as $\frac{10}{20}$ trials $= .5$ per 5 mmh. Obviously this process of estimation will be invalid for intervals greater than 10 mmh, since a probability of 1 or greater will be produced. It will be demonstrated later than the trial time interval should be chosen such that the probabilities generated will be .1 or less.

Two ways have been discussed for establishing probability estimates, one dealing with discrete trials, the other dealing with trials in time. The next section contains a more rigorous mathematical treatment of the concepts of probability.

4.4 Mathematical Calculations of Probability

Here, as with the reduction of Boolean expressions, the human mind is limited in what it can accomplish without the use of methods or techniques. The method of qualitatively going about predicting the probability with which an event will occur was shown above. Even there, however, the result was a numerical value that quantified this prediction. The tools and techniques of mathematics greatly aid the prediction of probability to the point where valid results can be obtained that may not be intuitively obvious.

For any Boolean event A, there is a corresponding probability, designated $P(A)$, which describes the relative frequency with which A occurs. The first axiom of probability states that for any event A,

$$0 \leq P(A) \leq 1 \tag{4.1}$$

If $P(A) = 1$, event A is certain to occur, and if $P(A) = 0$, event A is certain not to occur. These two concepts are important, although they may seem to be trivial. Examples will be given later. Generally the events of concern have probabilities between zero and 1.

It was demonstrated in Chapter 3 that the total of all possible events could be represented by mapping techniques. When this method of mapping was applied, it was necessary to represent each event (represented by a letter) in all combinations with all other events. Such a map is generally representative of the concept of a *sample space* in probability terminology. The *entire sample space*, or *universal set*, is the collection of all possible subsets or combinations of the events under consideration. Hence if all events in a sample space are ORed together to form the event S, its occurrence must be certain; that is,

$$P(S) = 1 \tag{4.2}$$

This is often referred to as a second axiom of probability.

A third axiom of probability requires the understanding of the term "mutually exclusive." Any two events* are mutually exlusive if and only if their mapping to the sample space results in no overlapping. That is, the two events have no common subset. If two events A_1 and A_2 are mutually exclusive, then

$$P(A_1 + A_2) = P(A_1) + P(A_2) \tag{4.3}$$

that is, the event formed by ORing events A_1 and A_2 together has a probability equal to the sum of the probability of A_1 plus the probability of A_2. Note that this holds *only* if the two events are mutually exlusive. This is an important axiom and one that will be used extensively in the calculation of systems reliability.

Let A be any Boolean event. A working definition of $P(A)$ can be obtained by applying the definition of relative frequency. It is given by

$$P(A) = \frac{N(A)}{N(S)} \tag{4.4}$$

where $N(A)$ and $N(S)$ are the number of occurrences of events A and S, respectively, S consisting of the entire sample space. Since S must occur in any

---

*As defined in Chapter 3, an event may be represented by any legitimate Boolean combination of Boolean variables.

trial, $N(S)$ is the number of trials. Remember that in order to predict future probability, the process is assumed to continue to perform as it has in the past. Equation (4.4) can be viewed as the empirically derived estimate of the probability of event A as indicated by the $N(S)$ past trials. Only when the assumption of purely random effects holds to a reasonable degree of accuracy can this be used as an estimate of future probability.

One further restriction: $N(S)$ must have occurred a sufficiently large number of times to form a representative sample. Naturally one trial, which could only yield a probability of zero or 1, would not be sufficient. On the other hand, an infinite number of trials (which is usually impossible) would yield a perfect picture of the probability that has governed the process in the past. All probabilistic estimates stand somewhere between these limits of obvious inaccuracy and perfect estimation.

It might be judged that, since there is some inaccuracy involved in estimating probabilities, such estimation is of little value. But consider two things. First, there is inaccuracy in all measurement. This does not render them useless, since a "close enough" approximation usually serves quite well. For example, in measuring the length of a board, the nearest $\frac{1}{8}$ or $\frac{1}{16}$ inch may be close enough. Second, consider the alternative. To throw probability away because it is not perfect is similar to throwing the tape measure away because it is not perfect. To guess at the likelihood of a future occurrence without using probabilistic estimation is similar to describing the length of a board in terms of short, long, middle-sized, not too long, and so on. This bit of philosophy is included to emphasize the true meaning of inaccuracy and uncertainty. The uncertainty of future events leads us to quantify their future occurrence in terms of probabilities. Because of changes in the state of randomness and limitations in sample size, these probabilities are, of necessity, inaccurate. Further discussion of these inaccuracies will be deferred until a broader understanding of the nature of probabilities is imparted.

In Chapter 3 the concept of a binary or binomial event (also called a Boolean event) was discussed. It should be clear that if event A is in the sample space under consideration, then

$$A + \bar{A} = S \tag{4.5}$$

In fact, this was stated as a rule with respect to the truth value in Table 3.4. Now since A and $\bar{A}$ cannot overlap, applying Equation (4.3)

$$P(A + \bar{A}) = P(A) + P(\bar{A})$$

and from Equation (4.5),

$$P(A + \bar{A}) = P(S)$$

which from Equation (4.2) was equal to 1. Hence

$$P(A) + P(\bar{A}) = 1$$

and

$$P(A) = 1 - P(\bar{A}) \tag{4.6}$$

Thus, if the probability of any Boolean variable, $P(A)$, is known, the probability of its complement $P(\bar{A})$ can be easily determined by subtracting from 1.

Unions of particular events can be formed in much more complicated ways, and it is of great benefit to know how to determine the probability of the union of any two events A and B. Consider the simple function $T = A + B$. This is mapped in Figure 4.1; and the event AB was marked twice to

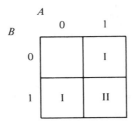

Figure 4.1. Mapping of $T = A + B$

demonstrate the contribution both from A and from B. Now consider Equation (4.3), which holds for mutually exclusive events. If A and B were mutually exclusive, then

$$P(A + B) = P(A) + P(B) \tag{4.7}$$

but, if in fact they overlap as in Figure 4.1, such an addition would add in the probability of event AB twice. A general expression may be derived for $P(A + B)$ from Equation (4.7) by making a correction for the extra addition of $P(AB)$. Hence this general expression is

$$P(A + B) = P(A) + P(B) - P(AB) \tag{4.8}$$

In other words, since $P(AB)$ was added twice, the expression may be corrected by subtracting it out once. Notice that this leads to a new concept of the meaning of the term "mutually exclusive." For Equation (4.8) only reduces to Equation (4.7) if $P(AB) = 0$. Hence A and B are mutually exclusive if and only if $P(AB) = 0$. From the meaning of AB (i.e., A and B) it should be clear that this means that the probability is zero that an event includes A and B simultaneously. And, of course, this is exactly what was originally implied in the initial definition of mutually exclusive (i.e., no over-laps in the two events).

Note that Figure 4.1 is still a correct representative of $T = A + B$ regardless of the fact that $P(AB) = 0$. This is important. Maps are used to impart the *Boolean* relationship and are in *no way affected by the probabilities of the individual events*.

But these maps are extremely useful in the determination of probabil-ities. This is due to the extension of Equation (4.3) to any number of events. This is easily done since A_2 may consist of two mutually exclusive events B_1 and B_2. Hence Equation (4.3) would become

$$P(A_1 + A_2) = P(A_1 + B_1 + B_2) = P(A_1) + P(B_1 + B_2)$$

but since B_1 and B_2 are also mutually exclusive,

$$P(A_1 + B_1 + B_2) = P(A_1) + P(B_1) + P(B_2)$$

This line of reasoning might be continued until a generalization of any n mutually exclusive events could be written as

$$P(A_1 + A_2 + \ldots + A_n) = \sum_{i=1}^{n} P(A_i) \tag{4.9}$$

Now the use of the map can be explained. As illustrated by the use of Figure 4.1, the map enables easy recognition of overlapping events. Equation (4.9) indicates that the probability of any event can be obtained by simply adding up *all* of the *mutually exclusive* events that make up that event. By taking the map and identifying the mutually exclusive events, it is easy to determine the probability of a given event. Since a Boolean expression is no more than an expansion of a Boolean event, the probability of *any* Boolean expression can be determined in the same manner.

To illustrate the procedure, consider Figure 4.1 again. Chapter 3 stated that there are many legitimate ways of writing $T = A + B$, which although not as simple, are still logically correct. For example, all of the following are logically acceptable:

$$T = A + B \tag{4.10}$$

$$T = A + \bar{A}B \tag{4.11}$$

$$T = A\bar{B} + B \tag{4.12}$$

$$T = \bar{A}B + AB + A\bar{B} \tag{4.13}$$

Now, although Equations (4.11 through 4.13) are not as simple as Equation (4.10), they do have one distinct advantage: *their terms are mutually exclusive*.

Hence it is a simple matter to sum up the probabilities of each to determine the probability of the expression. Thus, in addition to Equation (4.8), which is *always* legitimate, there are three other completely equivalent forms which will give precisely the same answer:

$$P(A + B) = P(A) + P(\bar{A}B) \tag{4.14}$$

$$= P(A\bar{B}) + P(B) \tag{4.15}$$

$$= P(\bar{A}B) + P(AB) + P(A\bar{B}) \tag{4.16}$$

respectively.

Note that the map is being used in a different way then when it was employed for the purpose of simplifications. Before, adjacent subsets of even powers of 2 were sought as large as possible; and generally overlaps were advantageous to aid simplification. Now, although the goal is still the largest adjacent subsets for combination as possible, it is essential that these combinations *do not overlap*. The only reason for desiring large combinations is to reduce computational rigor, not to improve the accuracy or validity. Hence Equations (4.14) and (4.15) are only superior to Equation (4.16) because they will result in fewer calculations to obtain the final answer. More examples of the use of maps will be given after the concept of independence is covered.

4.5 Independent and Mutually Exclusive Events

You may have noted above the assumption that $P(AB)$ could be easily obtained for any events A and B. This is not always the case, especially when the occurrence of A affects the occurrence of B, or vice versa. To illustrate how this might occur, consider two events A and B that could occur on any trial of a particular experiment. A and B are not mutually exclusive; that is, they may both occur simultaneously on any given trial. Consider 10 trials, the outcomes of which are recorded in Figure 4.2. Applying Equation (4.4), the

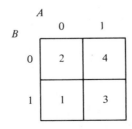

Figure 4.2. Record of ten trials of events A and B

following estimates of probability can be obtained:

$$P(A) = .7 \qquad P(\bar{A}) = .3$$
$$P(B) = .4 \qquad P(\bar{B}) = .6$$
$$P(\bar{A}\bar{B}) = .2 \qquad P(A\bar{B}) = .4$$
$$P(\bar{A}B) = .1 \qquad P(AB) = .3$$

Assume that these are direct empirical results obtained from past occurrences.

Now suppose that by some means it is known that event *B* had occurred. Or, the restriction of attention to only those events where *B* occurs might be desired. In both cases it is given (by assumption) that *B* occurs. The probability of *A* under these circumstances is denoted by $P(A|B)$ and is read "the probability of *A* given *B*." Obviously if *B* is given, the sample space is restricted, since now the probability of *B* is equal to 1. In other words, *B becomes the sample space*. Figure 4.3 depicts the limited-sample-space

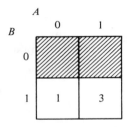

Figure 4.3. Limited sample space: given *B*

concept. The shaded area can be ignored, since interest is only in those events in which *B* has occurred. Now from Figure 4.3 and Equation (4.4), if the new sample space is the total number in *B*, the following can be estimated:

$$P(\bar{A}|B) = \tfrac{1}{4} \tag{4.17}$$
$$P(A|B) = \tfrac{3}{4} \tag{4.18}$$

To generalize this concept, an equation similar to Equation (4.4) can be written for conditional probabilities:

$$P(A|B) = \frac{N(AB)}{N(B)} \tag{4.19}$$

Again this reflects the fact that *B* is now considered to be the sample space, and therefore the only occurrence of *A* will be the mutual occurrence of *A* and *B*.

It is now possible to modify Equation (4.19) so that it is in terms of the probabilities of AB and B with respect to the sample space. This is done from Equation (4.19), as follows:

$$P(A\,|\,B) = \frac{N(AB)/N(S)}{N(B)/N(S)} = \frac{P(AB)}{P(B)} \qquad (4.20)$$

From Equation (4.20) a determination of $P(AB)$ follows by multiplying both sides of the equation by $P(B)$:

$$P(AB) = P(A\,|\,B)P(B) \qquad (4.21)$$

This equation yields a method for obtaining $P(AB)$ under any circumstances.

By definition, A is said to be independent of B if the occurrence of B in no way affects the occurrence of A and the occurrence of A in no way affects the occurrence of B. This can be stated mathematically very simply as

$$P(A\,|\,B) = P(A) \qquad (4.22)$$

and

$$P(B\,|\,A) = P(B) \qquad (4.23)$$

since the probabilities of the events are in no way influenced by the occurrence of the other. Obviously in the example given in Figure 4.2, A and B were not indicated to be independent [compare Equations (4.17) and (4.18) with the calculations for $P(A)$ and $P(B)$]. Figure 4.4 gives an example of 10 trials in which the relative frequency estimates of probability would indicate independence. Note that A and B are independent, by our definition in Equations (4.22) and (4.23), but are not mutually exclusive, since $P(AB) = .4$ and not zero.

The calculation of $P(AB)$ can be simplified when A and B are independent. Combining Equations (4.21) and (4.22), the following equation is obtained:

$$P(AB) = P(A)P(B) \qquad (4.24)$$

which only holds if A and B are independent.

By a similar argument of extension as was used to derive Equation (4.9), it can be shown that, if $A_1, A_2, \ldots, A_n$ are all independent events [i.e., they are pairwise independent as described by Equation (4.22) for any pair], then

$$P(A_1 A_2 \ldots A_n) = \prod_{i=1}^{n} P(A_i) \qquad (4.25)$$

On the other hand, if the events cannot be assumed independent, a more complicated expression is required as follows:

$$P(A_1 A_2 \ldots A_n) = P(A_1)P(A_2 \mid A_1)P(A_3 \mid A_1 A_2) \cdots$$
$$P(A_n \mid A_1 A_2 \ldots A_{n-1}) \tag{4.26}$$

The conditional probabilities given above are often quite difficult if not impossible, to determine empirically. For this reason, events are often defined in such a way as to take advantage of the simplicity of Equation (4.25).

Because of the confusion that often exists between the terms *independent* and *mutually exclusive*, an additional note is needed. It was stated above, in the process of resolving the differences between Equations (4.7) and (4.8), that if A and B were mutually exclusive, these events could not occur simultaneously or, mathematically speaking,

$$P(AB) = 0$$

Now contrast this with the definition of independence:

$$P(A \mid B) = P(A)$$

It can be shown that two occurring events, A and B, cannot be mutually exclusive and independent at the same time since that would require

$$P(AB) = P(A)P(B) = 0$$

[see Equations (4.24) and (4.26)]. The only exception would be if either or both A and B could not occur at all and therefore had a zero probability.

The fact that A and B are mutually exclusive infers a special kind of occurrence. The occurrence of A prevents the occurrence of B and vice versa. Although this is not the typical meaning of dependence, it is an important relationship and should be understood.

4.6 Reliability

Reliability will be discussed at this time for two reasons: (1) to provide examples for the use of the probabilistic concepts discussed above, and (2) to furnish the concepts of reliability which are essential to the safety engineer. These concepts have been applied rigorously by NASA and other organizations of government and industry in designing for a final product that will perform the mission for which it was designed. In the context of occupational safety and health it also has innumerable applications, although most are yet to be made.

The term "reliability" refers to the probability that a system will perform a previously defined mission. This mission must be defined in terms of quality, quantity, time, and so on, such that its performance can be evaluated. For our purposes here it will be assumed that such a definition has been specified to the extent that the occurrence of this mission can be described by a Boolean function with all of the assumptions of Boolean functions discussed in Chapter 3. Reliability is not always described by simple time-independent binary expressions. However, more sophisticated methods are beyond the scope of this text, and hence the reader interested in pursuing these methods is referred to the literature on reliability (see especially reference 3).

System reliability is the probability of system success; success being defined sufficiently for evaluation. The success of a system is highly dependent upon the success of the components. In most cases the relation between component success and system success can be determined to a sufficient degree of accuracy. This is important, for it is, comparatively speaking, very inexpensive to determine the reliability of a component as compared to a system.

To illustrate this point, the method by which the reliability of a component is determined will be demonstrated. Suppose that we have a certain component (e.g., a bearing, transistor, resistor, tire, etc.) and wish to determine its reliability. Success is defined for the specific item as the performance of a given set of specifications for a given period of time. Since reliability is probability of success, Equation (4.4) can be applied directly where event A is now the number of successes, and event S (the total sample space) is the total number that is tried. Hence the reliability of A, $R(A)$, is given by

$$R(A) = \frac{N(A)}{N(S)} = \frac{\text{number of successes}}{\text{total number tried}} \qquad (4.27)$$

Of course, in order that $R(A)$ be accurate, it is necessary that $N(S)$ be sufficiently large. In other words, a large number of each component will have to be tested in order to determine its reliability. Unless these tests are nondestructive, a great cost could result. However, most components are mass-produced, which results in a low per-unit cost. On the other hand, systems are assemblies of possibly a large number of components. These are relatively expensive mainly because of assembly costs. To construct and test a large number of complete systems would be economically infeasible. The economic savings in component testing over system testing is one of the primary motivations for the development of reliability techniques.

Given that the reliability of components can be determined experimentally, the next problem is to determine the reliability of the system. To do this, functional relationships must be established to express the system reliability as a function of the component reliabilities: henceforth called the *reliability*

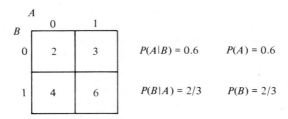

Figure 4.4. Relative frequencies of A and B indicate independence

function. For simple series and parallel redundant configurations, Boolean algebra provides a convenient tool for establishing the reliability function.

For convenience the terms "series" and "parallel" will be defined as they apply to system configurations. A pure series system is one in which every component must succeed for the system to succeed. The failure of any one component results in system failure. This is depicted by Figure 4.5 for a system of n series components. Note that Figure 4.5 is strictly the reliability relationship: it has nothing to do with the physical structure of the system (e.g., an

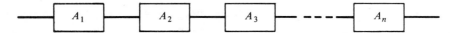

Figure 4.5. Symbolic representation of a pure series system

electronic system might have n components assembled in parallel and still be a series configuration as far as reliability is concerned). Since all components must work for the system to work, the Boolean AND functional relationship will be used to relate the components to the system; that is,

$$T = A_1 A_2 \ldots A_n \tag{4.28}$$

If these components fail independently (i.e., the failure of any one component in no way affects the failure of any other component), then Equation (4.25) can be applied and the reliability function is

$$R(T) = \prod_{i=1}^{n} R(A_i) \tag{4.29}$$

In the nonindependent case, of course, Equation (4.26) would be applied appropriately.

A pure parallel system is one in which the system will succeed if any one of the components succeeds. The success of any one of the components results in system success. Figure 4.6 depicts n components in parallel. Again, this representation is strictly for reliability purposes. The Boolean OR operator can be used to express the truth value of system success in terms of com-

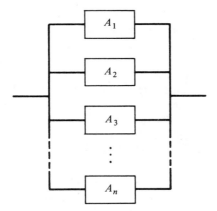

Figure 4.6. Symbolic representation of a pure parallel system

ponent success as follows:

$$T = A_1 + A_2 + \ldots + A_n \qquad (4.30)$$

Now $A_1, A_2, \ldots, A_n$ cannot be considered mutually exclusive in any but very exceptional cases, and therefore Equation (4.9) is not useful for simplification. Therefore, some mapping procedure, as explained immediately after Equation (4.9), or some other method of ensuring that the proper probabilities are calculated, is required. This will be explained below when the general solution for series–parallel configurations is presented.

Prior to this, consider another form of Equation (4.30). In order for a pure parallel system to fail, *all* its components must fail. Since $\bar{T}$ denotes system failure, and $\bar{A}_i$ denotes ith-component failure, the following relationship is obtained:

$$\bar{T} = \bar{A}_1 \bar{A}_2 \ldots \bar{A}_n \qquad (4.31)$$

Those familiar with Boolean complementation will recognize Equation (4.31) to be the complement of Equation (4.30).

Now if the probability of $\bar{T}$ can be determined, the probability of T (reliability) can also be determined, simply by subtracting from 1 [see Equation (4.6)]. Again, assuming independence, equation (4.31) can be transformed into a probability functional relationship, as follows:

$$P(\bar{T}) = P(\bar{A}_1)P(\bar{A}_2) \ldots P(\bar{A}_n) \qquad (4.32)$$

which can be expressed in terms of system and component reliabilities as follows:

$$1 - R(T) = [1 - R(A_1)][1 - R(A_2)] \ldots [1 - R(A_n)]$$

$$R(T) = 1 - \prod_{i=1}^{n} [1 - R(A_i)] \qquad (4.33)$$

The general discussion of reliability determination which follows will reveal that any reliability problem can be solved in two ways: (1) by calculating the reliability directly, or (2) by calculating the "unreliability" and subtracting from 1. The primary advantage of having an alternative approach is a savings in calculation time.

The reader at this point may wonder what the real significance of series and parallel systems is. Naturally one objective in most systems is increased reliability. There are basically two ways to increase reliability: (1) increased quality control of the components and the assembly of the system, and (2) redundancy. Briefly, the first includes advances in technology coupled with statistical sampling techniques, both of which are beyond the scope of the present discussion. The second, however, is of concern here. Given a fixed quality (reliability) of the components, the reliability of the system can still be improved by redundancy. This is of vital importance, since technology can go just so far toward the goal of perfection. And, although redundancy can never get the rest of the way, in many cases it gets as close as is necessary.

Except for systems that are designed for reliability, most systems contain no redundancy. The designer includes only those parts necessary for successful operation. If any component is removed, the system will not function properly; hence this is a series system. Later the designer may find that a particular component is causing the system to fail more than is desired. Although he has the two options for improving reliability given above, let us suppose that this critical component cannot be improved. His only recourse is to design into the system a backup, spare component that can come into action if and when the first component fails. This is a redundant component and will lead to parallelism in the configuration.

By studying Equations (4.29) and (4.33) it should be clear that the reliability of a system is always deteriorated by adding a component in series, and it is always improved by adding a component in parallel, regardless of the reliabilities of the components since they are strictly between zero and 1. Generally systems are not of the pure-series or pure-parallel type but are a mixture of the two. The general solution to these configurations is presented next.

4.7 Reliability Determination

A general approach to reliability determination has been given in terms of the two examples presented above. To generalize, the procedure consists of two steps:

1. Write a Boolean expression for the reliability configuration. This expression will relate the truth value (or success value) of the system to the success value of the components.

2. Translate this expression into an algebraic expression for the probability of the system success (system reliability) as a function of the reliabilities of the components.

The second step of the above procedure requires considerable enlargement. To do this, the foundation of the map method and the tabular method of simplification laid in Chapter 3 will be utilized. Before generalizing, consider the parallel example again, this time with four components each. For this system the Boolean expression is

$$T = A_1 + A_2 + A_3 + A_4$$

and step 1 is completed. This expression can be mapped as given by Figure 4.7a. Note that a mark has been placed for each covering. To determine the probability of event T, there are two alternatives. First, combinations of the covered events may be formed for simplification, and the probabilities of the events may be added provided that there is no overlapping in the combinations. This may be done in a large number of acceptable ways; one is presented in Figure 4.7b. Note that the number of times that a block was covered in Figure 4.7a is irrelevant. The new equivalent Boolean expression produced is

$$T = A_1 + \bar{A}_1 A_2 + \bar{A}_1 \bar{A}_2 A_3 + \bar{A}_1 \bar{A}_2 \bar{A}_3 A_4$$

Now this expression is certainly not in its simplest form, nor is it intended to be.* What is important is that each term be mutually exclusive of every other term. This is obvious from the map in Figure 4.7b; there are no overlaps. This can be translated into a probability function, or in this case a reliability function, as follows:

$$R(T) = R(A_1) + [1 - R(A_1)]R(A_2) + [1 - R(A_1)][1 - R(A_2)]R(A_3)$$
$$+ [1 - R(A_1)][1 - R(A_2)][1 - R(A_3)]R(A_4) \qquad (4.34)$$

This is an acceptable reliability function since now the success of the system can be determined by the success of the components. The expression is not unique since it depends upon judgment as far as making the combinations in Figure 4.7. However, the answer is unique: any legitimate *nonoverlapping* combination of events in Figure 4.7a will yield a Boolean expression that can be converted directly to a legitimate reliability function for the system. A wide variety of acceptable expressions can be written, but all will yield precisely the same reliability.

---

*In this case, $T = A_1 + A_2 + A_3 + A_4$ is the most simplified form of the expression.

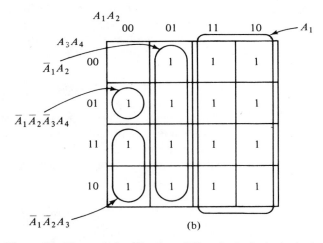

Figure 4.7. Map and simplification of $T = A_1 + A_2 + A_3 + A_4$

Before considering an alternative method of working this problem, the procedure will be generalized as follows:

1. From the system, write a Boolean expression that relates system success to component success.
2. Transfer this Boolean expression to a map.
3. Make combinations of all covered spaces such that there are no overlaps in combinations (i.e., all resulting events are mutually exclusive).
4. Since the resulting events of step 3 are mutually exclusive, apply Equation (4.9) to write the reliability function.

There are alternative methods of writing the reliability function from the maps, such as subtracting out overlapping areas and the like. However, the above is recommended for its simplicity of approach, even though it will not guarantee a minimum of calculational effort. Before leaving this example, an attempt should be made to explain the seeming discrepancy between Equation (4.33) and Equation (4.34), which, by the way, is not the result of choosing a different combination than Figure 4.7. Rather, the relationship expressed by Equation (4.6) was used. Figure 4.8 shows how this relationship can be

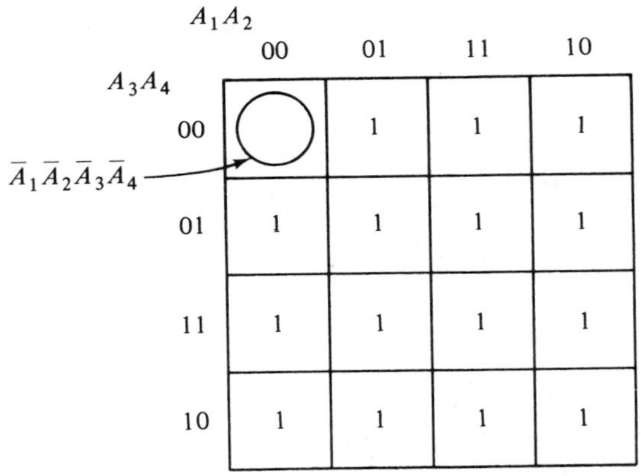

Figure 4.8. Graphical complement of Figure 4.7

utilized graphically. By writing an expression for those spaces which are *not* covered, the failure function ($\bar{T}$) can be obtained. This, in turn, when translated into probabilistic terms yields the unreliability function. This was explained independently of the map in connection with Equations (4.31)–(4.33).

4.8 Example Configurations

Some sample configurations will now be presented to illustrate the four-step procedure given above. As mentioned previously, configurations are generally not of a pure series or a pure parallel nature but are a mixture. Consider, for example, the configuration in Figure 4.9, which depicts two series subsystems AB and CD in parallel. This configuration will be solved in Example 4.1.

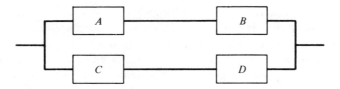

Figure 4.9. Two series subsystems in parallel

Example 4.1 Four-Stage Reliability Problem

Given a configuration as in Figure 4.9 with component reliabilities of $R(A) = .9$, $R(B) = .8$, $R(C) = .7$, and $R(D) = .6$, determine the reliability of the configuration.

Solution:

Step 1.   Write the success function for the configuration. This can be done by enumerating the paths to success. Either A and B are both working or C and D are both working; therefore,

$$T = AB + CD$$

Step 2.   Map the success function. This is done in Figure 4.10a.

Step 3.   Rearrange and combine the success function into mutually exclusive events. This is done in Figure 4.10b, resulting in the new success function

$$T = AB + \bar{A}CD + A\bar{B}CD$$

Step 4.   Apply Equation (4.9) to determine the reliability function for the system:

$$R(T) = R(AB) + R(\bar{A}CD) + R(A\bar{B}CD)$$

At this point Equation (4.25) or (4.26) could be applied term by term depending on whether the components failed independently or dependently, respectively. Assume independence and further adopt the following terminology:

$$p_X = R(X)$$

and

$$q_X = 1 - R(X)$$

Hence the reliability function can be written

$$R(T) = p_a p_b + q_a p_c p_d + p_a q_b p_c p_d$$

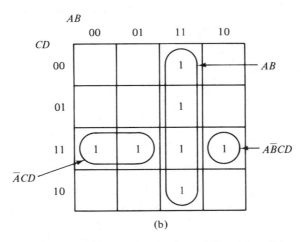

Figure 4.10. Map and combinations of $T = AB + CD$

Now the values of the component reliabilities and unreliabilities can be substituted into this equation to obtain

$$R(T) = (.9)(.8) + (.1)(.7)(.6) + (.9)(.2)(.7)(.6)$$
$$= .8376$$

Example 4.2 Application to Aircraft Reliability

Consider a four-engine aircraft that will fly as long as any two "balanced" engines are working. If the reliability of each engine is .9, what is the reliability of the aircraft?

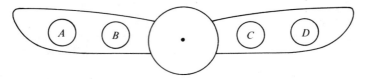

Figure 4.11. Four engine nomenclature

Step 1.   From Figure 4.11 and the statement of the problem, a Boolean success function can be written. This will include all engines working, any three engines working, and the balanced pairs AD and BC. AB, AC, BD, and CD are unbalanced. Hence the success function becomes

$$T = ABCD + ABC + ABD + ACD + BCD + AD + BC$$

Although this function is obviously redundant, this will have no effect on the result as long as the function is logically correct.

Step 2.   Figure 4.12 demonstrates the coverings of the map for the success function above. Note the symmetry and the potential simplification of the function.

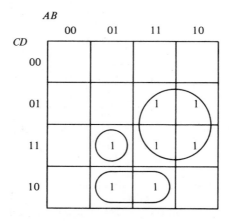

Figure 4.12. Map for Example 4.2

Step 3.   One set of mutually exclusive combinations is shown in Figure 4.12. Remember that any legitimate set of mutually exclusive combinations will render the correct answer. For combinations chosen, the new success function is

$$T = AD + BC\bar{D} + \bar{A}BCD$$

Step 4.   Assuming independence, since the terms of the above function are mutually exclusive, the reliability function may be written directly as

$$R(T) = p_a p_d + p_b p_c q_d + q_a p_b p_c p_d$$
$$= (.9)(.9) + (.9)(.9)(.1) + (.1)(.9)(.9)(.9)$$
$$= .9639$$

Note that the "equivalent" configuration in terms of series–parallel systems would be two series subsystems in parallel with the success function, $T = AD + BC$. This was not obvious from the word statement.

Example 4.3 Comparison of Configurations

It is often remarked that the configuration with the fewest components is the most reliable. This statement, of course, is only strictly true for a pure series system. In a mixed system it is sometimes difficult to determine the comparative reliability without a thorough analysis. Consider, for example, the two configurations depicted in Figure 4.13. Suppose that they both perform the same function and that $R(A) = .9$, $R(B) = .8$, $R(C) = .7$, and $R(D) = .6$. Which of the two is the superior configuration in terms of reliability?

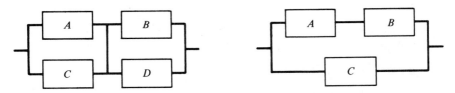

Figure 4.13. Two configurations compared in Example 4.3

Step 1.   The success functions for the two configurations are (respectively):

$$T_1 = AB + AD + CB + CD$$
$$T_2 = AB + C$$

Step 2.   The maps for these success functions are presented in Figure 4.14.

Step 3.   From Figure 4.14, the two new success functions are

$$T_1 = AB + A\bar{B}D + \bar{A}BC + \bar{A}\bar{B}CD$$
$$T_2 = C + AB\bar{C}$$

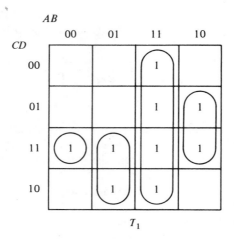

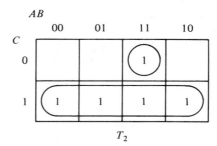

Figure 4.14. Maps for Example 4.3

Step 4. The corresponding reliability functions therefore are

$$R(T_1) = p_a p_b + p_a q_b p_d + q_a p_b p_c + q_a q_b p_c p_d$$
$$= (.9)(.8) + (.9)(.2)(.6) + (.1)(.8)(.7) + (.1)(.2)(.7)(.6)$$
$$= .8924$$

$$R(T_2) = p_c + p_a p_b q_c$$
$$= .7 + (.9)(.8)(.3)$$
$$= .916$$

Although the two configurations are very close, the second is more reliable. Note, however, that if the values of the probabilities were reversed (i.e., $p_a = .6$, $p_b = .7$, $p_c = .8$, and $p_d = .9$), then the outcome would be in favor of the first configuration. Obviously, no generalized conclusion should be made without a thorough analysis of the configurations of concern.

4.9 Application to Safety

The above examples demonstrate the use of probability theory as it applies to reliability. The relationship between reliability and safety is obvious since generally a more reliable system will be safer. As with other systems design objectives, however, there are limitations on the amount of money that can be spent to obtain a high degree of reliability. Example 4.3 presented a comparison between two configurations. In reality, this type of comparison would be performed for all alternatives proposed, and a configuration would be chosen based on economic as well as reliability factors.

To help crystallize the application of this theory to safety, consider the following generalizations:

1. Any two or more operations, events, or devices that perform the same (safety) function within a given system may be considered redundant (or parallel) operations, events, or devices.
2. Any two or more operations, events, or devices, all of which are required to prevent a given situation or accident, should be considered as series operations, events, or devices.
3. Generally, to increase the safety of a system, redundancy should be added at those points which contribute greatest to the unreliability of the system. Again, however, cost-effectiveness considerations should be included (see Chapter 5).

Although the quantification of component reliability is often difficult in safety applications, the realization of the above three generalizations is quite simple. A guard on a power saw may eliminate 95 % of potential accidents. Safety instruction and motivation may eliminate 98 % of potential accidents. In a sense these are parallel efforts on the same problem. Hence, if *independent*, the resultant probability of an accident in a given time frame could be calculated by parallel computations to be .001. The reliability would be .999 if defined purely in terms of safety.

Similarly, if a fire could start either from smoking carelessness or static electricity discharge, the prevention of these two events could be considered in series. It is important to distinguish between reliability and unreliability at this point. If the reliability of the system is dependent upon the nonoccurrence of a fire, the reliability of the components would be the probability that each prevented the fire. For example, if the probability of a fire *not* starting in a given time period by smoking carelessness and static electricity discharge were .99 and .98, respectively, the reliability of the system would be .9702. This assumes no other contributing factors and a stated time period.

The relationship between configuration reliability and system safety will be illustrated more completely in Chapter 5. When the safety of a system is dependent upon a mechanical device, the concepts of reliability should be thoroughly understood and applied. The results will provide an important input into subsequent analyses.

4.10 Time-Dependent Reliability; Maintenance

The definition of reliability assumed a well-defined mission in terms of quality, quantity, time, and so on, such that its performance could be evaluated. This section considers the element of time and discusses maintenance as an alternative to redundancy in improving reliability and safety.

In the discussion above, the reliability of a component was stated in terms of one probability of success. Hence the system reliability, for the time period under consideration, could be calculated. Very few systems, however, are used for one mission and then scrapped. Generally a system continues to be used until it becomes obsolete or until the cost of its maintenance exceeds the cost of its replacement (on a present-worth basis). Even if its service life is known, its reliability will fluctuate during its lifetime, owing to maintenance and wear. Thus a fixed time period of 1 month for a system might yield a reliability of .99 when new, but it might drop to .95 after 6 months of operation.

Two time measurements are being discussed here simultaneously and they should not be confused. The first is the fixed mission duration. This may be a definite mission time, as in the case of a lunar excursion mission. It may also be arbitrarily fixed, as in the case of most industrial systems, where reliability calculations are used for safety and maintenance evaluations. The second time measurement is of the physical age of the system. Because of maintenance the physical ages of all components may not be the same. Thus the reliability will depend upon the mission duration and the actual service age of each of the components.

For the present discussion, assume that the mission duration is fixed either by mission constraints or by other considerations. The variable of concern, then, is the time that each of the components have been in service. This, of course, can be obtained directly from the maintenance history of the system. From this, the reliability of the system for its next mission can be calculated.

To do this it is necessary to have entire component-life tests rather than component reliabilities as given by Equation (4.27). In other words, rather than $R(A)$, it is necessary to obtain $R_i(A)$, the reliability of component A at age i. Here age i will be measured in discrete mission durations (i.e., $i = 0$ is new, $i = 1$ is after 1 mission, $i = 2$ is after 2 missions, etc.). Table 4.2 gives examples of three possible time-dependent component reliabilities. The number in the body of the table is the probability of component success given that it has survived mission duration *i*.

The comparative descriptions at the bottom of Table 4.2 demonstrate the value of each component if it were to be used over several time periods. Note that the superiority of component *A* over *B* (and vice versa) depends upon the time period involved. Component *C* is superior by any measurement.

Table 4.2 EXAMPLE OF TIME-DEPENDENT RELIABILITIES

i	$R_i(A)$	$R_i(B)$	$R_i(C)$
0	.99	.90	.99
1	.95	.88	.98
2	.90	.86	.97
3	.84	.84	.96
4	.76	.80	.95
5	.65	.70	.93
6	.50	.60	.90
7	.35	.40	.85
8	.10	.30	.80
9	.05	.25	.75
Average	.609	.653	.911
Description	High initial, fast deterioration	Lower initial, lower rate of deterioration	High initial, low deterioration

Two ways have been discussed for improving system reliability. The first was the improvement of the quality of the component. This could be accomplished, by replacing component *B* (in Table 4.2) with component *C*, assuming that they are interchangeable in the system. The second method discussed, redundancy, provided backup component(s) to take the place of those which failed. A third method, maintenance, is now introduced. While optimum maintenance policies are extremely difficult to obtain, they are not particularly difficult to evaluate. The understanding of their evaluation will enable operational maintenance to be performed on a logical basis.

First, recognize what it means to maintain a system. For present purposes, assume that all maintenance involves the replacement of components, or its very close analogy. For example, oiling could be considered as replacing the component oil. Thus maintenance is merely the restoration of a component from a reliability $R_i(\cdot)$ to $R_j(\cdot)$, where $i < j$; generally *j* is 0, being a new component.

The following stepwise procedure will be used in evaluating maintenance policies:

1. Formulate a maintenance policy specifying the value of *i* at which each component is to be replaced.

2. Calculate the system reliability for the first time period and record it.
3. Make any changes in component reliabilities according to the policy. Calculate the reliability for the next time period.
4. Repeat step 3 until a cycling of reliabilities occurs, yielding the steady-state reliability of the system.
5. Repeat steps 2–4 for all alternative policies that may be given consideration.
6. Select the alternative policy that cost and reliability indicate to be optimal.

The shortcomings of this stepwise approach will be discussed prior to presenting an example. First, the procedure admitted no failures, which obviously there will be. Generally, when a component fails, it is immediately replaced, whether or not the policy calls for it. Thus the reliability of the system will be *higher* than that calculated above. The procedure above gives the reliability given that components are replaced before failure. If the probabilities of failure are low, this will be close to the actual reliability. More importantly, however, is the comparable value of the results. Since this assumption is made in all evaluations, all will be similarly affected, thus tending to cancel the effect of this oversimplification. However, if this is thought to be too large a shortcoming, methods of simulation discussed in Chapter 6 can be applied.

A second problem associated with this method, as well as more accurate simulation techniques, is the lack of a guaranteed optimal solution. However, this is only serious in very large systems where the sheer number of calculations prevents the finding of a good, if not optimal, solution. And, even the most sophisticated methods break down when confronted with very large systems. Fortunately, this method lends itself readily to the computerized techniques discussed in Section 4.11.

Example 4.4 Comparison of Maintenance Policies

Consider the example three-component configuration given in Figure 4.13. It had a success function of $T = C + AB\bar{C}$ when the mutually exclusive terms were identified by the map method in Example 4.3. Now suppose that the time-dependent reliabilities of Table 4.2 are in effect. The procedure will now be followed as outlined in the text.

Step 1. Policy I will be the replacement of all components after each time period.

Steps 2–4 (Policy I). $R(T) = R(C) + R(A)R(B)R(\bar{C})$, assuming independence. Hence $R(T) = .99891$. Since this will be the same for all time periods, this is the steady-state solution.

Step 5 (Policy II). A second policy will be evaluated which will consider replacing all components after three time periods. The following results are obtained:

i	$R(T)$
0, 3, . . .	.99891
1, 4, . . .	.99672
2, 5, . . .	.99322

Step 5 (Policy III). Suppose that policy I is considered to be too expensive and that policy II falls below acceptable reliability limits. Another policy is proposed: Assuming component 3 to be the most expensive, replace components 1 and 2 in every second time period, and component 3 in every third time period. The following results are obtained:

i	$R(T)$
0, 6, . . .	.99891
1, 7, . . .	.99672
2, 8, . . .	.99673
3, 9, . . .	.99871
4, 10, . . .	.99782
5, 11, . . .	.99508

Step 5 (Policy IV). According to the criteria of cost and reliability described above, policy III is probably optimum. However, to demonstrate the evaluation of another policy, and to provide a comparative example with the method of simulation discussed in Chapter 6, one additional policy will be evaluated. This will consist of the replacement of the entire system every five time periods. By the method given above, the following results are obtained:

i	$R(T)$
0, 5, . . .	.99891
1, 6, . . .	.99672
2, 7, . . .	.99322
3, 8, . . .	.988224
4, 9, . . .	.9804

It is obvious here that if replacements were made immediately after failure, the results would be different. When this becomes a factor, simulation methods should be used (see Example 6.13).

Step 6. Other policies could be evaluated. Judgment must be applied to balance the factors of cost and reliability. If .995 were the lowest acceptable reliability, policy 3 would satisfy this quite well at minimum cost.

In concluding this section on maintenance policies, some reflection should be given to the old cliche: "A well-maintained shop is a safe shop." The above methodology provides a means for evaluating the value of maintenance to safety. It is obvious that preventative maintenance, the task of replacing components before they break down, can greatly increase reliability. In turn, both productivity and safety should be favorably affected.

Once again, however, it is necessary to recognize that preventative maintenance must be evaluated in terms of cost/benefit, as was the case with component redundancy and component quality improvement. These, in turn, must be evaluated simultaneously in terms of other competing safety countermeasures. While models can be structured to evaluate component quality, redundancy, and maintenance policies simultaneously, the integration of this into the larger problem is quite complex. This is where systems engineering is required. The combination of these basic building blocks of reliability modeling, combined with sound experience, should go a long way toward greater total system safety.

4.11 Computerized Algorithm

We have given simple comparisons of alternative configurations and policies. Often large numbers of configurations must be compared, where each configuration has a large number of alternative components, each with a different reliability. In addition, the number of components in a given configuration is often greater than five or six, thus making the map method quite cumbersome. It is possible, in cases where independent subsystems may be formed as in Figure 4.15, to solve the subsystems and then to combine subsystems to solve the system. However, this still might require a great amount of computational effort.

With the use of an algorithm similar to that given in Chapter 3 for simplifying Boolean expressions, it is possible to program the computer to perform most of this work. In fact, the only input required is the Boolean success function and the reliabilities of the components. This is possible since again the problem lends itself to the use of binary numbers. The background reasons for using binary numbers were given in Chapter 3 and will not be repeated here.

The tabular method will be presented in five steps. Specific reference will be made to Table 4.3 in conjunction with the example given in Figure 4.16.

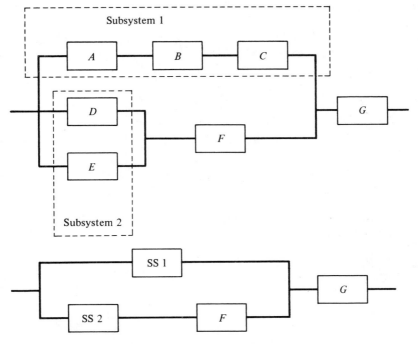

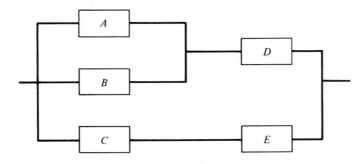

Figure 4.15. Reliability systems synthesis. Systems may be simplified by combining those components which are uniquely in series or uniquely in parallel with each other. Here the reliabilities of A, B, and C are combined to determine the reliability of subsystem 1. Similarly, D and E are combined to determine the reliability of subsystem 2. These are now used to determine the reliability of the system in the usual manner.

$T = AD + BD + CE$
Binary: 10010, 01010, 00101

Figure 4.16. Example configuration and success function

143

Table 4.3 Example of Computational Method for Generation of All Mutually Exclusive Terms

$$T = AD + BD + CE$$

Binary:	10010,	01010,	00101
Base 10:	18,	10,	5

	Binary Numbers	Transmission Function	Added Terms From: 5	10	18	Generated Transmission Function	Unsimplified Reliability Function
0	00000						
1	00001						
2	00010						
3	00011						$R =$
4	00100						
5	00101	5				$5 = \bar{A}\bar{B}C\bar{D}E$	$q_a q_b P_c q_d P_e$
6	00110						
7	00111		7			$7 = \bar{A}\bar{B}CDE$	$+q_a q_b P_c P_d P_e$
8	01000						
9	01001						
10	01010	10				$10 = \bar{A}B\bar{C}D\bar{E}$	$+q_a P_b q_c P_d q_e$
11	01011			11		$11 = \bar{A}B\bar{C}DE$	$+q_a P_b q_c P_d P_e$
12	01100						
13	01101		13			$13 = \bar{A}BC\bar{D}E$	$+q_a P_b P_c q_d P_e$
14	01110			14		$14 = \bar{A}BCD\bar{E}$	$+q_a P_b P_c P_d q_e$
15	01111		15	15		$15 = \bar{A}BCDE$	$+q_a P_b P_c P_d P_e$
16	10000						
17	10001						
18	10010	18				$18 = A\bar{B}\bar{C}D\bar{E}$	$+P_a q_b q_c P_d q_e$
19	10011				19	$19 = A\bar{B}\bar{C}DE$	$+P_a q_b q_c P_d P_e$
20	10100						
21	10101		21			$21 = A\bar{B}C\bar{D}E$	$+P_a q_b P_c q_d P_e$
22	10110			22		$22 = A\bar{B}CD\bar{E}$	$+P_a q_b P_c P_d q_e$
23	10111		23		23	$23 = A\bar{B}CDE$	$+P_a q_b P_c P_d P_e$
24	11000						
25	11001						
26	11010			26	26	$26 = AB\bar{C}D\bar{E}$	$+P_a P_b q_c P_d q_e$
27	11011			27	27	$27 = AB\bar{C}DE$	$+P_a P_b q_c P_d P_e$
28	11100						
29	11101		29			$29 = ABC\bar{D}E$	$+P_a P_b P_c q_d P_e$
30	11110			30	30	$30 = ABCD\bar{E}$	$+P_a P_b P_c P_d q_e$
31	11111		31	31	31	$31 = ABCDE$	$+P_a P_b P_c P_d P_e$

Step 1. Establish the success function in terms of Boolean algebra variables, as usual. As before, there should be no concern if the success function is not in minimum form, since any logically correct function will yield the correct answer.

Step 2. Assign to each variable in step 1 a successive power of 2 (starting with 0). Then transform the transmission function into a set of binary numbers, term by term, in the following way: if the variable appears in the term assign its position a 1; if not, assign its position a 0. For example, let

$T = AD + BD + CE$. Let E be 2^0, D be 2^1, C be 2^2, B be 2^3, and A be 2^4. Since there are five variables, there will be five positions in the binary representation. So AD becomes 10010, BD becomes 01010, and CE becomes 00101. Note that in reliability work, all variables appear in their affirmative mode, and hence there is no need for the use of the asterisk (*) as was used in the tabular method for simplification.

Step 3.   Take each transformed term and logically compare it with each succeeding binary number up to 2^{n-1}, where n is the number of variables. This logical comparison consists of checking the positions of the succeeding binary numbers that contain a 1 in the success-function terms. If in a succeeding binary number all these positions also contain 1s, then that number will be added to the success function. Once a number becomes a part of the success function, it will remain there and it can be ignored in other comparisons. It should not be added twice.

For example, consider the term AD, which was transformed to 10010. All the successive binary numbers are: 10011, 10100, 10101, 10110, 10111, 11000, 11001, 11010, 11011, 11100, 11101, 11110, 11111. Those numbers underlined above become part of the success function. Similarly, for the example of Table 4.3, BD (01010) and CE (00101) will also be used to generate additional terms. The next-to-last column in Table 4.3 contains all the generated terms.

Step 4.   Reconstruct the new success function using the terms generated by step 3. Each binary number that was added to the original set will form a new term in the success function. Owing to the nature of ordered binary numbers, all the mutually exclusive events that will result in system success have been generated if the binary representation corresponds to the truth value of the term (e.g., 00101 becomes $\bar{A}\bar{B}C\bar{D}E$, 00111 becomes $\bar{A}\bar{B}CDE$, as shown in Table 4.3).

Step 5.   Write the reliability function directly from the "complete" success function. Since all mutually exclusive events that lead to system success have been generated, it is a simple matter to add these probabilities to determine the reliability. If the events are statistically independent, the procedure used in the last column of Table 4.3 can be followed. Note that $q_i = 1 - p_i$ for all i.

The five-step solution presented has one major advantage over other methods—it is easily programmed on a digital computer. Input consists of the transmission function and the element probabilities of success. The computer easily generates the system reliability. The procedure is not without its disadvantages, however. One primary deficiency is noted when the number of elements becomes very large. On the average, the number of binary-number comparisons doubles with each additional component. A 10-element system would require 1,024 binary numbers to be generated and evaluated by the computer, whereas an 11-element system would require 2,048, and a 15-element system 32,768. Natually the problem gets very serious for systems of 20 to 30 elements. Yet, many design decisions might require a multitude of such configurations to be rapidly evaluated without decomposing the configuration into manageable subsystems. Therefore, a more efficient solution algorithm is presented in the next section.

4.11.1 Solution Algorithm

Once the problem is structured by the framework given above, it can be taken out of the area of reliability and restated. The problem is to generate all binary numbers from n_i to 2^N such that the unit digits in n_i match the unit digits in the generated number; all other digits are irrelevant. [Here N is the number of components and n_i the numerical representation for the ith term of the success function (see Table 4.3, column 3).] If this could be done quickly for all i, the problem would be solved, since, given this set of binary numbers, the reliability can be determined. An easy way to do this for small systems is actually to compare n_i digit by digit to each number between n_i and N, as in Table 4.3. However, for large systems this is computationally inefficient.

It is possible to generate the required set of numbers directly by again taking advantage of the nature of binary numbers. The following step-by-step procedure is recommended.

Step 1.   Follow steps 1 and 2 in the procedure given above to generate a binary representation for each term.

Step 2.   Determine the number of 0s in the first term. Set the variable z equal to this number.

Step 3.   Fix the positions of the 1s in the binary representation.

Step 4.   Generate all binary numbers from 0 to $2^z - 1$. Replace the unfixed positions of step 3 with the digits of each of these binary numbers, respectively, to generate the new terms in the transformed transmission function.

Step 5.   Repeat steps 2–4 for all terms. If one binary representation is generated more than once, consider it only once.

Table 4.4 demonstrates part of the example presented in Table 4.3. In this example the additional terms for the *CE* (i.e., 5, 00101) terms are generated. In the first column, all the "fixed" digits appear with the 0s blanked out. The "unfixed" digits are then replaced by the second column, the set of ordered binary numbers from 0 to $2^z - 1$, to form the new generated terms. This procedure has the advantage of being readily adaptable to computer programming as well as being computationally efficient.

Table 4.4 EXAMPLE OF DIRECT GENERATION OF ADDITIONAL TERMS IN THE TRANSMISSION FUNCTION FOR TERM *CE* OF TABLE 4.2

Term CE: 00101, 5		*Binary Generator:* $i = 0, 1, 2, \ldots, 7\ (2^z - 1)$; $z = $ *Three Positions*	*Generated Term*	*Base* 10
1 1	&	000	00101	5
1 1	&	001	00111	7
1 1	&	010	01101	13
1 1	&	011	01111	15
1 1	&	100	10101	21
1 1	&	101	10111	23
1 1	&	110	11101	29
1 1	&	111	11111	31

The procedure presented above has been programmed on a digital computer. The program, documented in Appendix 2, has been specifically designed for fault-tree analysis to be discussed in Chapter 5. However, its application to reliability is very straightforward. Input consists of the success function and the element reliabilities, and the output is the system reliability. With this rapid means of determining reliability, many alternative configurations can be compared by merely inputting various success functions. Maintenance policies can be evaluated by varying the input probabilities.

Example 4.5 Example of the Tabular Method

Perform Example 4.1 using the tabular method.

Step 1.   From Example 4.1:

$$T = AB + CD$$

which translated to the binary representation is

$$1100, 0011$$

According to the notation described above:

$$N = 4; \qquad n_1 = 12; \qquad n_2 = 3$$

For the first term, 1100:

Step 2.   The number of zeros is 2. Thus $z = 2$.

Step 3.   The first two digits will remain fixed.

Step 4. All binary numbers from 0 to $2^2 - 1$ are 00, 01, 10, and 11. Replacing the zeros in the term yields 1100, 1101, 1110, and 1111. For the second term, 0011:

Step 2. $z = 2$.

Step 3. The first two digits remain fixed.

Step 4. Again the binary numbers from 0 to 3 are 00, 01, 10, and 11. Thus the generated terms are 0011, 0111, 1011, and 1111.

Step 5. Steps 2, 3, and 4 are complete for all terms. Looking at all generated terms yields the following (order is irrelevant): 1100, 1101, 1110, 1111 (twice), 0011, 0111, and 1011. Hence the reliability function can be written:

$$R = p_a p_b q_c q_d + p_a p_b q_c p_d + p_a p_b p_c q_d + p_a p_b p_c p_d$$
$$+ q_a q_b p_c p_d + q_a p_b p_c p_d + p_a q_b p_c p_d$$
$$= .8376$$

Note that the 1111 term should be included only *once* in the final term. Obviously the above reliability function is extremely inefficient for hand calculations, even though it yields the same answer as that obtained in Example 4.1. The astute reader will recognize that the same results could have been obtained by the method employed in Example 4.1 if each individual space on the map were considered to be the combination of mutually exclusive events chosen. Fortunately, the entire procedure given above can be computerized; lengthy, error-prone computations are unnecessary.

4.12 Closure

This chapter has presented an overview of some of the basic laws of probability. These were then exemplified by applying them to problems of simple discrete reliability computations. Together this material forms a basic foundation for the techniques of fault-tree analysis that are presented in Chapter 5.

SELECTED REFERENCES

1. BROWN, D. B.; "A Computerized Algorithm for Determining the Reliability of Redundant Configurations,– *IEEE Trans. Reliability*, Vol. R-20, No. 3, 1971.

2. MILLER, I. and FREUND, J. E.; *Probability and Statistics for Engineers*, Prentice-Hall, Inc., Englewood Cliffs, N.J., 1965.

3. SHOOMAN, M. L.; *Probabilistic Reliability: An Engineering Approach*, McGraw-Hill Book Company, New York, 1968.

QUESTIONS AND PROBLEMS

1. Give three examples of events that (a) are and (b) are not mutually exclusive. Do the same for events that are and are not independent.

2. An urn contains 100 balls. Thirty are black and 70 are white. Half of the black balls have a red stripe. The probability of selecting a white ball with a stripe out of the urn is .1. Assuming random selection, what is the probability of
   (a) Selecting a white ball?
   (b) Selecting a white ball without a stripe?
   (c) Selecting a white ball with a stripe given that a white ball is selected?
   (d) Selecting a black ball with a stripe?

3. In the function $T = AB + ACD + BCD$, the probabilities of the events are $P(A) = .8$, $P(B) = .9$, $P(C) = .7$, and $P(D) = .6$. Assume independence and determine the probability of T.

4. Determine the reliability of a three-engine aircraft that can fly on any two engines given that the reliability of each engine is .9. Determine the change if the plane can also fly with only the center engine working.

5. A configuration will operate successfully if both components A and B work or if both components C and D work. In either event, however, component E must also be working. If the reliabilities of all components are .6, what is the reliability of the configuration?

6. A hazardous fire area is subject to a high concentration of fumes at any given time, with a probability of .6. Three things can ignite a fire: (1) a person smoking, (2) static electricity, and (3) sparks caused by maintenance. The probabilities of these three events are .001, .01, and .05, respectively. If there is a fire, the probability of it getting out of hand and burning the plant down is .1. What is the probability that the plant will be lost due to fire?

7. Two machines, comparable in price, are being considered from a safety point of view. The one utilizes three operators, each of which would have a .01 probability of removing its guard (in a given time period). If the guard is removed, any of the operators would have a .005 probability of getting injured, except that the one who removed it who would have a .001 chance. On the other hand, the second machine has only one operator. Although the guard is not removable, there is a safety hazard involved in maintenance. The probability of maintenance being required is .02, and, if it is required, the probability of an injury is .001. Which machine is better from the safety point of view?

8. Determine $P(T)$ for
   (a) $T = AB + AC + DE$
   (b) $T = A + ABC + DE$
   (c) $T = ABD + BC + E$
   (d) $T = A + B + CDE$
   (e) $T = ABC + BCD + CDE$

Use the probabilities

$$P(A) = .6 \qquad P(D) = .9$$
$$P(B) = .7 \qquad P(E) = .1$$
$$P(C) = .8$$

and assume independence.

9. Solve Problem 8 by use of "unreliability" [i.e., $1 - P(T)$], and check the answers with those determined in problem 8.

10. Solve Problem 8 by use of the tabular method. At the appropriate step, check the solution with the map method used in Problem 8.

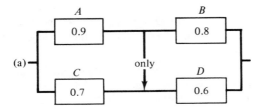

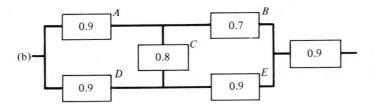

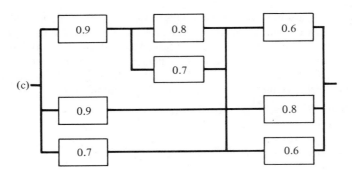

Figure P4.12.

11. Determine the reliability of a four-engine aircraft that will fly if any two engines operate. The reliability of each engine is .9.

12. Determine the reliability of the networks shown in Fig. P4.12.

13. The configuration in Problem 12(a) has the following time-dependent reliabilities for the components:

i	$R_i(A)$	$R_i(B)$	$R_i(C)$	$R_i(D)$
0	.90	.80	.70	.60
1	.85	.78	.69	.60
2	.80	.75	.68	.60
3	.75	.73	.67	.50
4	.65	.70	.65	.40
5	.50	.60	.66	.30

Test the following maintenance policies:
(a) Replace all components every two time periods.
(b) Replace A every time period, the rest every third time period.
(c) Replace A and B every two time periods, C and D every three time periods.
     What is the minimum cost alternative if the minimum acceptable reliability is .85? If it is .83? Assume equal cost among components.

FAULT-TREE ANALYSIS

5.1 Introduction

The application of probability theory to the determination of system reliability was demonstrated in Chapter 4. These concepts will now be generalized to include entire systems of men, machines, and materials. In place of reliability, the probability of an accident will be considered. This could be viewed as the "unreliability" of the system. However, since reliability is generally used to denote the probability of success, and since success can be structured to include or exclude many factors irrelevant to personnel or even system safety, we prefer to keep reliability and safety in two separate contexts. Nevertheless, the formulation and solution of these problems is basically the same.

In this chapter study will be centered on the evaluation of event probabilities in terms of cost/benefit. The objective will be to make decisions in an optimal manner such that the chance of accidents can be further reduced. Note that this is similar to the reliability objective, which was to increase the probability of system success. Instead of a schematic flow diagram, another type of diagram, called a *fault tree*, is used to aid this analysis.

Fault-tree analysis is a technique frequently discussed in safety literature

5

(e.g., see references 1 and 3). It is a logical approach to identify the areas in a system that are most critical to safe operation. According to Recht (3), fault-tree analysis was developed in 1962 by H. A. Watson of Bell Telephone Laboratories. Although it has been used extensively in the design of spacecraft by NASA (2), its application to occupational safety and health has been limited.

The implementation of more sophisticated methods, such as fault-tree analysis, has been gradual at best. In the past, many inexpensive efforts, such as safety campaigns, slogans, goals, and so on, have been used to attack the problem of industrial safety. When measured in terms of cost-effectiveness, the low cost of these activities probably renders them a favorable investment. The same is true of inexpensive equipment such as hard hats and fire extinguishers. No attempt should be made to deemphasize the effectiveness of these items. Also, in most cases sophisticated analyses are not required to determine that such investments should be made.

However, when the obvious expenditures are made and the accident level is still too high, a new level of study is required. A vast range of devices, protective equipment, motivational programs, and so on, are available, with a limited amount of funds available for their procurement. A most important decision must be made to allocate the total budget to those expenditures such that the greatest benefit will be obtained.

This chapter will present a working guide to the use of fault-tree analysis for the purposes of cost/benefit determination. The mere construction of fault trees for a system serves to reveal critical, although often hidden, faults of the system. In terms of understanding safety aspects of the system and focusing on the most critical areas, the qualitative use of fault-tree analysis is invaluable. However, quantitative aspects of fault-tree analysis greatly multiply its power. This chapter proceeds by presenting the methodology of fault-tree analysis, including an approach to constructing the fault tree. This is followed by the uses of the fault tree for accident classification, and finally for cost/benefit determination.

5.2 Methodology

The methodology developed here will not be a rigorous and comprehensive review of all work performed in fault-tree analysis. Rather, it is specifically designed to guide the reader into the use of those concepts consistent with the scope of this book.

In order to discuss the methodology of fault-tree analysis, some basic definitions must be understood. These are presented below.

1. Event: a previously defined and specified occurrence within the system. It may be defined at the system or component level. Examples: (a) switch A fails to perform properly, (b) switch A fails due to temperature extreme, and (c) system mode change by switch A. *Note:* For our purposes here, events are defined such that they are binary in nature; they either occur or fail to occur—two states only. An event does not have to cause failure; it can be a normal functioning of the system. Thus there are two classifications of events, as given below.

2. Fault event: an event in which one of the two states is an abnormal occurrence, resulting in some type of failure or fault.

3. Normal event: an event in which both states are expected and intended to occur at a specified time. A normal event could possibly become a fault event due to timing, and therefore it is important that the timing of a normal event be specified if necessary. (*Note:* Events defined by definitions 2 and 3 are mutually exclusive and all-encompassing.)

4. Basic event: an event (fault or normal) that occurs at the element level. An element here refers to the smallest subdivision of the analysis of the system. For example, if failure rates are available for certain components of the system, these components would form the elements,

inasmuch as any further subdivision of these components would be unnecessary. Thus a basic event would be the failure of one of these components. This would also be a fault event.

5. Primary event: an event caused by a characteristic inherent in the component itself, such as the failure of a light bulb due to a worn filament.

6. Secondary event: an event caused by an external source, such as the failure of a light bulb due to excessive voltage.

7. Head event: the event at the top of the fault tree which is analyzed by the remainder of the fault tree. This is generally a resultant failure which removes the system (man or machine) from normal operation.

The symbology of fault-tree analysis will be presented in the next section in terms of examples. To obtain an understanding of the procedure, it is helpful to get an overview of the general fault-tree approach prior to proceeding. The construction of a fault tree originates by a process of synthesis and analysis. This can be observed by the following stepwise outline.

A. SYNTHESIS

1. Determine at the most general level, *all* events that are to be considered as undesirable for the normal operation of the system under study.

2. Separate these events into mutually exclusive groups, forming groups according to some common relationships (such as similar causes).

3. Utilizing the common relationship, establish one event that encompasses all events in each group. This event will form the head event and will be considered by a separate fault tree.

B. ANALYSIS: TOP-DOWN APPROACH

1. Select one head event, the event that is to be prevented. One system might have many head events, as indicated in A.3.

2. Determine all primary and secondary events that could cause the head event.

3. Determine the relationship between the causal events and the head event in terms of the AND and OR Boolean operators.

4. Determine the value of further analysis of each event determined in steps 2 and 3. For each causal event that is to be developed further, repeat steps 2 and 3 where the term "head event" is now replaced by the particular causal event that is to be further developed.

5. Continue to reiterate steps 2, 3, and 4 until all events are either in terms of basic events, or else it is not desirable to develop the event further, owing to insignificance, lack of data, and so on.

6. Diagram the events using the symbols discussed in the next section.

7. Perform qualitative and quantititive analyses as discussed further below.

This stepwise procedure is an oversimplification of the actual process, which may be reiterated many times. The common relationship mentioned in synthesis steps 2 and 3 is the key to proper organization. Another factor that must be considered, as will be obvious when the quantitative methods are discussed, is the severity associated with the head event. All head events will have variability as far as severity is concerned, since the same accident can result in a wide range of possible damage and injuries. However, when a given class of accidents is generally more severe than another class of accidents, the two should be organized under separate head events. For example, in setting up fault trees for forest harvesting there was a clear distinction in severities between two types of "struck by" accidents: (1) limb and (2) logs or trees. The former generally resulted in superficial wounds, whereas the latter, at least on the average, resulted in a significantly higher severity. Therefore, separate fault trees were set up for the two accident classifications. The purpose for this in evaluating cost/benefit will be discussed below.

Generally there will be more than one fault tree developed for each system. Although they may later be linked together, it is beneficial to keep them separate at the development stage. Also, if a system operates in more than one mode, it may be necessary to develop a different set of fault trees for each mode of the system.

Example 5.1 Synthesis Steps of Fault-Tree Analysis

Suppose that the system under consideration is a grinding operation, and according to synthesis step 1, the following list of fault events must be prevented:

1. Hand or fingers touch wheel.
2. Arm contacts wheel.
3. Clothing gets caught in machine.
4. Speck of metal contacts eye.
5. Improper grounding causes electrical shock.
6. Motor overheating produces fire.

Of course, this is a partial list of possible undesirable events, but it is sufficient to illustrate the procedure. A more complete list could be obtained from past accident reports. Also, for the present example it will be assumed that there is no significant difference between the average severities of the above classifications.

The second step under Synthesis is to separate these events into mutually exclusive groups according to some classification scheme. This is, in a sense, an art rather than a science and depends highly upon the organizational ability of the analyst. However, certain relationships should be immediately apparent. Events 1 and 2 are very closely related and could be analyzed concurrently. The same is true of events 5 and 6. Event 4 is rather unique and should be analyzed separately. Event 3 might be part of the 1 and 2 group; however, it refers to clothing rather than direct contact, so we prefer to analyze it separately.

With the four groups, four head events can now be created:

Event	Head Event
1 and 2	Wheel contact with person
3	Clothing caught in machine
4	Chip in eye
5 and 6	Motor-related accident

Each of these head events should be used to begin a top-down approach to constructing a fault tree, as indicated above under ANALYSIS. Prior to continuing this example, however, the symbols and Boolean techniques of fault-tree analysis must be understood.

5.3 Symbols and Boolean Techniques

Fault-tree diagrams are very useful, especially in qualitative analyses of a system for safety. In order to use the digrams to their greatest advantage, it is helpful if the standard symbols are used. The standard meanings of the Boolean AND and OR operators, which were discussed in Chapter 3, are used here. The following symbols are recommended:

1. Rectangle: used to identify events that will generally be developed further in the analysis.
2. Circle: used to identify basic events (i.e., events that are in no need of further development because of the sufficiency of empirical data).
3. House: identifies events that are expected to occur in the normal operation of the system.
4. Diamond: used for an event that will not be developed further in the logic diagram either because of insufficient data or because the event itself is inconsequential. Figure 5.1 illustrates the first four symbols.
5. AND Gate: the symbol in Figure 5.2 is used for an AND gate. Two or more inputs (lines from symbols below) and one output characterize this symbol. In order for the event immediately above the symbol to occur, *all* the input events must occur.

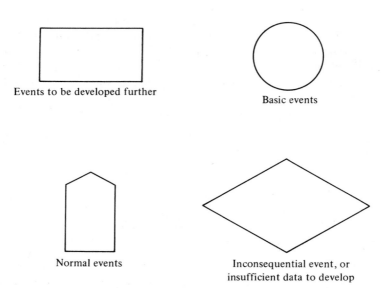

Events to be developed further

Basic events

Normal events

Inconsequential event, or insufficient data to develop

Figure 5.1. Event symbols for fault tree analysis

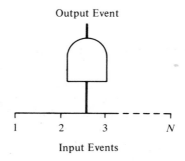

Output Event

1   2   3   *N*

Input Events

Figure 5.2. "AND" gate

6. OR Gate: the symbol in Figure 5.3 is used for the OR gate. As with the AND gate, multiple inputs and one output are required. In order for the output event to occur, *at least one* of the input events must occur.

7. Conditional AND and OR Gates: in many situations, strict Boolean AND and OR functions are not appropriate. For example, in some AND situations, the sequence of events is important. Also, there may be an EXCLUSIVE-OR situation where one or the other input events would cause the output event, but if both occurred simultaneously, the output event would not occur. In situations such as these, the AND and OR gates are modified by another symbol attached to the

Output Event

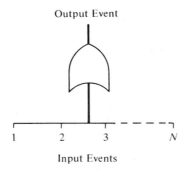

Input Events

Figure 5.3. "OR" gate

side. Generally, an oval is used to qualify the AND and OR gates; however, a rectangle has also been used in aerospace applications, where the particular time relationships between input and output events are of importance. Care should be used with these symbols and relationships, since the quantitative analysis given below is predicated upon the strict AND and OR relationships. (This will be further illustrated in Figures 5.6 and 5.7.)

8. Transfer Symbol: the triangle is used for convenience to transfer to another diagram or to another part of the same diagram.

Example 5.2 Example Analysis

In Figure 5.4 the event "Person Gets Chip in Eye" could be considered as a head event or the result of the analysis of another event. In either case, assume that it is not basic and requires further development. Two things are

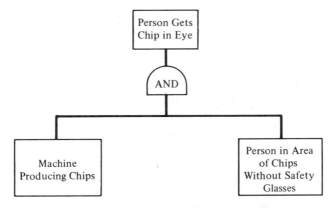

Figure 5.4. "AND" gate example. Both of the input events must occur for the output event to occur

required for this event to occur. First, the machine must be producing chips, and second, a person must be in the area of the machine without being adequately protected by safety glasses. This breakdown is given in Figure 5.4, where the two events are in rectangles indicating the possibility of further analysis.

The particular analysis of the above event is not unique, but neither is it arbitrary. The objective is to resolve the complex event into simple events that can be evaluated easier. In this case the probability of the "Machine Producing Chips" can be obtained by direct observation. The same is true of the other event, "Person in Area of Chips Without Safety Glasses." Now these event probabilities alone will not be sufficient to determine the probability of the event "Person Gets Chip in Eye." The reason for this is important. Refer to Figure 5.4 again and note that the two events, *as stated*, could occur without the occurrence of an accident (i.e., the resultant event).

This problem could be rectified by changing the second event to "Person in Area of Chips Without Safety Glasses and Gets Chip in Eye Given That Machine Is Producing Chips." This is now technically correct, although overly cumbersome. Rather than insisting upon a technically correct labeling of events, it greatly facilitates the process if a general rule is established regarding the AND gate.

5.3.1 Rule 5.1: AND-Gate Event Rule

The input events to an AND gate must be defined such that the second is conditioned upon the first, the third is conditioned upon the first and second, . . , and the last is conditioned on all the others. In addition, at least one of the events must include the occurrence of the output event.

Note that it is not necessary to label each event completely. In fact, such labeling would be detrimental to graphical clarity. As long as the events are interpreted in light of rule 5.1, lengthy descriptions are unnecessary. For consistency the events should be ordered from left to right such that those on the right are conditioned on all preceding events to the left (see Figure 5.2). Also, if not otherwise labeled, the occurrence of the output event will be assumed to be implied in the last event (the Nth event of Figure 5.2).

Although this level of sophistication in implied conditional events may seem trivial at this point, it is essential for the quantitative analysis to follow. Also important is a clear definition of each event, which is seldom obtained in the brief label. For example, in Figure 5.4 the event "Machine Producing Chips" will probably include any action of the machine that would make chips come into proximity with the eye. This might not be restricted to the actual machining operation if the machine could propel chips under other circumstances (i.e., the air draft caused by the normal grinding machine whether or not the wheel is in contact with the workpiece). This expanded definition should be understood by all analysts involved, and hence the formalizing of definitions in separate written form is recommended.

Prior to giving an example of analysis using the OR gate, Rule 5.2 will be given.

5.3.2 Rule 5.2 OR-Gate Event Rule

The input events to an OR gate must be defined such that together they constitute all the possible ways that the output event can occur. In addition, each one of the events must include the occurrence of the output event.

Again, the explicit label of the events need not obey Rule 5.2. However, Rule 5.2 must be understood by the analyst in applying fault-tree analysis.

Example 5.3 Illustration of the OR-Gate Event Rule

Rule 5.2 can be illustrated by consideration of Figure 5.5. Here the objective is to analyze the event "Person Without Safety Glasses Enters Area of Chips" by the use of an OR gate. The first part of Rule 5.2 is satisfied by virtue of the third event covering all other possibilities. Unless it is obviously not required, it is safe to create an event such as this to cover those events of very low probability which could cause the output event. Notice further that the second part of Rule 5.2 is satisfied since each input event label contains the

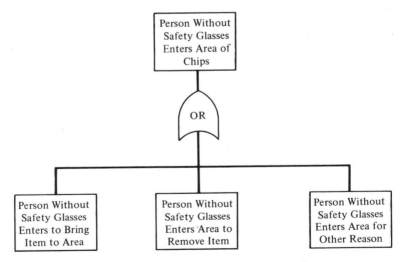

Figure 5.5. "OR" gate example. At least one of the input events must occur for the output event to occur

clause "Person Without Safety Glasses Enters . . . ." An appropriate abbreviation would consist of eliminating this clause since, by Rule 5.2, this must be understood.

Before leaving the use of AND and OR gates, some consideration should be given to the two collectively. Analysis step 2 in the procedure for constructing the fault tree indicates that the analysis is the result of a search for causes of the event being analyzed. If there are several causes, all of which must occur simultaneously, then obviously an AND gate will be used. In other words, the AND gate inputs answer the question: What *must* occur for the output event to take place? On the other hand, if there are several causes any one of which could by itself create the result, an OR gate should be used. Therefore, the OR gate inputs answer the question: What events could possibly occur (each by itself) to bring about the output event?

Now obviously some combinations of ANDs and ORs will be necessary. The problem is to determine which are to be developed first, since alternatives may arise for any given event. Here the solution will be given by Rule 5.3, which will not be thoroughly discussed until the next section. It is sufficient to say at this point that Rule 5.3 was developed strictly from data-collection and probability-calculation considerations.

5.3.3 Rule 5.3 Fault-Tree Development Rule

For any event that requires further development, analyze all possible OR input events first prior to analysis for AND inputs. This will hold for the head event or any subsequent events that need further development.

The remaining techniques of fault-tree construction can best be obtained by studying the uses of the fault tree as well as examples. Prior to this, two examples of the use of conditional gates will be discussed.

Example 5.4 Conditional AND

Figure 5.6 presents the analysis of the event "Wheel Contact Injures Person." Two events are required to cause this event. First, the machine must be operating and, second, some part of the body must come in contact with the wheel. The qualification attached to the AND gate indicates that if the machine is not operating when the wheel contact is made, the accident will not occur.

The situation depicted in Figure 5.6 generally does not occur; that is, it is generally possible to get into an accident by contacting the wheel first and then inadvertently turning on the machine. Here we assume some clutch mechanism, or a very slow starting wheel, such that this would be no problem. Without this assumption a two-part analysis might be in order; one for the

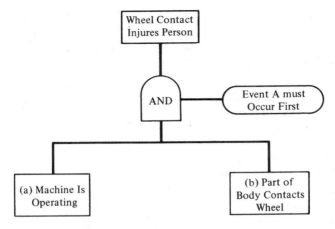

Figure 5.6. Conditional "AND" example

events as depicted in Figure 5.6, and one for the case where the machine is started after body contact with the wheel.

Example 5.5 Conditional OR

Figure 5.7 demonstrates the use of a conditional OR gate. Note that the above assumption of the slow start or clutch mechanism is not in effect here. This type of accident is usually prevented by having two switches sufficiently

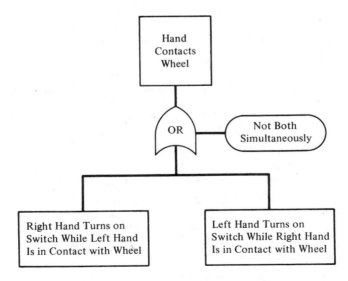

Figure 5.7. Conditional "OR" example

apart that both hands are required to start the machine operating. In any event, although this is clearly an OR situation, it is obvious that both events cannot occur simultaneously. Therefore, it is necessary to have the conditional OR, as given in Figure 5.7.

Example 5.6 General Fault-Tree Development

To demonstrate the entire synthesis and analysis procedure in terms of a total industrial system would require a great deal of time and space. Therefore, one head event will be developed here to moderate detail so that it is still useful for illustrative purposes. Figure 5.8 incorporates some of the examples given above in developing the head event "Chip in Eye (Grinding)." This particular fault tree is to analyze the specific type of eye injury that might be caused by the grinding operation. Many of the events now in diamonds could be placed in rectangles and developed further. Also, the fault tree developed here is not unique. A comparison of Figure 5.5 with the corresponding branches of Figure 5.8 illustrates this point.

Assume that the synthesis part of the fault-tree development is completed and that one of the head events is "Chip in Eye (Grinding)." By Rule 5.3 the first analysis is performed in terms of the OR operator. Those who might have this accident fall into two mutually exclusive and all-encompassing categories: (1) operators and (2) nonoperators. Further, assume that the accident will not occur if adequate eye protection is worn. Therefore, the two events shown illustrate the first breakdown.

Past accident records should be consulted before the analysis and brought to bear at this point. Suppose that these records indicate that very few operators are injured in this way since they wear their protective equipment a great proportion of the time. Therefore, this event can be placed in a diamond, since, although it could be developed further, there is no need for further effort here. Before leaving this event, a reminder is in order with respect to Rule 5.2. The event "Operator Fails to Wear Safety Glasses" has an abbreviated label which, if spelled out in detail, would read "Operator Fails to Wear Safety Glasses and Is Injured by Chip in Eye."

Assume that the records show that nonoperators are involved in the vast majority of accidents of this sort. It is now desirable to further analyze this event to determine causes. It is correct to apply Rule 5.3 again and attempt to further subdivide the class of nonoperators. Assuming no significant differences between nonoperators or between various causal sequences, it is necessary to proceed with an AND functional relationship. Remember that the AND relationship asks the question: "What *must* happen?" not "What *could* happen?"

Four things must occur in order for the nonoperator to be injured in this way. These four are listed appropriately under the AND gate in Figure 5.8.

Note that Rule 5.1 applies here so that the events are conditional by implication. Now in the first three events this is inconsequential. However, the fourth
is obviously conditional, since the operator could not stop the machine unless

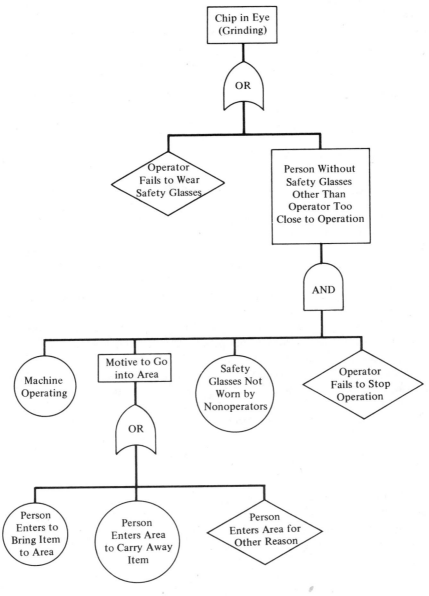

Figure 5.8. Example fault tree

it was first operating. This may seem a trivial point in this stage of the development. However, when assigning probabilities, the necessity for this detail will become apparent.

The event "Motive to Go into Area" is analyzed into the specific reasons. Assume here that the large number of nonoperators is caused by a tool rack close to the operation. Past causes of this location for the tool rack are irrelevant at this point, and as in the three-level approach of Chapter 1, time should not be expended in enumerating all details in this phase of the analysis. The breakdown in Figure 5.8 illustrates Rule 5.2, in that all possibilities are covered, and each of the input events includes the output event. The example in Figure 5.8 will be used as an example of the quantitative techniques discussed below.

5.4 Quantitative Analysis*

Although the development of the fault tree gives the analyst an insight into the problem otherwise unobtainable, the full potential of this technique cannot be realized without further quantification. The purpose here in quantifying the analysis is to effectively allocate the safety budget. To do this, the various alternative safety investments are considered in light of their effect upon the fault tree and the resulting head event. A measure of cost/benefit is then determined for use in decision making.

The term *cost/benefit* is a vague term used in describing a variety of applications. Here it is defined to be the dollars spent per negative utility reduction. This can be better understood if we separate, for the time being, the measure of cost from the measure of benefit. Cost will be defined as the dollar outlay to pay for the incorporation of a device, method, procedure and so on (henceforth called a *countermeasure*) into the industrial system for a given unit period of exposure.† Thus the cost of devices that must be periodically recharged and/or replaced is based on average costs for a given unit (e.g., a 1,000,000-man-hour exposure period). Permanent fixtures, such as machine guards, can be prorated on the basis of the life of the machine. The cost of educational programs can be prorated, based upon their frequency. All countermeasures must, for comparison purposes, have a common denominator.

The measure of benefit, expected negative utility reduction, requires a thorough explanation. There is a negative utility (or cost in terms of dollars and personal well-being) associated with every accident. This negative utility

---

*Some of the material that follows was originally published in "Cost/Benefit of Safety Investments Using Fault Tree Analysis," *J. Safety Res.*, Vol. 5, No. 2, 1973.

†Any unit of time that satisfies the conditions described below would be appropriate.

depends directly upon the severity of the accident. Now, each head event chosen to be analyzed by a fault tree will consist of (or result in) an accident, the severity of which may not be constant. Historical records can be used to determine the relative frequency of the head event causing First Aid, Temporary Total, Permanent Partial, Permanent Total, or Fatal injuries. For each of these classifications, a negative utility can be established as in Table 5.1.

Table 5.1 Example of Negative Utility Schedule

Severity Classification	Severity	Negative Utility,* u_i
1	First Aid	20
2	Temporary Total	345
3	Permanent Partial	2,500
4	Permanent Total (including fatalities)	21,000

*This value need not be a dollar figure if other intangibles, such as social costs, are to be considered. For this example, however, First Aid was a dollar value per case estimated by the author. All other figures are average costs per case given by the National Safety Council, *Accident Facts*, 1971. Generally, management judgment will be required to formulate the negative utility values consistently with company policy.

The expected negative utility of the head event *if it occurs* can now be calculated by the following:

$$E = \sum_{i=1}^{N} p_i u_i \tag{5.1}$$

where p_i is the probability of occurrence of the ith severity class given that the head event occurs, N the number of severity classes, and u_i the negative utility associated with the ith severity class. Note that this gives the expected negative utility *if the head event occurs*, so it is not an absolute measure of criticality. The value of E, therefore, can be viewed as the expected cost of the accident if it happens to occur. The values of u_i might be expressed in dollars, lost workdays, and so on.

An alternative method for calculating E would be more appropriate if the values of negative utility from a large number of past occurrences of the head event were measured directly. For example, suppose that the negative utility unit chosen was "lost workdays." Further, suppose that in the past n occurrences of the head event, the measures of negative utility were u_1, $u_2, \ldots, u_n$, respectively. The expected negative utility associated with the head event would be obtained from the arithmetic mean of these measures:

$$E = \frac{\sum_{i=1}^{n} u_i}{n} \tag{5.2}$$

In fact, it can be shown that Equations (5.1) and (5.2) are equivalent under the conditions that there are n severity classes ($N = n$) and that the probability of each severity class is equivalent ($p_i = 1/n$). This occurs when we consider each accident as a unique situation.

Although E, given by Equation (5.1) or (5.2), tells how bad the accident is expected to be if it occurs, it does not entirely tell how important the accident is as far as safety investments are concerned. Several other important factors must be considered. One of these is the frequency with which the accident occurs. This will be measured in terms of probability through the use of the fault trees themselves. Prior to discussing the technique for obtaining the head event probability, P, the use of this probability will be discussed.

Given the values of E and P, an absolute measure of "criticality" associated with the head event can be obtained as follows:

$$C = P \cdot E \tag{5.3}$$

where C is the expected negative utility associated with the head event in the given time or production unit. This can best be illustrated by an example. Suppose that a certain head event has been observed to occur five times in the last 100 million man-hours (mmh). Thus the probability of occurrence, P, in any mmh would be .05, or 5%. Now further suppose that the average severity of this event, when it occurred, was calculated from measurements [using Equation (5.1) or (5.2)] to be 8 lost workdays. Then the absolute criticality associated with this head event would be

$$C = (.05 \text{ occurrence/mmh})(8 \text{ lost workdays/occurrence})$$

$$= .4 \text{ workday/mmh}$$

This is an independent measure of criticality, since it takes into consideration both the frequency and the severity. Hence it is valid to use this measure in comparing the importance of head events. Note that any unit of time or production, and any unit of severity, could be substituted for mmh and lost workdays, respectively.

This last statement must be qualified somewhat before proceeding. Although it is true that any unit of production or time may be used, it should be remembered that the number of occurrences per this unit is being used as a *probability*. Hence the laws of probability must be applicable and the unit chosen accordingly. This does not affect the type of unit, but the magnitude of the unit. In the example above the type of unit is exposure time, specifically 1 million man-hours. Other units of exposure time are 1,000 man-hours and 1 hundred million man-hours. The determination of P for Equation (5.3) was accomplished in the example above by considering the duration of one unit as one *trial* (see Section 4.2). Therefore, since the event occurred five times in 100 trials, its probability of occurrence in any one given trial was .05. If the unit had been 1,000 man-hours, the number of trials would have been 100,000 and the probability would have been .00005. On the other hand, if a unit of 100 million man-hours had been chosen, the number of trials would have been 1, and the probability calculation would yield a value of 5.

This may seem confusing, but it is not if the basic definition of a trial is understood. A trial must be so defined that the event in question either occurs or fails to occur. The repetitive occurrence of an event within a trial is not within the binary assumption (see definition of "event"). Therefore, a rule for establishing the *size* of the trial must be formulated such that the probability of multiple occurrences is small and hence need not be considered. At the same time, it is needless to choose a unit that is too small, for this does nothing but increase computational complexity. Rule 5.4 provides an adequate guide to the determination of unit magnitude.

5.4.1 Rule 5.4 Production-Time-Unit Scaling Rule

Once a unit of production or time is chosen to represent one trial, it should be scaled such that the probabilities of head events will generally fall between zero and 0.1.

Rule 5.4 ensures that the probability of the head event occurring twice in any given trial is small (maximum of .01 assuming independence). Further, it ensures against undue computational complexity. It contains the word "generally," since no one unit may be able to satisfy both requirements under all conditions. To what extent head-event probabilities can be allowed to exceed 0.1 will depend upon the accuracy required of the computations.

5.4.2 Determination of Head-Event Probabilities

The value of P can be obtained assuming that a proper unit of time or production has been determined to adequately define one trial. Of course, a method for determining P has already been described in the example above. This method follows directly from the concepts of relative frequency and probability discussed in Chapter 4. the formula is

$$P = \frac{n_h}{n_u} \tag{5.4}$$

where n_h is the number of occurrences of the head event in n_u trials given by the chosen time or production unit. It should not be alarming that this probability is dependent upon the choice of the trial size. An analogous situation is the probability of choosing an ace from a poker deck as opposed to a bridge deck. In both cases the sample space is being changed.

An alternative way to determine P is by using the fault tree end branch probabilities. This is necessary if the effect of alternative countermeasures is to be determined. It will be seen in the next section how end branch probabilities are obtained. For the present it will be assumed that they are available for use.

In the OR situation, any of the events will cause the subsequent event to occur and, therefore, assuming independence, the probability of occurrence of the subsequent event is given by

$$P_0 = 1 - \prod_{i=1}^{n} (1 - q_i) \tag{5.5}$$

where q_i is the probability of the ith causal event and n is the number of parallel branches. In the AND situation, all the events must occur for the subsequent event to occur and, therefore, assuming independence, the probability of occurrence of the subsequent event is given by

$$P_A = \prod_{i=1}^{n} q_i \tag{5.6}$$

The astute reader here will notice the similarity between Equations (5.5) and (4.33) as well as between (5.6) and (4.29). Indeed, one of the reasons that reliability was discussed in Chapter 4 was to demonstrate the concepts applied here. Hence the basis for Equations (5.5) and (5.6) will not be reviewed here.

Through a reiterative process, as illustrated in the example below, the probability of the head event can be determined from a knowledge of the probabilities of the branch events. This is the value of P that is used in Equation (5.3) to obtain the absolute expected negative utility associated with the head event. A system modification will produce a change in this value of expected negative utility, thus providing the measure of benefit.

Example 5.7 Evaluating Alternatives Within a Fault Tree

For an example, consider the fault tree developed in Figure 5.8. In Figure 5.9 the probabilities of occurrence are given for the end branch events for any million-man-hour period. Suppose that records show that in the past there have been 10 accidents of this type, of which 7 were First Aid, 2 were Temporary Total (man had to leave job), and one resulted in a Permanent Partial (caused permanent eye damage). Using Table 5.1 and

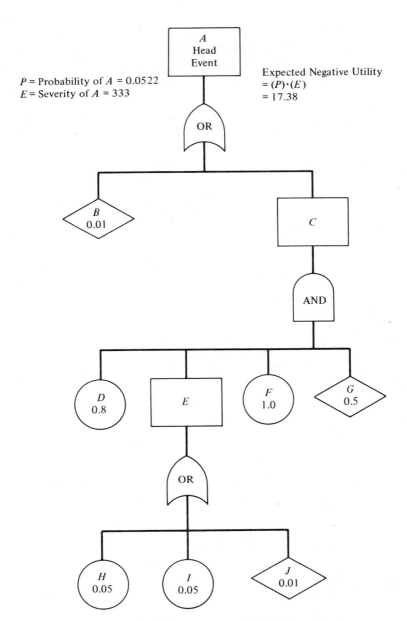

Figure 5.9. Example fault tree with probabilities assigned

Equation (5.1), the following expected negative utility is obtained given that such an accident occurs:

$$E = .7(20) + .2(345) + .1(2500) = 333 \tag{5.7}$$

To determine the probability of occurrence for the head event, Figure 5.9 is used along with Equations (5.5) and (5.6). Those sets of end branches which have no further development should be considered first. For example, in Figure 5.9 the very bottom events have no further development. Thus these three can be combined to determine the probability of the subsequent event. Since they are related by an OR function, Equation (5.5) will be used as follows:

$$P_0 = 1 - (1 - .05)(1 - .05)(1 - .01) = 1 - .8935 = .1065 \tag{5.8}$$

Thus, probabilistically speaking, Figure 5.9 reduces to Figure 5.10.

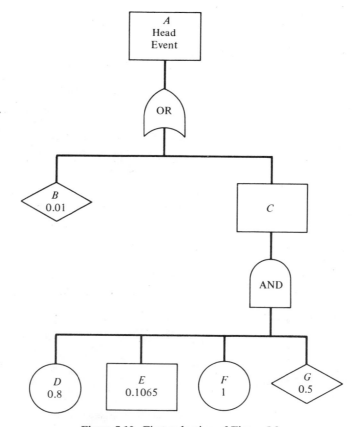

Figure 5.10. First reduction of Figure 5.9

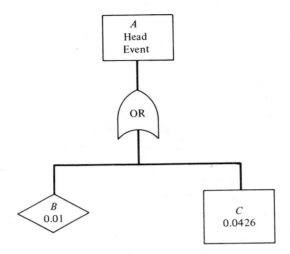

Figure 5.11. Reduction of Figure 5.10

Similarly, by using Equation (5.6) as follows,

$$P_A = (.8)(.1065)(1)(.5) = .0426 \qquad (5.9)$$

Figure 5.10 reduces to Figure 5.11. Finally, the probability of the head event can be determined again using Equation (5.5):

$$P = 1 - (1 - .01)(1 - .0426) = 1 - (.99)(.9574) = .0522 \qquad (5.10)$$

This is the probability of occurrence of the head event in any million man-hours of exposure. To obtain a measure of the criticality associated with the head event, Equation (5.3) is used:

$$C = P \cdot E = (.0522)(333) = 17.38 \qquad (5.11)$$

where P and E were calculated in Equations (5.10) and (5.7), respectively. Note that if Table 5.1 were in units of dollars, this value would represent the expected dollar cost per million man-hours of exposure from this particular hazard. Generally, Table 5.1 is not restricted to dollar costs, however.

This example will be pursued a bit further to determine how modifications on the fault tree are handled. If money is spent to improve the safety of this system, one or more of the basic event probabilities in the fault tree should be reduced or else the expected severity should be reduced. If not, either the expenditure should not be made, or else the fault tree is incorrect. A reduction in the basic event probabilities will always reduce the probability of the head event, P, and therefore it will also reduce the criticality, C, of the event. The amount by which the criticality is reduced will provide a measure

Table 5.2 ALTERNATIVES FOR EXAMPLE 5.7

Alternative	Description	Prorated Cost/mmh	Effect
1	Ensure that operator stops operation whenever anyone enters area	$25	Reduce probability of event G to .05
2	Move storage area away from grinding area	$15	Reduce probability of events H and I to zero
3	Both 1 and 2	$30	Same effects as both 1 and 2

of benefit for the change that was made. Hence a measure of benefit can be estimated for any safety investment.

This procedure will now be illustrated by a continuation of the example. Consider three proposed countermeasures to reduce the probability of the head event "Grinding Chip in Eye" originally presented in Figure 5.8 and reduced in Figures 5.9, 5.10, and 5.11. These three alternatives are given in Table 5.2 in terms of their costs prorated to 1 million man-hours of operation and the estimated or measured effect on the fault tree (see Figure 5.9). There are many other obvious alternatives; these three are given for simplicity. The first alternative is to have the operator shut down the grinding operation whenever someone comes around. It has been noted in the past that he has done this about 50% of the time without instruction. Therefore, if allowed to continue, .5 provides an estimate of the relative frequency with which he will shut the machine down. It is estimated that, even if he is instructed to stop the operation when someone approaches, he will not in 5% of the cases because either the operation is at a critical point or else he does not see the approaching person. The incurred cost results from delays caused by stopping the process.

The second alternative is to eliminate the major portion of the need for other persons to be in the area by removing a tool storage shelf from the area. Since these tools are used primarily for grinding, a cost will be incurred for the inconvenience of their leaving the area as well as for the initial relocation cost.

Alternative 3 is a combination of alternatives 1 and 2. The cost is not the sum of costs, however, since fewer people will be entering the area, resulting in fewer shutdowns. Other alternatives, such as the use of safety glasses by all persons, or the purchase of a duplicate set of tools, could also be evaluated.

Once the costs and returns are stated quantitatively, the evaluation can be removed from the subjectivity often caused by lengthy descriptions. Figure 5.12 presents the evaluation of alternative 1. The reduction of event G from .5 to .05 reduces the probability of the head event from .0522 to .0142. This is obtained by reducing the fault tree in the same method as in the example

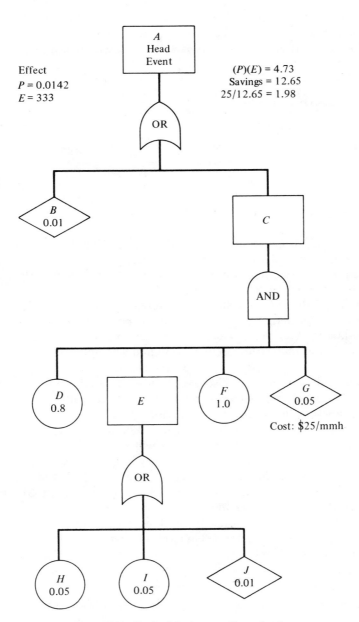

Effect
$P = 0.0142$
$E = 333$

$(P)(E) = 4.73$
Savings = 12.65
25/12.65 = 1.98

Figure 5.12. Revised fault tree, alternative 1

above. Since the probability of the head event is reduced, so is the criticality, since $C = PE$. The new criticality is 4.73, a savings of 12.65 units (in this case lost workdays). Hence a benefit of 12.65 lost workdays has been purchased for a cost of $25, and hence the cost/benefit is $1.98 per lost workday.*

Similar evaluations for alternatives 2 and 3 are performed in Figures 5.13 and 5.14, respectively. Table 5.3 is established based upon these calculations

Table 5.3 SUMMARY OF ALTERNATIVES FOR EXAMPLE 5.7

Alternative	Cost	Original Criticality	New Criticality	Benefit	Cost/Benefit
1	$25	17.38	4.73	12.65	1.98
2	$15	17.38	4.65	12.73	1.18
3	$30	17.38	3.46	13.92	2.16

in order to aid decision making. At this point, subjective reasoning must again be applied to choose between alternatives. Now, however, the evidence has been crystallized into useful information. Remember that the costs are expressed in terms of the unit period of time chosen, in this case million man-hours. Table 5.3 indicates that alternative 2 is the best in terms of cost/benefit. The actual alternative to choose, however, will depend upon the circumstances of application and, in particular, the money available for this purpose. For example, suppose that $25/mmh were the *maximum* that could be spent. This eliminates alternative 3 from consideration. Alternative 2 has a greater benefit, and since it is also less expensive, it also has a low cost/benefit. Barring any other unforeseen circumstances, such as a major labor walkout or the like, alternative 2 is obviously superior.

Now consider the decision given that $30/mmh is available to spend. Alternative 1 can be eliminated since it will not be an improvement over alternative 2 just because there is more money to spend. The choice is between alternatives 2 and 3, where alternative 2 has an excess of $15/mmh that can be spent elsewhere. Looking at it a different way, the added $15/mmh of alternative 3 buys only 1.19 additional benefit units, a marginal cost/benefit of $15/1.19 = 12.60$. Comparatively speaking, this is no bargain. But again evaluations can only be performed in terms of alternatives. If increased safety is required and this is the only way to get it, by all means invest in alternative 3.

---

*Remember that any unit of time and negative utility could also be used.

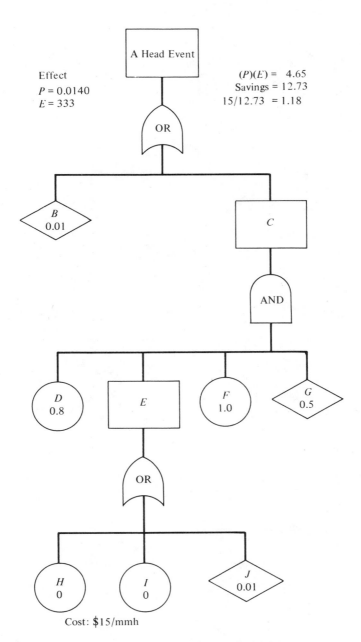

Figure 5.13. Revised fault tree, alternative 2

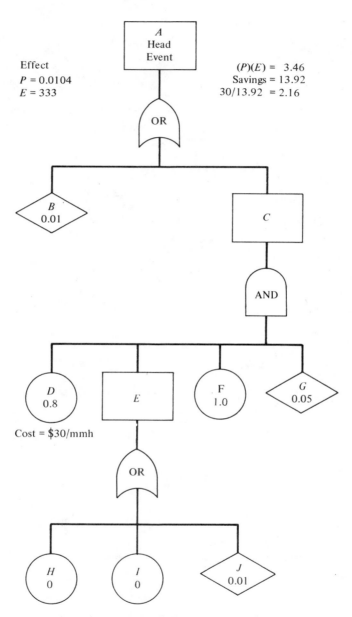

Effect
P = 0.0104
E = 333

(P)(E) = 3.46
Savings = 13.92
30/13.92 = 2.16

Cost = $30/mmh

Figure 5.14. Revised fault tree, alternative 3

On the other hand, suppose that there were another, more urgent use for the $15/mmh. For example, the company may own another facility identical to the first, in which case it would be possible to make two applications of alternative 2. Another possibility is given in Example 5.8.

Example 5.8 Evaluating Alternatives Among Fault Trees

Example 5.7 will be continued, but now suppose that there are two other head events that are competing for funds with the head event discussed in Example 5.7. This is the usual situation, and it is the reason that negative utility had to be brought into the picture. For this example, three head events will be considered, although generally there are many more. Further, assume that the preliminary evaluation of alternatives is available so that Table 5.4

Table 5.4 SUMMARY OF ALTERNATIVES FOR EXAMPLE 5.8

Head Event	Alter-native	Cost	Original Criticality	New Criticality	Benefit	Cost/Benefit
1	1	$25	17.38	4.73	12.65	1.98
	2	$15	17.38	4.65	12.73	1.18
	3	$30	17.38	3.46	13.92	2.16
2	1	$50	5.75	3.00	2.75	18.18
	2	$100	5.75	2.00	3.75	26.66
	3	$75	5.75	1.00	4.75	15.79
3	1	$5	80.00	60.00	20.00	.25
	2	$90	80.00	40.00	40.00	2.25

can be prepared. Note that the original criticality for each head event will generally be different, since both the probability of occurrence and the expected negative utility as used in Equation (5.3) will vary. The "New Criticality" is obtained by repeating the calculations and substituting the new probabilities for the old. As before, benefit is the difference between the new and the original criticality.

Generally the best investments are those with the lowest cost/benefit figures, and these should be made first. This is especially true with comparisons between head events. For example, alternative 1 under head event 3 is obviously superior to the others, since a cost of $5 buys a reduction of 20.00 criticality units. Under head event 1, which was discussed in detail above, alternative 2 is superior to alternative 1 in terms of cost/benefit. Alternative 1 only provides .08 additional reduction for a cost of an additional $10. Clearly there are other, superior ways to spend this $10.

5.4.3 Summary of Method

Before going on to other uses of the fault trees, a step-by-step procedure will be given to guide its implementation for cost/benefit calculations. Assuming that all fault trees are constructed, it is suggested that the following plan be used:

1. Determine from past accident records the expected negative utility (e.g., lost workdays) *if* the head event occurs.
2. Using Equations (5.5) and (5.6), determine the probability of occurrence of the head event.
3. From steps 1 and 2, determine the expected criticality associated with the head event, using Equation (5.3).
4. Repeat steps 2 and 3, substituting another alternative into the fault tree.
5. Determine the reduction of criticality obtained by applying the alternative of step 4.
6. Divide the cost of the alternative of step 4 by the reduction of criticality of step 5 to get the measure of cost/benefit.
7. Repeat steps 2 through 6 for all alternatives of the fault tree.
8. Repeat steps 1 through 7 for all fault trees.
9. To aid decision making, set up a table, such as Table 5.4, to compare alternatives and allocate the safety budget.

This procedure quantifies the decision process to provide the decision maker with the facts required to choose among alternatives. The actual decisions themselves will depend upon these figures as well as other intangible considerations. The most important of these, of course, is the total safety budget. Given management support for a very large safety effort, all the alternatives might be funded. On the other hand, if funds are limited, hard decison making will be required.

Great care and judgment should be used so that this technique is effectively applied. A thorough understanding of both the results and the techniques is essential. No hard-and-fast rules should be placed on choosing the lowest cost/benefit alternative first. Rather, it should be understood that this technique is designed to improve the ability of the decision maker and not to relieve him of his prerogatives. When viewed in this perspective it becomes a powerful tool for improving the quality of safety investments. Methods to further process the cost/benefit data will be discussed in Chapter 7.

5.5 Computerized Methods

One complaint about the use of fault trees in determining head-event probabilities may come from the large number of computations involved in applying Equations (5.5) and (5.6). In Appendix 2 a program is documented that performs the calculations of cost/benefit for any fault tree of 10 or fewer events. It uses the same techniques as described in Section 4.9.1. Some considerations concerning the use of this algorithm and program are presented here to demonstrate further the similarity between reliability and fault-tree analysis.

Any fault tree that uses strict AND and OR gates only can be expressed in terms of a Boolean algebra expression for the occurrence of the head event. The algorithm presented at the end of Chapter 4 will work with any Boolean algebra function, regardless of whether this function was obtained from a reliability configuration or a fault tree. The other requirements were knowledge of the probabilities of the elemental events and the assumption that the events were independent.

The general distinctions between using this algorithm for fault trees and for reliability problems should be clarified before proceeding. In reliability determination the probability of "success" is being sought. The elemental probabilities used in the algorithm are the probabilities that each of the elements will succeed. In calculating reliability, the objective is to alter the configuration or otherwise affect the components such that the highest possible value is obtained. This value is generally close to 1.

On the other hand, in applying this algorithm to fault-tree analysis, the probabilities of the basic events are relatively small. And, since the head-event probability represents an undesirable occurrence, the objective is to reduce it as much as possible. Generally, the probability of the head event should be close to zero.

Another differentiation involves the assumption of independence. In the reliability problems the strict assumption of independence was maintained throughout. According to Rule 5.1, the AND-gate event rule, the fault-tree events must be defined such that all input events to the AND gate must be conditioned upon each other. However, if the probabilities are defined in this way, the algorithm will still be useful, since the effect of multiplying these conditionally defined probabilities will yield the same result as in the analogous independent case. Further ramifications of Rule 5.1 will be discussed in Section 5.6.

Given that the Boolean truth function and the basic event probabilities are all that are needed to determine the probability of the head event using the algorithm of Chapter 4, the next step is to determine methods by which the Boolean function can be obtained from any fault tree.

The first step in translating a fault tree to a Boolean function is to assign the value of a Boolean variable (i.e., a letter) to each *basic* event. Only the basic events need be assigned a variable, since all other events can be represented in terms of the basic events. This is illustrated by Figure 5.15 for both

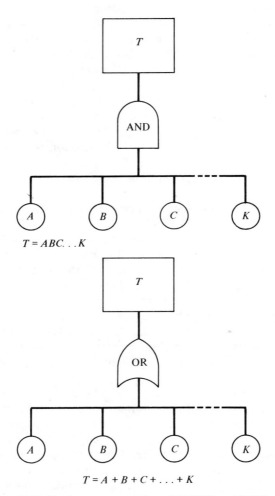

Figure 5.15. Boolean representation of "AND" and "OR" gates

the AND and the OR gates. In fact, these two examples can be regarded as simple fault trees.

Determining the Boolean function for more complicated fault trees is equally as simple. Generally the fault trees will lead to a product of sums or to a sum of products. Consider the continued development of the pure AND fault tree in Figure 5.15, where each of the events is developed with an OR gate. Such a tree with three events is given in Figure 5.16. As shown, the resulting Boolean function is a product of sums.

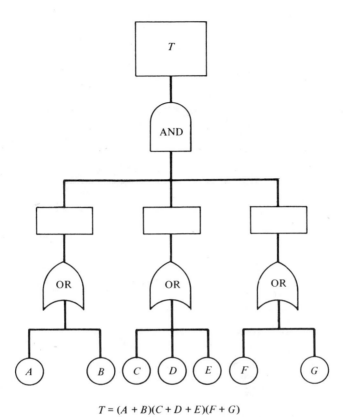

$$T = (A + B)(C + D + E)(F + G)$$

Figure 5.16. Boolean representation of an example fault tree (1)

On the other hand, the pure OR fault tree of Figure 5.15 can be developed with AND gates. The result will be similar to that given in Figure 5.17. As shown, the resulting Boolean function is a sum of products.

In the most general case there will be a variety of "mixed" fault trees. These can all be reduced to one of the two basic forms given in Figures 5.16 and 5.17. An example of such a general case is given in Figure 5.18. Note that *only* the basic events are lettered. Figure 5.18 also illustrates the translation of the Boolean representation into a pure sum of products. In order to use the algorithm of Chapter 4 or the program in Appendix 2, the Boolean expression must be expressed as a sum of products and coded in binary as indicated. This employs the use of *Boolean multiplication*, an example of which is given next.

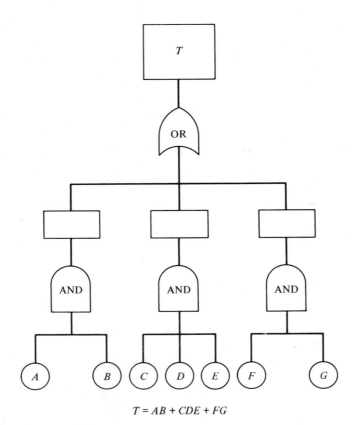

$$T = AB + CDE + FG$$

Figure 5.17. Boolean representation of an example fault tree (2)

Example 5.9 Preprocessing to Use Algorithm

Consider the example fault tree given in Figure 5.19. Since this fault tree has two OR gates connected by an AND gate, a product of sums is generated. This is given under the fault tree. The sum of products is obtained as follows:

$$
\begin{aligned}
T &= A + (B + C + D)(E + F + G)H \\
&= A + (BE + BF + BG + CE + CF + CG + DE + DF + DG)H \\
&= A + BEH + BFH + BGH + CEH + CFH + CGH \\
&\quad + DEH + DFH + DGH
\end{aligned}
$$

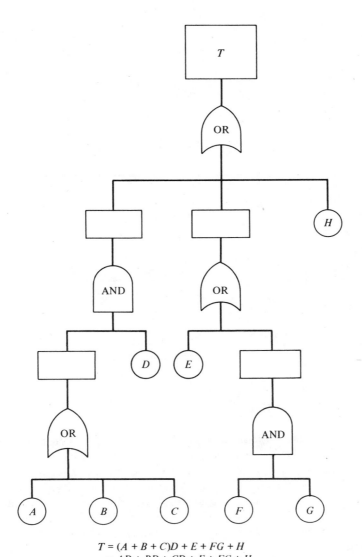

$$T = (A + B + C)D + E + FG + H$$
$$= AD + BD + CD + E + FG + H$$

Figure 5.18. Boolean representation of a general example fault tree

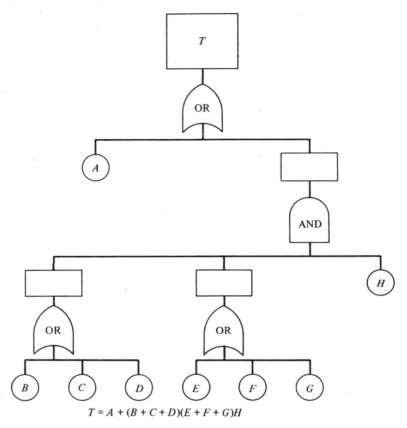

$$T = A + (B + C + D)(E + F + G)H$$

Figure 5.19. Fault tree for Example 5.9

5.6 Data Collection

In order to obtain useful information out of the cost/benefit procedure discussed above, it is essential that the input data be as accurate as possible. To accomplish this, sound data-collection procedures are required. In this section the use of data-collection fault trees will be discussed in terms of obtaining accurate estimates of frequency and severity of basic and head events, respectively. The other two factors required to perform the cost/benefit procedure are the costs of alternatives and the effects of the alternatives upon the basic events. Generally the first of these can be measured directly, and therefore no formal data-collection techniques are required for costs. As far as the projected effect of an alternative on the basic event probabilities, this factor must be estimated. Although it is true that controlled before–after studies may be effective in this determination, this type of study is beyond the scope of the present discussion. For methods of conducting such studies, those interested are referred to Chapter 6.

A distinction is required between the normal fault trees that have been discussed in the sections above and what will be called data-collection fault trees. Suppose that a complete set of fault trees is available for a given industry. It would be a simple matter to go back through the accident records and determine for each accident the basic event(s) that occurred. From this, an estimate of basic event probabilities could be determined.

To do this easily, a coding scheme for the basic events should be established. However, all the basic events need not be coded. In fact, wherever an AND gate appears, the input events should not be coded, since, for the output event to occur, *all* the input events must occur. Thus the recording of an *output* event for any AND gate is sufficient.

Any effective coding scheme devised by the user will be sufficient for the recording of event occurrences. One good scheme consists of using a decimal number, where the number to the left of the decimal represents the fault-tree number, and the number to the right represents the basic event within the fault tree. Figure 5.20 demonstrates this concept as well as the facts illustrated in the paragraph above. This is assumed to be the fourth of a number of

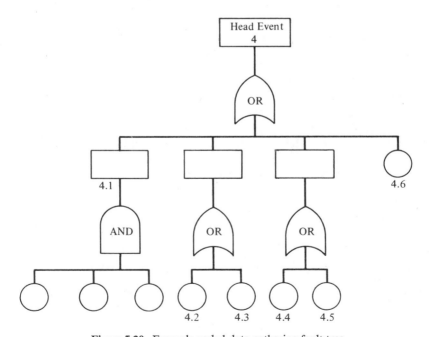

Figure 5.20. Example coded data gathering fault tree

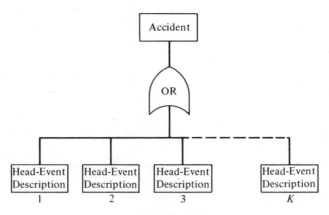

Figure 5.21. Index fault tree

similar fault trees. When a large number of such fault trees are being used simultaneously, an index fault tree, as given in Figure 5.21, can be used.

Now, given that all fault trees have been developed and properly coded, each accident (past, present, or future) can be assigned a number which thoroughly describes the event or events that caused the accident. The procedure is to first go to the index fault tree and determine the head event that was involved. Then, proceed to that head event and trace down the fault tree to the particular event that was involved. This number can be recorded for permanent reference.

Using past records, this can be performed for a large number of accidents over a long period of time. If the period is long enough, and if environmental conditions have generally remained constant, the frequency of occurrence of each event can be used to determine its probability of occurrence in some unit of time or production.

Assume that the total system has been divided with respect to safety into head events $E_1, E_2, \ldots, E_N$. Further assume that each of these events has been analyzed by fault-tree analysis such that the basic events of E_i may be designated $e_{i1}, e_{i2}, \ldots, e_{iM(i)}$, where $M(i)$ is the number of basic events into which E_i has been divided. Then the probability of any event e_{ij} can be determined as follows [see Equation (4.4)]:

$$P(e_{ij}) = \frac{N(e_{ij})}{N(U)} \qquad (5.12)$$

where $N(e_{ij})$ is the number of occurrences of the basic event e_{ij}, and $N(U)$ is the number of occurrences of the chosen time or production unit chosen for the study. For example, if the given unit chosen were 10,000 units of production, and the basic event e_{ij} occurred 9 times in 1,000,000 units, then $N(e_{ij}) =$

9, $N(U) = 100$, and $P(e_{ij}) = .09$ occurrence/10,000 units. That is, the probability of an occurrence in any 10,000 units of production time is .09.

The severity associated with any head event can be determined by calculating the average number of days lost (or a similar measure of severity). This is indicated by Equations (5.1) and (5.2). Since it is easy to record the severity during data collection of the basic events, this is strongly recommended.

5.7 Environmental Factors

Environmental factors can be handled very effectively by the use of AND gates. It was stated above that accident data will not provide information for basic events below an AND gate, since, if the accident occurred, of necessity *all* inputs to the AND gate must have occurred. The most simple case will result in a situation indicated by Figure 5.22. Here an event A is analyzed by means of an AND gate into (1) an environmental factor and (2) the occurrence of the accident given this environmental factor. Assume that all gates above event A are OR gates. Therefore, from accident data the probability of event A can be determined. The analysis must go further, however, since the proposed countermeasures may affect the environmental factor.

Before demonstrating the method by which this is done, it is helpful to present an example. Suppose that event A consists of a man getting hit by a

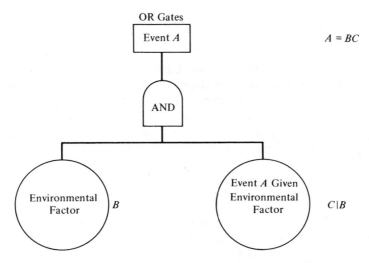

Figure 5.22. Simple environmental factor consideration

tree during a forest-harvesting operation. More specifically, the tree is one that was leaning upon another prior to the harvesting operation. This is a particularly dangerous situation in forest harvesting, and the name given to this environmental factor is "natural leaner." Obviously, a man cannot be hit by a "natural leaner" unless it first exists. Second, it must for some reason become dislodged and fall. Hence the diagram of the situation as given in Figure 5.23 is in effect. Now the probability of the event "Injury Due to Natural Leaner" may be determined by the accident records. But, since this probability cannot be affected directly, a further step is required.

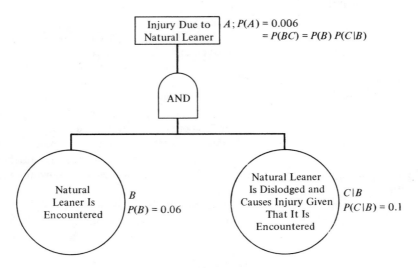

Figure 5.23. Example from forest harvesting

In order to determine the probability of the two input events, it is necessary to obtain data on the environmental factor. For this example it is a simple matter to go into the work environment, the forest, and determine the average number of leaners per some unit of production. However, great care must be taken at this point to ensure that the proper units are used. Since events B and $C|B$ in Figure 5.23 are inputs to an AND gate whose output is event A,

$$P(A) = P(B)P(C|B) \qquad (5.13)$$

In other words, the two input events must multiply to give the probability of the output event A. Now assume that $P(A) = .006$ in any 10,000-cord production period; the units of $P(B)$ and $P(C|B)$ must multiply to give a probability in a 10,000-cord period. Suppose that data have been collected on a representative 10,000-cord area (i.e., over the number of trees that would make up 10,000 cords). Suppose further that 600 natural leaners that could cause injury were found. Clearly the probability of occurrence of a natural leaner in 10,000 cords is not 600/1, since probabilities must be between zero and 1 inclusive. No doubt this probability will be very close to 1 since a 10,000-cord production area would have a very good chance of having one or more natural leaners. However, the probability cannot be greater than 1.

The solution to this problem lies in the realization that it is permissible to change the definition of a "trial" and thereby obtain a ratio that is suitably called a *probability*. Remember that the probability of two occurrences on any given trial must be close enough to zero that its probability can be neglected. It is obvious that 600 occurrences (oc.) per 10,000 cords is equivalent to

$$\frac{600 \text{ oc.}}{10,000 \text{ cords}} = \frac{.6 \text{ oc.}}{10 \text{ cords}} = \frac{.06 \text{ oc.}}{1 \text{ cord}} = \frac{.006 \text{ oc.}}{.1 \text{ cord}}$$

The latter three are suitable for use as probabilities. However, the probability of two occurrences in any 10-cord production period is roughly .36, which is relatively high. The last probability is unnecessarily small, and it will produce an unrealistic estimate of $P(C|B)$. Hence .06/cord is chosen to be a good estimate of $P(B)$ in Figure 5.23.

To determine $P(C|B)$, equation (5.13) can be used, $P(C|B) = P(A)/P(B) = .1$, which is the probability of this accident *in 10,000 cords* given the occurrence of natural leaners in every *cord* of production. This can be seen by examining the units of $P(B)$ and $P(A)$. Another way of viewing $P(C|B)$ is 10,000 times the probability of this accident in one cord of production given a natural leaner in one cord of production.

In essence the probabilities are "forced" to work out by choosing units that will keep the probabilities within good boundaries, generally between zero and .1. This can always be done since the two must multiply to give the resultant output probability, which also is generally between 0 and .1. *This manipulation does not affect the validity of the data.* It merely allows the assumptions to prevail under which probability theory can be effectively used.

When an alternative is applied to remedy the problem, the units used must be borne in mind. For example, one remedy (although not always a controllable factor) might be to work in an area with fewer natural leaners. This will alter the probability of event *B*. In fact, this is one way that a given study can be easily modified to apply to many different environmental conditions. Hence the application of alternatives to remedy the "natural leaner problem" will not obtain as high a priority if $P(B)$ is reduced. Alternatives that do not alter the environment but act within a given environment to prevent the accident would reduce $P(C|B)$ but would not reduce $P(B)$ in Figure 5.23. This was illustrated without explanation in Example 5.7.

To summarize, the procedure for considering environmental factors involves the following steps:

1. From accident data determine the probability of the event immediately above the AND gate.
2. Determine the probability of B from environmental factors.
3. Adjust the units of $P(B)$ so that both $P(B)$ and $P(C|B)$ will be acceptably small (generally less than .1).
4. Compute $P(C|B) = P(A)/P(B)$.
5. When estimating the effect of alternatives (modifications), modify $P(B)$ for changes in the environmental factor, and $P(C|B)$ for all other changes.

5.8 Conclusion

In this chapter the use of quantitative techniques to determine priorities on alternative investments has been discussed. The benefits of the methods speak for themselves in terms of greater understanding of the system and of the decision-making process. The procedure given above quantifies the decision process to provide the decision maker with the facts required to choose among alternatives. The actual decisions depend most heavily upon the outside budget appropriated for safety. This will be further considered in Chapter 8.

Quantitative tools of this nature can provide safety professionals with improved ability to make proper decisions. This technique is not without limitations, however. One occurs in the necessity of the analyst to anticipate the head events. If he fails to anticipate a particular head event, obviously it cannot be evaluated. Similarly, the method itself forces a modeling of reality that will yield only an approximation of the true answer. Specifically, the assumption that each basic event is an independent binary event can lead to inaccuracy. But the realist will not reject improvement because it is not perfect; rather, he will realize that each step of improvement brings him that much closer to perfection.

In conclusion, it should be stressed that great care and judgment must be applied if the results are to be useful and meaningful. The output results can be of little value if the input data are not reliable. Thus the analyst is greatly encouraged to be familiar and modify his thinking by each assumption made.

SELECTED REFERENCES

1. KOLODNER, HERBERT H., "The Fault Tree Technique of System Safety Analysis as Applied to the Occupational Safety Situation," *ASSE Monograph 1*, June 1971.

2. NOLAND, C. A.; CANNIZO, W. M., and JOHNSTON, J. T., *NASA Technical Memorandum, NASA TM-X*-000, Feb. 23, 1972.

3. RECHT, J. L., "Systems Safety Analysis: The Fault Tree," *National Safety News*, Apr. 1966.

QUESTIONS AND PROBLEMS

1. A sidewalk has irregular cracks and bumps and could cause an accident if someone is walking too fast, not paying attention, carrying an object to block his view, suffering from sight deficiency, or is otherwise distracted. Draw a fault tree that analyzes this hazard.

2. Home fires are caused by many sources, among them: (1) spontaneous combustion, (2) faulty wiring, (3) gas leaks, and (4) children playing with matches. For spontaneous combustion to occur, there must be a substance that is subject to oxidation (such as oily rags) and an environment wherein oxidation can take place and heat can build up. Faulty wiring will only cause trouble if it is undetected and is not caught by a breaker or a fuse. Gas leaks also must go undetected for some time to cause a buildup, and then they must be ignited. Finally, children must have access to matches and their discipline must be such that they will play with matches if they are available. Draw a fault tree including the above hazards which could lead to a house fire.

3. Draw a fault tree of a home fire-alarm system. Assume the event "fire within the house" to be basic in this case. Develop the various failures that could occur in the alarm system itself. Assume the following: (1) an initial and secondary power source, (2) periodic tests on the alarm and sensor systems, and (3) the possibility of failure caused by man, animals, insects, decay, and so on.

4. Develop the head event in a forest-harvesting operation "limb falls from tree, injuring worker." Consider the various ways that the limb could be broken off and dislodged.

5. Consider a car on the highway as the system under consideration. The undesirable event to be analyzed is "car wreck," and the only failure modes to be considered are: (1) blowout, (2) loss of steering, and (3) brakes locked by driver. Draw the fault tree down to the basic level.

6. Construct a fault tree with the following Boolean function: $T = A + B(C + D) + E$. If $p_a = .01$, $p_b = .03$, $p_c = p_d = .4$, and $p_e = .001$, determine $P(T)$.

7. Construct a fault tree with the following Boolean function: $T = AB + AC + DE$. Assuming the same probabilities as given in Problem 6, determine the reduction of the probability of the head event if p_c and p_d are reduced to .2. Explain the results.

8. Construct a fault tree that has the following Boolean function: $T = A(B + C)(D + E)$. The initial probabilities of the basic events are .01, .05, .05, .02, and .02 for A, B, C, D, and E, respectively, per year. In the past the head event has occurred 20 times; 5 resulted in permanent totals, 10 in permanent par-

tials, and the rest in temporary totals. The negative utilities assigned by management to these types of injuries are 30,000, 3,000, and 500, respectively.
Determine the following:

(a) The expected negative utility if the head event occurs.

(b) The absolute negative utility associated with the head event.

(c) The cost/benefit associated with an investment of $100/year to decrease the
probability of occurrence of event B to .01.

9. In Problem 8, part (c), what would be the cost/benefit if this investment also
reduced the severity of all permanent totals to permanent partials?

10. Perform a cost/benefit analysis using fault-tree analysis on the nearest stairway.
Assume that in the past five years there have been three accidents caused by
slippery surfaces, five caused by inadequate railings, two by inattention, and
one by someone negligently placing obstacles on the steps. The negative utility
for each accident was an average cost of $200. Assume that you have been
given $1,000 to spend on improving the safety of the stairway. There are three
alternatives, each costing $500, which could be employed: (1) new surfaces,
which will reduce accidents caused by slippery surfaces by 70%; (2) new railings,
which will reduce accidents caused by inadequate railings only by 50%, since it
has been observed that a large number of pedestrians do not use the railings;
and (3) signs and educational programs, which are estimated to reduce railing-
related accidents and obstacle-related accidents both by 20%. The effects on
railing related accidents are considered to be cumulative.

(a) Draw a fault tree of the situation.

(b) Evaluate all alternatives to determine the best allocation of the $1,000.

(c) Would an alternative investment of $500 in another area that would yield a
cost/benefit of 50.00 be justified? (Use a 60-month denominator in calculating basic event probabilities.)

11. Consider the fault tree given by the Boolean function $T = AB + CDE + F$.
The severity of T is 100 workdays/accident. The probabilities of the basic events
in 1 million man-hours of operation are .02, .03, .01, .5, .04, and .05 for A, B, C,
D, E, and F, respectively.

(a) Construct the fault tree.

(b) What is the expected loss associated with the head event as is?

(c) Compare the two alternatives from a cost/benefit standpoint:

    (1) $100/mmh to reduce C and D to .005.

    (2) $200/mmh to reduce F to .01.

12. Over the past few years a certain critical machine has caused injury and property damage by failing in the middle of an operation. Draw a fault tree for this
operation and evaluate the probability of the head event. The following might
be helpful:

(1) Pr {power failure primary and emergency} = .005/mmh

(2) Pr {operator inadvertently causing failure} = .001/mmh

(3) Pr {other failure} = .001/mmh

(*Note:* Items 2 and 3 are not related to item 1.)

(4) An emergency power supply is available. To provide protection against primary power failure, two main components must operate. These are (a) a switch, with reliability .02 per mmh, and (b) a starter, with reliability .5/mmh

13. In Problem 12, what is the probability of a primary power failure? Suppose that the installation is moved to an area where primary failure is .01. What is the new head-event probability?

14. Using Table 5.4, determine the way to invest $50, $75, $100, and $200 such that a maximum *benefit* will be obtained. (*Hint:* Ignore cost/benefit.)

STATISTICAL ANALYSIS

6.1 INTRODUCTION

In the discussions in Chapter 5, the effect of safety alternatives upon the basic events was assumed to be known. Often the changes in the basic event probabilities can be determined or estimated very closely from a knowledge of the physical situation. On the other hand, it would be beneficial if some of the alternatives could be tested, possibly on a small scale, prior to their implementation. Some alternatives might have unseen harmful effects which would be uncovered. Also, it would be useful to test an alternative whenever it is implemented for purposes of evaluation, if for no other reason.

For these reasons, some basic techniques of statistical analyses are presented here. The methods and tests recommended are not intended to be a rigorous coverage of statistics. Rather, this is intended to reveal the potential and necessity of using statistical techniques. Therefore, great care should be exercised to ensure that the assumptions mentioned are valid for the situation being analyzed.

6

6.2 Random Variables and Distributions

For our purposes here, a *random variable* is defined to be a measurement that when repeated is subject to variation due only to chance. This concept needs qualification, since obviously there are causes for all occurrences and therefore no event is strictly due to chance. However, when these causal events can neither be measured nor controlled, and when they act in an inconsistent, though relatively stable way, they will be called *chance occurrences*.

Any measurement when repeated is subject to a certain amount of variation, if from no other source than human error. Safety measurements (e.g., number of lost workdays from a given type of accident) will have considerable vaiiation, owing to the undeterministic nature of accidents. Thus procedures are required for comparing measurements to determine if the differences between measurements are due to causal effects or merely to random variations. This is the goal of statistical analysis.

Associated with any random variable is a function called its *probability density function*, which relates to the probability of the random variable taking on a given value or a set of values. Since the random variable cannot be adequately described by one number, it must be described in terms of its distribution in the real numbering system.

There are two types of distributions: continuous and discrete. A *discrete distribution* describes a measurement that can take on only specific incremental values. The number of persons injured in a given way would have a discrete distribution, since it could take on values 0, 1, 2, . . . , N, where N is the total number in the population under consideration. A *continuous distribution* has the characteristic that it can take on an infinite number of values. Measurements of length, volume, and time are continuous in the sense that the measurement is truncated or rounded as a function of the measurement process and not because the occurrence exists in discrete entities.

The distinction between continuous and discrete distributions is nebulous at times. However, the distinction is made for expedience, and hence technical exceptions are of more interest to the mathematician than to the engineer. Many discrete distributions are considered to be continuous because of the large number of possible occurrences. Similarly, at times it is convenient to express a continuous measurement as a discrete random variable by using an appropriate rounding process. The distinction is important, however, as will be demonstrated below.

In a discrete distribution it is reasonable to refer to the probability with which any given point occurs. Hence the probability density function may be represented merely by plotting the probability versus the random variable for all possible value. Consider, for example, the random variable: number of accidents in a given plant for one month. Suppose that over the course of two years, the following measurements (counts of accidents) were made: 2, 4, 8, 3, 2, 1, 0, 0, 3, 4, 5, 6, 11, 5, 4, 3, 5, 6, 7, 3, 2, 0, 4, and 4. Table 6.1 presents this distribution in a more meaningful form. Using the concept of probability (relative frequency projected into the future) given in Chapter 4, the third column presents the probability of each of the events (numbers of accidents) based upon these empirical data. Note the assumption that if an event did not occur it will not occur, which is probably not valid, especially for the 9-accident event. It should be obvious that the larger the number of samples (number of months), the more reliable will be the probability estimates. However, no matter how large the sample size, there will still be some doubt as to the perfect validity of the estimates. Therefore, there exists a trade-off between the cost of uncertainty and the cost of acquiring additional data which must be balanced in acquiring empirical data. Figure 6.1 illustrates the probability distribution of the number-of-accidents random variable of Table 6.1. Here it is possible to determine the probability with which any value of the random variable will occur. Since one of these values will have to occur in any given month, the sum of the probabilities is equal to 1. This diagram gives an insight into the distribution of accidents per month. It would be beneficial to determine if a given remedy has a significant effect upon this distribution. To

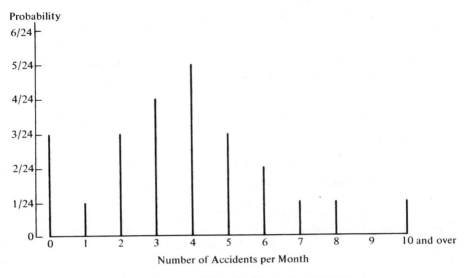

Figure 6.1. Probability distribution for Table 6.1

Table 6.1 DISTRIBUTION OF MONTHLY ACCIDENTS

Event (Number of Accidents)	Frequency of Occurrence	Probability
0	3	$\frac{3}{24}$
1	1	$\frac{1}{24}$
2	3	$\frac{3}{24}$
3	4	$\frac{4}{24}$
4	5	$\frac{5}{24}$
5	3	$\frac{3}{24}$
6	2	$\frac{2}{24}$
7	1	$\frac{1}{24}$
8	1	$\frac{1}{24}$
9	0	0
10 or over	1	$\frac{1}{24}$

do this, more quantitative techniques must be introduced. Since most of these are based upon the assumptions of continuous distributions, it is helpful if these are understood first.

For any continuous distribution the probability of occurrence of a given point has no meaning. First, it is impossible to define a point without the specification of an infinite number of digits. Recognizing this, it should be obvious that the probability is virtually zero for any point in the distribution to occur. Hence a *probability* distribution such as that given in Figure 6.1 is not possible to draw. However, an analogous concept proves very useful in practice—that of the probability distribution function of a random variable x, $f(x)$.

For any random variable x there exists a function called the probability distribution of x, $f(x)$, which has the following three properties:

1. $f(x)$ is positive for all x.
2. The area under the curve of $f(x)$ between any two points is equal to the probability of the random variable falling between those two points.
3. The total area under the curve of $f(x)$ must equal 1.

To illustrate this concept consider the continuous random variable, lost time due to a given type of injury. Suppose that the time lost due to this type of injury was anywhere from 0 to 100 work hours. Further, assume that it has been observed in the past that no one interval of time has a higher probability of occurrence than any other. From the first characteristic given above, the curve of $f(x)$ must always be positive. Since the random variable is contained between zero and 100, property 2 says that it must be positive in these areas. Also, since all intervals of equal width are equally likely, the curve must be flat, forming a rectangle of length $100 - 0 = 100$. Finally, property 3 states that the area under this curve must be 1, hence the height of this curve must be $\frac{1}{100}$. Figure 6.2 gives the graph of this curve whose probability density function is

$$f(x) = \begin{cases} \frac{1}{100} & 0 \le x \le 100 \\ 0 & \text{elsewhere} \end{cases} \tag{6.1}$$

Remember that the height of this curve does not represent the probability of a given point, since this probability is essentially zero for any point. However, by property 2, the probability of the random variable occurring in any given interval can be calculated. In this example the probability of lost work

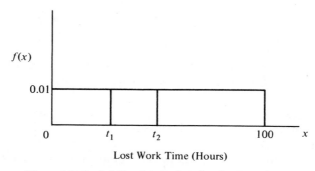

Lost Work Time (Hours)

Figure 6.2. Probability density function for Equation (6.1)

time falling between t_1 and t_2 is just $(.01)(t_2 - t_1)$. For example, the probability of an accident causing lost work hours between 40 and 50 is 0.1.

Of course, the form of the distribution many not be level. Many random variables exhibit a concentration of probability toward some midpoint. Distributions of this nature often fall into a class of distributions called *normal* or *Gaussian*. The general form of this distribution is found in Figure 6.3. Tables of probabilities for the normal distribution have been fully tabulated and their use will be described more fully below.

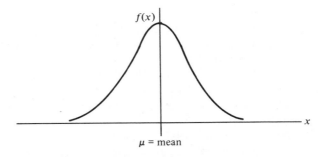

Figure 6.3. General shape of the normal distribution

6.3 Mean of a Probability Distribution

To compare two probability distributions and to perform other statistical tests, it is necessary to extract from the distributions estimates of certain numbers which characterize them, called *parameters*. A parameter is a unique number for a given distribution, and hence it may be viewed as a constant. Generally, the exact value of any parameter is unknown, since this would require a total knowledge of all outcomes (past, present, and future). Therefore, parameters must be estimated. But, since these estimates are functions of random variables, they are subject to random fluctuation themselves, and hence the parameter estimates have distributions.

The most used, and probably the most useful, parameter is called the mean. It provides a useful measure of where the *center* of the distribution lies. By definition, the mean is given by μ, where

$$\mu = \sum_{\text{all } x} xf(x) \qquad (6.2)$$

where x ranges over all the values that the random variable may take and $f(x)$ is the probability density function discussed above. For a discrete distribution (only), $f(x)$ can be viewed as the probability with which the specific value of x occurs.

As mentioned in the general discussion of parameters, typically some values of x are unknown, especially those that are yet to occur. Hence it is necessary to estimate the value of μ using the sample measurements that we have available. This is done for a discrete distribution by calculating $\bar{x}$, where

$$\bar{x} = \sum_{i=1}^{n} x_i P(x_i) \tag{6.3}$$

where x_i is the ith value that the discrete random variable can take, the number of values is n, and the estimated probability of occurrence of the ith value is $P(x_i)$. $P(x_i)$ corresponds to the third column of Table 6.1. Since there are a finite number of x_i's in a discrete distribution, it might be concluded that $\bar{x} = \mu$. However, the example displayed by Table 6.1 illustrates that $P(x_i)$ is an estimate subject to error. To perfect $P(x_i)$, *all* the measurements within the population would have to be taken. Hence Equation (6.3) must stand as an *estimate* of the true population mean.*

For a continuous random variable, Equation (6.2) may be used directly by substituting the appropriate function and integrating over all x. A discussion of calculus is outside of the scope of this book, and therefore the interested reader is referred to any standard statistics text. For our purposes here, the exact functional form of the distribution, as given in Equation (6.1), is rarely known. It is still possible to calculate the estimated mean of the distribution, however, by taking a random sample of n measurements of the variable from its population and calculating

$$\bar{x} = \sum_{i=1}^{n} \frac{x_i}{n} \tag{6.4}$$

This should be immediately recognized as the *arithmetic mean* or *average* of the n measurements. Equation (6.4) follows from Equation (6.3) if the probabilities of choosing any one of the n values is equally likely and hence have probability $1/n$. Thus the n values must be chosen at random from the distribution; no bias or favoritism must be exercised in selection.

It follows then that Equation (6.4) is valid for a random sample of n measurements taken from any distribution. This is a general formula for computing an effective measure of the location of the center of a distribution. This is essential if random variables are to be quantitatively compared.

---

*Note that this is not the case in situations where probabilities are *known* a priori, such as in the rolling of a die or the toss of a coin. Also, if all measurements within a population were taken, direct comparisons could be made with no need for statistical inference.

6.4 Standard Deviation of a Probability Distribution

It is not enough to know the location of the center of the distribution when comparing random variables. Two random variable will almost always differ by some amount even if they are taken from identical populations. This is due to the nature of random variation. The question is not: Do they differ? but rather, is their difference *significant*? Now to know whether two random variables are significantly different, it is necessary to know how much variation could be expected due to chance or normal causes (i.e., random variation). Hence a measure of spread, or variation, is required to determine to what extent the distribution is concentrated about its central point.

To measure the distance from each point in the distribution to the central point, or mean, it is necessary to subtract $x - \mu$, for all x. A first thought might be to measure the spread of the distribution by summing all these individual measures of distance. However, the resultant will be zero, since half the distribution is above the mean and half is below it. Somehow it is necessary to get rid of the cancellation effect caused by the sign of $x - \mu$. One thought is to take the absolute value of $x - \mu$ and form a parameter called the *absolute deviation*. However, a more useful statistic is formed by squaring the difference.

The variance of a probability function* with mean μ and density function $f(x)$ is defined to be

$$\sigma^2 = \sum_{\text{all } x} (x - \mu)^2 f(x) \tag{6.5}$$

Again since not all x are known on an empirical basis, it is necessary to estimate the true parameter σ^2 with an estimator, s^2. The value of this can be calculated for any distribution given a sample of n measurements of the random variable by using the following formula:

$$s^2 = \frac{\sum_{i=1}^{n} (x_i - \bar{x})^2}{n - 1} \tag{6.6}$$

An even more useful statistic is the standard deviation, which is defined to be the square root of the variance:

$$s = \sqrt{\frac{\sum_{i=1}^{n} (x_i - \bar{x})^2}{n - 1}} = \sqrt{\frac{\sum_{i=1}^{n} x_i^2 - \left[\left(\sum_{i=1}^{n} x_i\right)^2 / n\right]}{n - 1}} \tag{6.7}$$

---

*As in Equation (6.2), the summation sign will be replaced by an integral for a continuous distribution.

the latter representation being an easier computation formula of the estimate of the standard deviation.

Given above are estimates for two parameters of random-variable distributions, the mean and the standard deviation. These will measure the central location and the spread of the distribution, respectively. Remaining is a technique whereby these measurements can be used to determine if causal factors are in reality altering the value of a random variable.

6.5 Distribution of the Mean

The following theorem will be stated without proof. However, the rationale for the theorem is given below.

Theorem 6.1

If n samples are taken at random from a population with mean μ and variance σ^2, then a new random variable can be formed, $\bar{x}$, as given by Equation (6.4), whose distribution also has mean μ. If the population from which the samples are taken is infinite, then the variance of $\bar{x}$ is given by σ^2/n.

To demonstrate the validity of this theorem, suppose that a sample of n measurements is taken over and over again from an infinite population. For each sample a value of $\bar{x}$, the sample mean, is calculated. Now since the sample taken from the population is generally different from sample to sample, the values of $\bar{x}$ calculated will also vary. In fact, $\bar{x}$ can be considered a random variable, since it is a function of random variables. It should be reasonable that the mean of the distribution of $\bar{x}$'s will be the population mean μ. Since an average of the sample measurements is being taken, the largest $\bar{x}$ cannot exceed the largest individual measurement, and the smallest $\bar{x}$ obtained in this way cannot be smaller than the smallest measurement. In fact, generally the higher measurements will tend to cancel or at least buffer the low measurements, and vice versa, in the determination of the sample means. Hence the variation of the $\bar{x}$'s will be smaller than the variation of individual values, depending upon the sample size taken. The larger the sample size, the smaller the variation. The sample size could be enlarged to such a point that, if the sample size equaled the population size, the variance would be zero. Assuming this to be possible, a value of μ would be obtained on each sample. Statistical derivations have shown the variance of $\bar{x}$ for samples of size n to be σ^2/n.

The reason for discussing the distribution of $\bar{x}$'s is that it can be shown that as n gets large, this distribution is close enough to the normal distribution to give excellent results. In fact, for as low a value as $n = 10$ there is a close approximation, depending upon the underlying distribution of the x_i's. But for $n = 25$ or 30, the approximation gives excellent results with most distributions encountered in practice.

6.6 The Normal Distribution

With the above references to the normal distribution, the reader may wonder why this particular distribution is so important. It is important because its properties are well known and therefore many statistical tests can be performed if the assumption of normality holds. Further, since so many random variables fall into the normal distribution, it is possible to draw probabilistic conclusions about these variables. In particular, if a random variable has a normal distribution and its mean and standard deviation are known, any measurement from that distribution will lie within ± 1 standard deviation of the mean approximately 68% of the time (see Figure 6.4). For two times the standard deviation (σ) the probability is approximately 95%, and for 3σ units, the probability is over 99%. In practice this means that the probability that certain measurements come from a given normal distribution can be determined. If this probability is very small, it will probably be advantageous to determine the reason for the shift in the distribution.

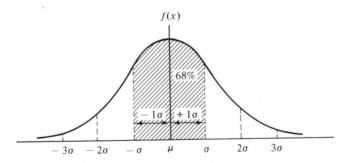

Figure 6.4. Probability structure of the normal distribution

Conversely, it may be necessary to determine if a given remedy or treatment has an effect upon the random variable in question. For example, the effect that the enforced wearing of safety glasses has upon eye-injury frequency might be studied. If measurements of current eye-accident frequency fall "significantly" below those before implementation of safety glasses, the conclusion would be that safety glasses do have a favorable effect. "Significantly" here means below a prechosen level (i.e., 2σ or 3σ), as will be shown below with an example.

Table 6.2 can be used to determine the probability of a normally distributed random variable departing from this mean by more than z standard

Table 6.2 VALUES OF THE STANDARD NORMAL DISTRIBUTION FUNCTION*

z_f	0	1	2	3	4	5	6	7	8	9
−3.0	.0013	.0010	.0007	.0005	.0003	.0002	.0002	.0001	.0001	.0001
−2.9	.0019	.0018	.0017	.0017	.0016	.0016	.0015	.0015	.0014	.0014
−2.8	.0026	.0025	.0024	.0023	.0023	.0022	.0021	.0021	.0020	.0019
−2.7	.0035	.0034	.0033	.0032	.0031	.0030	.0029	.0028	.0027	.0026
−2.6	.0047	.0045	.0044	.0043	.0041	.0040	.0039	.0038	.0037	.0036
−2.5	.0062	.0060	.0059	.0057	.0055	.0054	.0052	.0051	.0049	.0048
−2.4	.0082	.0080	.0078	.0075	.0073	.0071	.0069	.0068	.0066	.0064
−2.3	.0107	.0104	.0102	.0099	.0096	.0094	.0091	.0089	.0087	.0084
−2.2	.0139	.0136	.0132	.0129	.0126	.0122	.0119	.0116	.0113	.0110
−2.1	.0179	.0174	.0170	.0166	.0162	.0158	.0154	.0150	.0146	.0143
−2.0	.0228	.0222	.0217	.0212	.0207	.0202	.0197	.0192	.0188	.0183
−1.9	.0287	.0281	.0274	.0268	.0262	.0256	.0250	.0244	.0238	.0233
−1.8	.0359	.0352	.0344	.0336	.0329	.0322	.0314	.0307	.0300	.0294
−1.7	.0446	.0436	.0427	.0418	.0409	.0401	.0392	.0384	.0375	.0367
−1.6	.0548	.0537	.0526	.0516	.0505	.0495	.0485	.0475	.0465	.0455
−1.5	.0668	.0655	.0643	.0630	.0618	.0606	.0594	.0582	.0570	.0559
−1.4	.0808	.0793	.0778	.0764	.0749	.0735	.0722	.0708	.0694	.0681
−1.3	.0968	.0951	.0934	.0918	.0901	.0885	.0869	.0853	.0838	.0823
−1.2	.1151	.1131	.1112	.1093	.1075	.1056	.1038	.1020	.1003	.0985
−1.1	.1357	.1335	.1314	.1292	.1271	.1251	.1230	.1210	.1190	.1170
−1.0	.1587	.1562	.1539	.1515	.1492	.1469	.1446	.1423	.1401	.1379
− .9	.1841	.1814	.1788	.1762	.1736	.1711	.1685	.1660	.1635	.1611
−.8	.2119	.2090	.2061	.2033	.2005	.1977	.1949	.1922	.1894	.1867
−.7	.2420	.2389	.2358	.2327	.2297	.2266	.2236	.2206	.2177	.2148
−.6	.2743	.2709	.2676	.2643	.2611	.2578	.2546	.2514	.2483	.2451
−.5	.3085	.3050	.3015	.2981	.2946	.2912	.2877	.2843	.2810	.2776
−.4	.3446	.3409	.3372	.3336	.3300	.3264	.3228	.3192	.3156	.3121
−.3	.3821	.3783	.3745	.3707	.3669	.3632	.3594	.3557	.3520	.3483
−.2	.4207	.4168	.4129	.4090	.4052	.4013	.3974	.3936	.3897	.3859
−.1	.4602	.4562	.4522	.4483	.4443	.4404	.4364	.4325	.4286	.4247
−.0	.5000	.4960	.4920	.4880	.4840	.4801	.4761	.4721	.4681	.4641
.0	.5000	.5040	.5080	.5120	.5160	.5199	.5239	.5279	.5319	.5359
.1	.5398	.5438	.5478	.5517	.5557	.5596	.5636	.5675	.5714	.5753
.2	.5793	.5832	.5871	.5910	.5948	.5987	.6026	.6064	.6103	.6141
.3	.6179	.6217	.6255	.6293	.6331	.6368	.6406	.6443	.6480	.6517
.4	.6554	.6591	.6628	.6664	.6700	.6736	.6772	.6808	.6844	.6879
.5	.6915	.6950	.6985	.7019	.7054	.7088	.7123	.7157	.7190	.7224
.6	.7257	.7291	.7324	.7357	.7389	.7422	.7454	.7486	.7517	.7549
.7	.7580	.7611	.7642	.7673	.7703	.7734	.7764	.7794	.7823	.7852
.8	.7881	.7910	.7939	.7967	.7995	.8023	.8051	.8078	.8106	.8133
.9	.8159	.8186	.8212	.8238	.8264	.8289	.8315	.8340	.8365	.8389
1.0	.8413	.8438	.8461	.8485	.8508	.8531	.8554	.8577	.8599	.8621
1.1	.8643	.8665	.8686	.8708	.8729	.8749	.8770	.8790	.8810	.8830
1.2	.8849	.8869	.8888	.8907	.8925	.8944	.8962	.8980	.8997	.9015
1.3	.9032	.9049	.9066	.9082	.9099	.9115	.9131	.9147	.9162	.9177
1.4	.9192	.9207	.9222	.9236	.9251	.9265	.9278	.9292	.9306	.9319
1.5	.9332	.9345	.9357	.9370	.9382	.9394	.9406	.9418	.9430	.9441
1.6	.9452	.9463	.9474	.9484	.9495	.9505	.9515	.9525	.9535	.9545
1.7	.9554	.9564	.9573	.9582	.9591	.9599	.9608	.9616	.9625	.9633
1.8	.9641	.9648	.9656	.9664	.9671	.9678	.9686	.9693	.9700	.9706
1.9	.9713	.9719	.9726	.9732	.9738	.9744	.9750	.9756	.9762	.9767
2.0	.9772	.9778	.9783	.9788	.9793	.9798	.9803	.9808	.9812	.9817
2.1	.9821	.9826	.9830	.9834	.9838	.9842	.9846	.9850	.9854	.9857
2.2	.9861	.9864	.9868	.9871	.9874	.9878	.9881	.9884	.9887	.9890
2.3	.9893	.9896	.9898	.9901	.9904	.9906	.9909	.9911	.9913	.9916
2.4	.9918	.9920	.9922	.9925	.9927	.9929	.9931	.9932	.9934	.9936
2.5	.9938	.9940	.9941	.9943	.9945	.9946	.9948	.9949	.9951	.9952
2.6	.9953	.9955	.9956	.9957	.9959	.9960	.9961	.9962	.9963	.9964
2.7	.9965	.9966	.9967	.9968	.9969	.9970	.9971	.9972	.9973	.9974
2.8	.9974	.9975	.9976	.9977	.9977	.9978	.9979	.9979	.9980	.9981
2.9	.9981	.9982	.9982	.9983	.9984	.9984	.9985	.9985	.9986	.9986
3.0	.9987	.9990	.9993	.9995	.9997	.9998	.9998	.9999	.9999	.9999

*B. W. Lindgren, *Statistical Theory*, Macmillan Publishing Co., Inc., New York, 1960.

deviation units. The number of standard deviation units for any value of x can be determined by subtracting the distribution mean from the value of x in question and dividing by the distribution stanard deviation:

$$z = \frac{x - \mu}{\sigma} \tag{6.8}$$

where x is any specified value, μ is the population mean, and σ is the population standard deviation. Since x in Equation (6.8) is a random variable, then z is also a random variable. Here z is said to have the standard normal distribution with a mean of 0 and a variance and standard deviation of 1. By using Equation (6.8) to "standardize" any normally distributed random variable, the standard normal table can be used for all problems involving the normal distribution.

Table 6.2 can thus be used to determine the probability that a value chosen from a normal distribution (at random) will have the value x *or less*. This, in turn, can be used to determine the probability of occurrence of more than x, merely by subtracting the table value from 1.

Example 6.1 Use of Table 6.2

Given that the number of accidents in a given installation over a monthly period is normally distributed with mean of 10 and a standard deviation of 2, what is the probability that 16 accidents in a given month are due solely to chance?

Solution: Interpret "solely to chance" to mean that the 16 accident sample came from the same distribution as the previous measurements. Hence, using Equation (6.8),

$$z = \frac{16 - 10}{2} = 3$$

or 3σ units. Table 6.1 states that when $z = 3$, the probability of z or less is .9987. Hence there is only a .0013 chance of 3 or more σ units, or 16 or more accidents. This is the probability that 16 or more accidents were due solely to chance.

6.7 The *t* Distribution

One problem in applying the preceding theory is that the underlying distribution must be normal. If, instead of testing an individual value from the underlying population, the value of a sample mean of size n were being tested, this value would have an approximately normal distribution if the value of n is large. The random variable becomes $\bar{x}$, and, by Theorem 6.1, its standard deviation is $\sigma/\sqrt{n}$, where σ is the population standard deviation. Hence, for this type of statistical analysis, Equation (6.8) can be rewritten as

$$z = \frac{\bar{x} - \mu}{\sigma/\sqrt{n}} \tag{6.9}$$

So for any test involving a sample mean of size n (n large), where the population mean and standard deviation are known, Equation (6.9) may be used in conjunction with Table 6.1. Remember that $\bar{x}$ is being viewed as a random variable with mean μ and variance σ^2/n.

It is sometimes desirable to relax two assumptions regarding the use of Equation (6.9). The first is the necessity for a large value of n, and the second is the assumed knowledge of the population standard deviation, σ. In order to do this, however, it is necessary to go back to the assumption that the underlying distribution is at least roughly normal. In practice this means that the distribution must be roughly bell-shaped and symmetrical. Methods for verifying these assumptions are given in most elementary texts. Many natural phenomena fall into this category, and most others can be transformed by logarithms, exponentiation, and so on.

Given the above conditions—that the underlying distribution is roughly normal—the particular distribution that the sample means have is a function of their sample size. For example, means of a large sample size will obviously fall into a tighter distribution than means of a small sample size. In the extreme, means of sample size 1 would have the greatest variance, whereas means of the population size would have zero variance, each sample being the population mean. Thus to consider samples of small n ($n < 30$), it will be necessary to consider a different distribution for each sample size n.

To relax the second assumption, the knowledge of σ, consider the work performed in 1908 by W. S. Gosset (who used "Student" as a pseudonym when he published these results). He defined the t statistic to be

$$t = \frac{\bar{x} - \mu}{s/\sqrt{n}} \tag{6.10}$$

where t is the value of a random variable said to have a Student's t distribution, $\bar{x}$ is the mean of a sample of size n, μ is the population mean, s is the sample standard deviation given by Equation (6.7), and n is the sample size. The t statistic is a random variable that has a different distribution for each value of n. For $n = 2$ to 30, the values of t are given in Table 6.3 for certain cumulative probabilities. Figure 6.5 illustrates the use of Table 6.3 with four examples

Table 6.3 CRITICAL VALUES FOR STUDENT'S *t* DISTRIBUTION*

			Pr{Student's $t \leq$ tabled value}			
n	0.75	0.90	0.95	0.975	0.99	0.995
2	1.0000	3.0777	6.3138	12.7062	31.8207	63.6574
3	.8165	1.8856	2.9200	4.3027	6.9646	9.9248
4	.7649	1.6377	2.3534	3.1824	4.5407	5.8409
5	.7407	1.5332	2.1318	2.7764	3.7469	4.6041
6	.7267	1.4759	2.0150	2.5706	3.3649	4.0322
7	.7176	1.4398	1.9432	2.4469	3.1427	3.7074
8	.7111	1.4149	1.8946	2.3646	2.9980	3.4995
9	.7064	1.3968	1.8595	2.3060	2.8965	3.3554
10	.7027	1.3830	1.8331	2.2622	2.8214	3.2498
11	.6998	1.3722	1.8125	2.2281	2.7638	3.1693
12	.6974	1.3634	1.7959	2.2010	2.7181	3.1058
13	.6955	1.3562	1.7823	2.1788	2.6810	3.0545
14	.6938	1.3502	1.7709	2.1604	2.6503	3.0123
15	.6924	1.3450	1.7613	2.1448	2.6245	2.9768
16	.6912	1.3406	1.7531	2.1315	2.6025	2.9467
17	.6901	1.3368	1.7459	2.1199	2.5835	2.9208
18	.6892	1.3334	1.7396	2.1098	2.5669	2.8982
19	.6884	1.3304	1.7341	2.1009	2.5524	2.8784
20	.6876	1.3277	1.7291	2.0930	2.5395	2.8609
21	.6870	1.3253	1.7247	2.0860	2.5280	2.8453
22	.6864	1.3232	1.7207	2.0796	2.5177	2.8314
23	.6858	1.3212	1.7171	2.0739	2.5083	2.8188
24	.6853	1.3195	1.7139	2.0687	2.4999	2.8073
25	.6848	1.3178	1.7109	2.0639	2.4922	2.7969
26	.6844	1.3163	1.7081	2.0595	2.4851	2.7874
27	.6840	1.3150	1.7056	2.0555	2.4786	2.7787
28	.6837	1.3137	1.7033	2.0518	2.4727	2.7707
29	.6834	1.3125	1.7011	2.0484	2.4671	2.7633
30	.6830	1.3114	1.6991	2.0452	2.4620	2.7564
∞	.6755	1.2817	1.6450	1.9600	2.3267	2.5760

*D. E. Owen, *Handbook of Statistical Tables*, Addison-Wesley Publishing Company, Inc., Reading, Mass., 1962.

in which $n = 12$. It should be recognized that Table 6.3 is analogous to Table 6.1 except that only six points are given for each distribution, and these are given in the body of the table. In fact, when $n = \infty$, the *t* statistic is normally distributed, and hence these particular entries in the table are identical. Hence the use of the *t* statistic is very similar to the use of the *z* statistic, only now the population standard deviation need not be known.

Example 6.2 Use of the t Statistic

Suppose that the average noise level in a given factory has been 90 dB(A) in the past. A series of nine measurements are taken at random during working

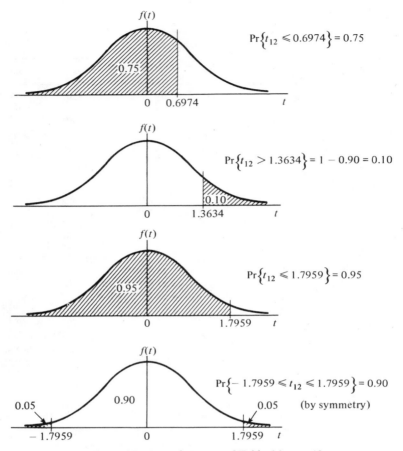

Figure 6.5. Example usages of Table 6.3, $n = 12$

hours, yielding the following results: 95, 98, 92, 84, 105, 92, 110, 86, and 98 dB(A). What conclusions can be drawn concerning the shift in the population?

Solution: If the population has not shifted, it will still have a mean of 90 dB(A). The mean of the nine random samples will be a random variable drawn from a population with a Student's t distribution whose mean is 90 dB(A), assuming no population change. The population standard deviation is estimated by s in Equation (6.7). For this example the value of the random variable is

$$\bar{x} = \frac{95 + 98 + 92 + 84 + 105 + 92 + 110 + 86 + 98}{9}$$

$$= 95.56$$

The population standard deviation is estimated by

$$s = \sqrt{\frac{\sum x_i^2 - \frac{(\sum x_i)^2}{n}}{n-1}} = \sqrt{\frac{82{,}738 - \frac{739{,}600}{9}}{8}}$$

$$= \sqrt{70.03} = 8.37$$

Thus the standard deviation of the random variable $\bar{x}$ is

$$s_{\bar{x}} = \frac{s}{\sqrt{n}} = \frac{8.37}{\sqrt{9}} = \frac{8.37}{3} = 2.79$$

The value of the t statistic is

$$t = \frac{\bar{x} - \mu}{s/\sqrt{n}} = \frac{95.56 - 90}{2.79} = 1.99$$

standard deviation units. Hence the sample value deviates from the theoretical value ($\mu = 90$) by 1.99 σ units. According to Table 6.3, the probability of this occurring (i.e., the value of a random variable being 1.99 *or less* σ units from the population mean when $n = 9$) is between .95 and .975. Interpolation yields a value of .9573, about 96%. Thus the probability of obtaining the mean value of 95.56 or greater in a sample of 9 if $\mu = 90$ is, in fact, only about 4%. One of two conclusions can be drawn: (1) the analysts were "lucky," inasmuch as they accomplished something that has less than a 1 in 20 chance of occurring, or (2) the population mean has, in fact, shifted to somewhat higher than $\mu = 90$. If there is good reason to believe the latter, then appropriate actions should be taken. However, if there is no apparent reason for the increase, further sampling might be justified to further verify the results. If a repetition of the experiment again obtained a value at or above 95.56, this would provide more conclusive evidence that the noise level had increased.

6.8 Tests of Hypotheses

The example above illustrates the use of statistical estimation to draw conclusions. It is helpful to formulate a more precise structure for performing tests. In statistical terminology this is referred to as *testing hypotheses*. Before continuing to discuss this procedure, some thought should be given to the types of errors that can be incurred.

Testing a hypothesis involves first the formulation of a hypothesis and then the specification of definite outcomes of an experiment. These outcomes are asserted to either cause an acceptance or a rejectance of the hypothesis.

This process is structured in such a way that a deterministic result (i.e., to accept or to reject) is obtained based upon probabilistic data (i.e., the outcome of an experiment which is a random variable). Therefore, one of two possible errors could be made, which are named as follows:

α error: to *reject* the hypothesis when, in fact, it is correct.

β error: to *accept* the hypothesis when, in fact, it is false.

Usually the α error will be chosen before the test is begun. After the test the β error will be studied, as part of the interpretation of the results.

The general procedure that will be followed for testing hypothesis is as follows:

1. State the hypothesis.
2. State the alternative hypothesis.
3. Specify a value of α.
4. Determine the statistic that is to be used.
5. Determine the sample size of the experiment to be performed, n.
6. Determine the acceptance and rejection regions.
7. Perform the experiment and draw the conclusions based upon step 6.
8. Study the conclusions in light of β-error calculations.

This procedure will be followed in specifying the particular tests to be performed under the following conditions: (1) tests of one mean when (a) σ is known or $n \geq 30$ or (b) σ is unknown and $n < 30$; and (2) tests of two means under conditions (a) and (b).

6.8.1 Hypothesis Concerning One Mean (σ Known or $n > 30$)

The general rule of thumb for testing one mean when n is large (greater than 30) is that s calculated by Equation (6.7) will closely approximate σ, and hence the test can proceed under the assumption that σ is known. There are exceptions to this rule when the underlying distribution varies greatly from normal. However, if a general bell-shaped symmetrical distribution is in effect, this assumption should give good results. The general procedure given above will now be applied to these circumstances.

1. State the hypothesis. In this case the hypothesis will be $\mu = \mu_0$ (i.e., the mean of the distribution in question is equal to a specified value μ_0).
2. State the alternatives. Two possible alternatives* will be considered

---

*In most elementary texts, a third $\mu \neq \mu_0$ is considered, which leads to what is called a "two-tailed test." For simplicity this is excluded here, since it is primarily of theoretical value. The interested reader is referred to any elementary statistical textbook.

for simplicity, $\mu > \mu_0$ and $\mu < \mu_0$. Generally, one of these will be chosen and specified depending upon the objective of the test.

3. Specify a value of α. As an example, suppose that $\alpha = .05$ (i.e., we are willing to tolerate a 5% probability of rejecting the hypothesis $\mu = \mu_0$ even though it is true).

4. Determine the statistic to be used. Since n is large, assume that σ is known. Thus the z statistic may be used, assuming that the means of the samples have normal distributions (which is generally a valid assumption). Hence the statistic $z = (\bar{x} - \mu_0)/(\sigma/\sqrt{n})$ will be used according to Equation (6.9).

5. Determine n, the sample size. Generally, n should be as large as possible. However, there is a restriction by practical as well as economic considerations. Only a certain number of samples may be available. To obtain more measurements may cost more than the verification of the hypothesis is worth. Formulas are given in many texts to determine n, given the α and β errors desired. Given a particular α error, the value of the β error is reduced by increases in the sample size. Judgment is usually applied in choosing n as large as possible subject to practical considerations.

6. Determine the acceptance and rejectance regions. Here assume that the hypothesis is true and use the value of α to determine the "cutoff" point for acceptance or rejection. Figure 6.6 illustrates the rationale upon which this point is determined. For $H_0: \mu = \mu_0$ and $H_a: \mu >$

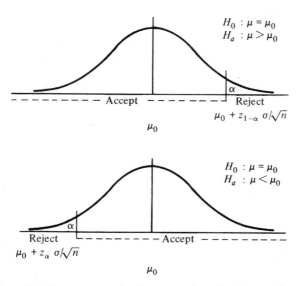

Figure 6.6. Acceptance and rejectance regions for the two alternative hypotheses

μ_0, assuming that H_0 is true, the distribution is centered at μ_0. Now according to the definition of α we will accept α probability of rejecting H_0 even though it is true. Given a population mean of μ_0 and a population standard deviation of σ, there is a probability of α of the random variable being $z_{1-\alpha}$ or more standard deviation units above the mean. Hence the upper acceptance limit is

$$\text{U.L.} = \mu_0 + z_{1-\alpha}\frac{\sigma}{\sqrt{n}} \tag{6.11}$$

where $z_{1-\alpha}$ is the value of z for which there is a probability of $1 - \alpha$ of not exceeding. The value of $z_{1-\alpha}$ for $\alpha = .05$ is given as 1.645 in Table 6.2. Since μ_0 is specified, n is determined in step 5, and σ is estimated by s for large samples, U.L. can now be determined. There is no lower limit since, if $\mu < \mu_0$, this is not of concern under this particular test.

If the second alternative is in effect (i.e., $H_0: \mu = \mu_0$ and $H_a: \mu < \mu_0$), then the bottom half of Figure 6.6 is used. That is, assuming that the distribution is centered at μ_0, if H_0 is rejected and H_a is accepted, our sample must be *significantly* less than μ_0. Hence the rejectance region is below μ_0, the lower limit for acceptance being

$$\text{L.L.} = \mu_0 + z_\alpha\frac{\sigma}{\sqrt{n}} \tag{6.12}$$

where z_α for $\alpha = .05$ is given as -1.645 in Table 6.1. Hence the lower limit of acceptance for this example would be

$$\text{L.L.} = \mu_0 - 1.645\frac{\sigma}{\sqrt{n}}$$

as indicated in Figure 6.6.

7. The experiment is performed to test the hypothesis by taking a sample of n and calculating the mean $\bar{x}$. In step 6 it was assumed that σ is known or could be closely estimated by s, which might come from this or a series of prior samples. The conclusions are based upon the following:
   (a) Reject H_0 and accept $H_a: \mu > \mu_0$ if $\bar{x} > \text{U.L.}$, as given by Equation (6.11).
   (b) Reject H_0 and accept $H_a: \mu < \mu_0$ if $\bar{x} < \text{L.L.}$, as given by Equation (6.12).
   (c) Otherwise, accept H_0.

8. The conclusions should be weighed in the light of practical considerations. Was there, in fact, a nonrandom influence that caused a shift in the population mean? If circumstantial evidence tends to validate the

statistical conclusions, the analyst should be secure in his findings. However, additional testing may be required if this is not the case.

Another way of evaluating the results is by calculating the β error associated with the test. This is fairly easy to do with the z statistic. However, it must be realized from the outset that there is more than one value of β for any given test. There was only one value for α, since α is the probability of rejection *given that the hypothesis H_0 is true*. Since there is only one true value, $\mu = \mu_0$, the α value could be easily represented as in Figure 6.6 and, once specified, it leads to a unique determination of U.L. and L.L.

The β error is defined as the probability of accepting H_0 when it is false. Since there is an infinite number of ways that H_0 may be false, there is an infinite number of β errors. However, if there is a specification of *how false* the hypothesis is, one of the β-error points can be generated. By calculating a sufficient number of points, a curve, called the *operating characteristic* (O.C.) *curve*, can be drawn of β versus "degree of falseness." The general method of calculating β will be given here, and then the entire curve for an example will be drawn.

The top distribution of Figure 6.7 depicts the distribution when the hypothesis $H_0: \mu = \mu_0$ is true. It is centered at μ_0; and α, the area above the upper acceptance limit (U.L.), is the probability of rejecting H_0 even though it is perfectly true. Now assume that H_0 is not true (i.e., $\mu \neq \mu_0$), and further specify that the degree of untruth is $\mu = \mu_1$, where $\mu_1 \neq \mu_0$. Thus the distribution has shifted to be centered at μ_1, as depicted by the lower distribution of Figure 6.7. But acceptance or rejectance of H_0 is based solely upon U.L. no matter what the shift may be. U.L. does not change; it is a

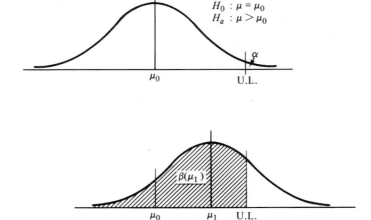

Figure 6.7. Calculation of β for $\mu = \mu_1$

function of α, σ, and μ_0, not of the sample mean or the population mean. Hence U.L. is the same value in the bottom distribution as it is in the original in Figure 6.7.

Now β is just the probability of *accepting* H_0 when it is false. Obviously this depends upon "how false" H_0 is. But given that $\mu = \mu_1$, it is possible to calculate β, which at μ_1 will be designated $\beta(\mu_1)$. It is just the area below U.L. in the distribution centered about μ_1, since this is the probability that H_0 will be accepted even though $\mu = \mu_1$ and not μ_0. The probability of this is given by $\beta(\mu_i)$ where, for any μ_i,

$$z_\beta(\mu_i) = \frac{\text{U.L.} - \mu_i}{\sigma/\sqrt{n}} \tag{6.13}$$

and the probability is obtained by using Table 6.1. The same principle applies if H_a: $\mu < \mu_0$ and an analogous formula to Equation (6.13) becomes effective. The student is urged to become acquainted with the concept of β error, since the mere memorization of formulas has never proved successful. In the example below, the entire O.C. curve is drawn for H_a: $\mu < \mu_0$.

Example 6.3 Test of Hypothesis with Operating Characteristic Curve

Past experience over the years has revealed that the noise level in a given process is 85.0 dB(A), with a standard deviation of 4.3 dB(A). Test the hypothesis H_0: $\mu = 85.0$ against the alternative $\mu < 85.0$ based upon a sample of 64 measurements which yields a mean of $\bar{x} = 83.9$ at a .01 level of significance.

Solution: Use the stepwise procedure given above:

1. H_0: $\mu = 85.0$.
2. H_a: $\mu < 85.0$.
3. $\alpha = .01$ (often called level of significance).
4. z statistic, since σ is known and n is large.
5. $n = 64$.
6. The top distribution in Figure 6.8 depicts the distribution when H_0 is true. The .01 significance level requires that the lower limit be 2.326 standard deviation units below the mean since $z_{.01} = -2.326$. The standard deviation unit is $\sigma/\sqrt{n}$ since the random variable is the mean. Equivalently, using Equation (6.12),

$$\text{L.L} = 85 - 2.326\frac{4.3}{\sqrt{64}}$$
$$= 83.75$$

7. Accept H_0, since the sample value, $\bar{x} = 83.9$, is not below 83.75.

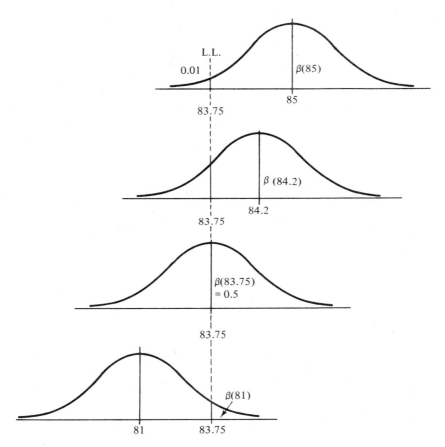

Figure 6.8. Diagrams for Example 6.3

8. Clearly the burden of proof is upon the alternative hypothesis. However, once α is chosen and the test is set up, it is unethical, if not
downright dishonest, to vary α to prove a point. It is interesting to
note, however, that had α been chosen as .05, H_0 would have been
rejected and H_a would have been accepted. Quite often the verdict is
not as close, and these complications are eliminated. The purpose of
this example was to illustrate this point: possibly more experimentation should be performed before the decision is cast in bronze.

Another way to see this is to calculate β for some of the points in
question. $\beta(85)$ is the probability of rejecting H_0 when it is true. Thus
$\beta(\mu_0) = 1 - \alpha$, since we must either accept or reject and α is the probability
of accepting given $\mu = \mu_0$. $\beta(84.2)$ can be calculated by determining the area

above the 83.75 line if the mean is 84.2 (see Figure 6.8). That is, using Equation (6.13),

$$z_{1-\beta(84.2)} = \frac{83.75 - 84.2}{4.3/\sqrt{64}} = -.84$$

So, from Table 6.1, $1 - \beta(84.2) = .2005$ and $\beta = .80$. Similarly, $\beta(83.75) = .5$ since half of the distribution is above the lower limit. Likewise,

$$z_{1-\beta(81)} = \frac{83.75 - 82}{4.3/\sqrt{64}} = 3.26$$

which yields a probability of $\beta = .0006$. A series of calculations would reveal the operating characteristic curve given in Figure 6.9. Note that $\mu > 85$ is not considered, since this is not of concern in this test.

From Figure 6.9, it can be seen that even if the true distribution mean shifted to 84.2, H_0 would still be accepted 80% of the time. When the shift gets down to 83.75, it is still accepted 50% of the time. Certainly the burden of proof is on the shift. That is, it must shift *significantly*, with an $\alpha = .01$ significance, before we will reject H_0 and accept H_a.

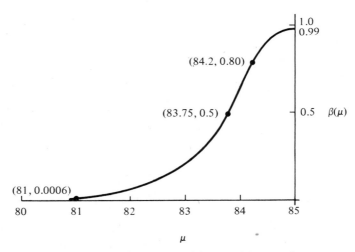

Figure 6.9. Operating characterisic curve for Example 6.3

6.8.2 Hypothesis Concerning One Mean (σ Unknown and $n < 30$)

It n is small and a past knowledge of σ has not been constructed, Student's t distribution can be used. The underlying distribution must be fairly bell-shaped and symmetrical for this approximation to hold. This is usually not a severe restriction in natural situations, into which most safety work falls. The general step-by-step procedure is applicable:

1. State the hypothesis. Here, as above, the hypothesis will be $\mu = \mu_0$, where μ_0 represents an assigned value.
2. State the alternative. Again consider either of two possibilities: $\mu > \mu_0$ or $\mu < \mu_0$, as specified by the analyst.
3. Specify a value of α depending upon how severe a burden of proof is required on the alternative.
4. Determine the statistic to be used. In this case $(\bar{x} - \mu)/(s/\sqrt{n})$ has a t distribution, as discussed in conjunction with Equation (6.10). Since n is small and σ is unknown, s will be used, calculated from the same sample from which $\bar{x}$ is obtained.
5. Determine n, the sample size. It is assumed for this test that n is small (i.e., less than 30). It bears repeating that the accuracy of the test will increase (practically, not theoretically) with a larger value of n. If n can be increased above 30, the test given in Section 6.8.1 should be used.
6. Determine the acceptance and rejectance regions. Again assuming that the hypothesis is true, α will be used to determine the acceptance and rejectance regions. This test proceeds exactly as the one above except that the t statistic will be used instead of the z statistic. Figure 6.6 can be used provided that the limit formulas are modified accordingly. Hence if $H_0: \mu = \mu_0$ and $H_a: \mu > \mu_0$, the upper acceptance limit is

$$\text{U.L.} = \mu_0 + t_{n,1-\alpha}\frac{s}{\sqrt{n}} \qquad (6.14a)$$

corresponding to Equation (6.11). Also, if $H_0: \mu = \mu_0$ and $H_a: \mu < \mu_0$, the lower limit of acceptance is

$$\text{L.L.} = \mu_0 - t_{n,1-\alpha}\frac{s}{\sqrt{n}} \qquad (6.14b)$$

which corresponds to Equation (6.12). Note, however, that since Table 6.3 is not as complete as is Table 6.2, it is necessary to rely upon symmetry to obtain the negative portion of the t values. Since the table values are all positive, compensation is made by subtracting in Equation (6.14b).

7. In many cases the experiment will have been performed already. However, sound statistical practice requires that the testing procedure be established (e.g., α chosen, etc.) prior to the experiment, or at least prior to processing the data. Assume that the experiment (e.g., the random sampling) is completed and that $\bar{x}$ and s are calculated so that

either Equation (6.14a) or (6.14b) can be evaluated. The conclusions will be based upon the following:

(a) Reject H_0 and accept H_a: $\mu > \mu_0$ if $\bar{x} >$ U.L., as given by Equation (6.14a).

(b) Reject H_0 and accept H_a: $\mu < \mu_0$ if $\bar{x} <$ L.L., as given by Equation (6.14b).

(c) Otherwise, accept H_0.

8. The evaluation of the test using the O.C. curve is much the same as given for the large-sample-size case discussed above. It will not be repeated here. The only difficulty that might be incurred is the lack of intermediate values given in Table 6.3.

Example 6.4 Use of the t Statistic to Test Hypotheses

The noise level of a certain process is supposed to be reduced by a given procedure. The level before this procedure is applied was an average of 85.0 dB(A). A sample of nine measurements is taken, as follows: 84.0, 86.0, 82.2, 81.3, 85.2, 84.3, 86.3, 85.3, and 82.1. Has the sound level been significantly reduced?

Solution: Assuming that individual values are roughly from a normal distribution use, the test given above is as follows:

1. H_0: $\mu = 85$.
2. H_a: $\mu < 85.0$.
3. α can be chosen according to the degree of burden placed upon H_a. It helps to choose a value that can be obtained from Table 6.3. Let us choose a moderate degree of significance, $\alpha = .10$.
4. t statistic since n is small and σ is unknown.
5. $n = 9$.
6. Since to disprove H_0, the sample must be *significantly lower* than 85.0, Equation (6.14b) will be used to establish the lower cutoff point. To use this equation it is necessary to have a knowledge of $\bar{x}$ and s. These are obtained by the standard formulas:

$$\bar{x} = \sum_{i=1}^{n} \frac{x_i}{n} = \frac{756.7}{9} = 84.08$$

$$s = \sqrt{\frac{\sum x_i^2 - (\sum x_i)^2/n}{n-1}} = 1.82$$

Thus, using Equation (6.14b) and Table 6.3,

$$\text{L.L.} = 85.0 - 1.3968\frac{1.82}{\sqrt{9}}$$

$$= 84.15$$

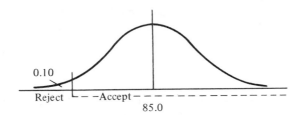

Figure 6.10. Figure for Example 6.4

7. Since $\bar{x} = 84.08 < \text{L.L.} = 84.15$, reject H_0 and accept H_a; that is, conclude that the mean of the population from which the nine samples were taken has a mean less than 85.0 (see Figure 6.10).

6.8.3 Hypothesis Concerning Two Means

In many practical situations it is not possible to obtain a knowledge of μ. For example, in many before–after types of studies, sampling must be performed both before and after, necessitating the test between $\bar{x}_1$ and $\bar{x}_2$. Now, it is possible to assume that one of these (e.g., the "before" sample, $\bar{x}_1$) is equal to the previous population mean μ and proceed by using one of the two tests above. However, it is more accurate to perform the test in light of the following theorem:

If a third distribution is formed by summing random variables from two distributions, one which has mean μ_1 and variance σ_1^2, and the other which has mean μ_2 and variance σ_2^2, then the mean of the third distribution is $\mu_1 + \mu_2$, and its variance is $\sigma_1^2 + \sigma_2^2$. If the third distribution is formed by taking the difference of the second from the first, then the mean of the third distribution is $\mu_1 - \mu_2$ but its variance is still $\sigma_1^2 + \sigma_2^2$.

To apply this theorem to test whether two distributions differ, take the mean of a sample from the first which would have a standard deviation of $\sigma_{\bar{x}_1} = \sigma_1/\sqrt{n_1}$; and take the mean of a sample from the second which would have a standard deviation of $\sigma_{\bar{x}_2} = \sigma_2/\sqrt{n_2}$. Then the standard deviation of the differences between the sample means would be

$$\sigma_{\bar{x}_1 - \bar{x}_2} = \sqrt{\frac{\sigma_1^2}{n_1} + \frac{\sigma_2^2}{n_2}}$$

Now, the same assumptions regarding normality that were discussed above still apply, even though the concern is now with sums and differences of means instead of the means themselves. Hence the test statistics that are in effect are the z and t, depending upon the sample size. If the samples are large (or theoretically if σ_1 and σ_2 are known), then the test statistic is

$$z = \frac{\bar{x}_1 - \bar{x}_2}{\sqrt{\dfrac{\sigma_1^2}{n_1} + \dfrac{\sigma_2^2}{n_2}}} \tag{6.15}$$

where σ_1 and σ_2 may be estimated by s_1 and s_2 provided that n_1 and n_2, the sample sizes, are relatively large (i.e., greater than 30).

If one or both of the sample sizes are small, then the test statistic* is

$$t = \frac{(\bar{x}_1 - \bar{x}_2)}{\sqrt{(n_1 - 1)s_1^2 + (n_2 - 1)s_2^2}} \sqrt{\frac{n_1 n_2 (n_1 + n_2 - 2)}{n_1 + n_2}} \tag{6.16}$$

where the value that will be used in Table 6.3 can be considered to have a sample size of $n_1 + n_2 - 1$.

Rather than enumerating the general procedure for tests between two means, an example that illustrates the use of these two statistics will be given. The tests are performed similarly to those given above. For the z statistic, if $H_0: \mu_1 = \mu_2$ and $H_a: \mu_1 < \mu_2$, then a lower acceptance limit (L.L) is developed from Equation (6.15) as follows:

$$\text{L.L.} = z_{1-\alpha}\sqrt{\frac{\sigma_1^2}{n_1} + \frac{\sigma_2^2}{n_2}} \tag{6.17}$$

On the other hand, if $H_0: \mu_1 = \mu_2$ and $H_a: \mu_1 > \mu_2$, an upper acceptance limit (U.L.) is developed from Equation (6.15) as follows:

$$\text{U.L.} = z_{\alpha}\sqrt{\frac{\sigma_1^2}{n_1} + \frac{\sigma_2^2}{n_2}} \tag{6.18}$$

Notice that the L.L. will always be negative and the U.L. will always be positive. The following rules are used to draw conclusions:

(a) If $(\bar{x}_1 - \bar{x}_2) < $ L.L., then reject H_0 and accept $H_a: \mu_1 < \mu_2$.
(b) If $(\bar{x}_1 - \bar{x}_2) > $ U.L., then reject H_0 and accept $H_a: \mu_1 > \mu_2$.
(c) Otherwise, accept H_0.

Now the same acceptance and rejectance rules as those given above will hold if the sample sizes n_1 and n_2 are small. Now, however, the upper and lower acceptance limits are calculated using Equation (6.16). These are:

$$\text{L.L.} = t_{n_1+n_2-1,\alpha}\sqrt{\frac{(n_1 + n_2)[(n_1 - 1)s_1^2 + (n_2 - 1)s_2^2]}{n_1 n_2 (n_1 + n_2 - 2)}} \tag{6.19}$$

---

*For the development of this formula, see I. Miller and J. E. Freund, *Probability and Statistics for Engineers*, Prentice-Hall, Inc., Englewood Cliffs, N.J., 1965.

and

$$U.L. = t_{n_1+n_2-1,\alpha}\sqrt{\frac{(n_1 + n_2)[(n_1 - 1)s_1^2 + (n_2 - 1)s_2^2]}{n_1 n_2(n_1 + n_2 - 2)}} \qquad (6.20)$$

Two examples will now be given to illustrate the use of each of these statistics.

Example 6.5 Test of Means, Large Sample Size

It is desired to compare the accident records of two similar plants over the past 3 years. Monthly measurements have been taken for both. However, one of the plants started collecting data 4 months later than the other. The monthly number of lost time accidents was recorded and the results were processed to obtain the following:

Plant 1	Plant 2
$n_1 = 36$	$n_2 = 32$
$\bar{x}_1 = 16.1$	$\bar{x}_2 = 14.3$
$s_1 = 4.3$	$s_2 = 5.1$

Solution: Since n is large, assume that $\sigma_1 \simeq s_1 = 4.3$ and $\sigma_2 \simeq s_2 = 5.1$. It appears that plant 1 might be slightly higher in accident frequency than plant 2. From the outset there is no way to show μ_2 to be significantly higher than μ_1. Hence any meaningful comparison necessitates the following analysis, using the step-by-step general procedure given above.

1. $H_0: \mu_1 = \mu_2$.
2. $H_a: \mu_1 > \mu_2$.
3. Assume that $\alpha = .05$, again at the discretion of the analyst.
4. Since n_1 and n_2 are large, the z statistic may be used.
5. $n_1 = 36$ and $n_2 = 32$, of necessity.
6. To disprove H_0, $\bar{x}_1 - \bar{x}_2$ must be significantly greater than zero. Hence an upper limit is required, which is obtained from Equation (6.18):

$$U.L. = 1.645\sqrt{\frac{18.49}{36} + \frac{26.01}{32}} = 1.895$$

7. Now $\bar{x}_1 - \bar{x}_2 = 1.8$, which is less than 1.895. Hence conclude at the .05 significance level that plant 2 is not significantly different from plant 1.

Example 6.6 Test of Means, Small Sample Size

Suppose that monthly accident frequencies are available for the past 30 months. Ten months ago a new accident-prevention system was implemented, and it is desirable to compare before and after. Preliminary processing reveals the following information:

After Implementation	Before Implementation
$n_1 = 10$	$n_2 = 20$
$\bar{x}_1 = 14.3$	$\bar{x}_2 = 18.1$
$s_1 = 3.8$	$s_2 = 4.1$

Solution: The calculated averages indicated a drop in accidents after implementation. The question is: Was this due to chance or has the population in fact shifted? Proceed according to the standard outline:

1. $H_0: \mu_1 = \mu_2$.
2. $H_a: \mu_1 < \mu_2$.
3. Let us choose $\alpha = .05$.
4. Since n_1 and n_2 are small, the t statistic must be used.
5. Of necessity, $n_1 = 10$, $n_2 = 20$.
6. To disprove H_0, $\bar{x}_1 - \bar{x}_2$ must be significantly lower than zero. Thus a lower limit will be required using the t statistic. This is given by Equation (6.19):

$$\text{L.L.} = -1.7011\sqrt{\frac{30[9(14.44) + 19(16.81)]}{200(28)}}$$

$$= -2.639$$

7. Since $x_1 - x_2 = -3.80$, which is less than -2.639, we conclude at the .05 level of significance that the population has, in fact, shifted and that the implementation of the safety procedures has been effective.

6.8.4 Tests of Proportions

In many situations the absolute number of occurrences is not of as much concern as the relative frequency. For example, in comparing the safety of two plants that produce at different levels, it may be more meaningful to test the number of accidents per million man-hours or per some other production or time unit. Hence quite often a fraction between zero and 1 is generated which can be interpreted as the probability of an accident in the given time or production unit specified. In establishing this unit, it is possible to scale so that appropriately sized fractions are generated to fit the assumptions that are required below.

Suppose that in a given time or production interval there is a probability p of an accident occurring. This could be determined by past records (i.e., $p = x/n$, where x is the number of occurrences and n is a relatively large number of time intervals over which the accidents occurred). If the occurrences of accidents are independent (i.e., p is the same for any time interval), then an estimate of the population parameter proportion of occurrences p' is $p' = x/n$. This is called a *binomial distribution*, and it can be shown that the variance of p is

$$\sigma^2 = \frac{p(1-p)}{n} \tag{6.21}$$

If p is not too small or too large (i.e., $.1 \leq p \leq .9$), and n is large ($n > 30$), then the normal distribution yields a good approximation to the binomial.* This results from the fact the proportions are, in fact, a type of average.

In testing one proportion against a determined value, it can be shown that the statistic

$$z = \frac{p - p_0}{\sqrt{\dfrac{p_0(1-p_0)}{n}}} \tag{6.22}$$

has a standard normal distribution, where p is a proportion calculated from a sample size n taken from a population whose proportion is p_0. Thus an upper and lower limit can be formed using Equation (6.22). If $H_0: p = p_0$ and $H_a: p < p_0$, then the lower acceptance limit is

$$\text{L.L.} = p_0 + z_\alpha \sqrt{\frac{p_0(1-p_0)}{n}} \tag{6.23}$$

and if $H_0: p = p_0$ but $H_a: p > p_0$, then the upper acceptance limit is

$$\text{U.L.} = p_0 + z_{1-\alpha} \sqrt{\frac{p_0(1-p_0)}{n}} \tag{6.24}$$

---

*Although it is true that there are other good approximations to the binomial, we may effectively use our knowledge of the normal distribution by scaling the time or production unit in question so that the given assumptions hold. However, if this is not possible, reference should be made in any standard statistics text to the Poisson approximation to the binomial.

The following acceptance rules apply:

(a) If $p <$ L.L., reject H_0 and accept $H_a: p < p_0$.
(b) If $p >$ U.L., reject H_0 and accept $H_a: p > p_0$.
(c) Otherwise, accept H_0.

A slightly different approach must be applied in the event that two proportions are being tested. Given two samples from the same population, the statistic*

$$z = \frac{p_1 - p_2}{\sqrt{\bar{p}(1 - \bar{p})\left(\frac{1}{n_1} + \frac{1}{n_2}\right)}} \tag{6.25}$$

has a standard normal distribution. Here p_1 and p_2 are two sample proportions calculated on the basis of samples of size n_1 and n_2, respectively. The value of $\bar{p}$ is the weighted average of p_1 and p_2:

$$\bar{p} = \frac{n_1 p_1 + n_2 p_2}{n_1 + n_2}$$

Again, upper and lower acceptance limits for the *difference* between p_1 and p_2 can be derived. The hypothesis H_0 is $p_1 = p_2$, and if this holds, $p_1 - p_2 = 0$. Hence the upper and lower acceptance limits are

$$\text{L.L.} = z_\alpha \sqrt{\bar{p}(1 - \bar{p})\left(\frac{1}{n_1} + \frac{1}{n_2}\right)} \tag{6.26}$$

and

$$\text{U.L.} = z_{1-\alpha} \sqrt{\bar{p}(1 - \bar{p})\left(\frac{1}{n_1} + \frac{1}{n_2}\right)} \tag{6.27}$$

where z_α is obtained from Table 6.1.

The acceptance rules are:

(a) If $p_1 - p_2 <$ L.L., reject $H_0: p_1 = p_2$ and accept $H_a: p_1 < p_2$.
(b) If $p_1 - p_2 >$ U.L., reject $H_0: p_1 = p_2$ and accept $H_a: p_1 > p_2$.
(c) Otherwise, accept $H_0: p_1 = p_2$.

Two examples will now be given to illustrate tests of hypothesis that involve proportions.

Example 6.7 Test of One Sample Proportion

Over the past 5 years the number of disabling injuries within a particular industry has been established to be 5.11 per 1 million man-hours. A particular

---

*For the development of this formula, see I. Miller and J. E. Freund, *Probability and Statistics for Engineers*, Prentice-Hall, Inc., Englewood Cliffs, N.J., 1965.

company in this industry has operated for 3 million man-hours and has had 13 disabling injuries. Is this company significantly below the value established for the industry?

Solution: Comparing individual means has little value here, since there is no measure of the variance. As an approximation, this problem can be fitted to the methods for testing proportions mentioned above. Since this is a comparison of a sample value with an established value, Equations (6.23) or (6.24) will apply. In this case the established value is 5.11 per 1 million man-hours = .511/100,000 man-hours = .0511/10,000 man-hours. For the normality assumptions to hold, $.1 \leq p_0 \leq .9$ and hence the value of $p_0 = .511$ is most appropriate. The denominator of this proportion is 100,000 man-hours. Note that multiple occurrences are being ignored. Converting the sample value of 13 to the same units yields 13/3,000,000 man-hours = .433/100,000 man-hours, with a sample of $n = 30$ trials of 100,000 man-hours in the 3,000,000 man-hours. Hence the step-by-step procedure for testing this hypothesis is as follows:

1. $H_0: p = .511$.
2. $H_a: p < .511$.
3. Suppose that $\alpha = .01$ this time, to place a heavier burden of proof on H_a. Note that this means that H_0 will only be rejected with a probability of 1 out of 100 if H_0 is true.
4. Since p is neither too small nor too large, and n is fairly large, the normal approximation to the binomial is appropriate.*
5. The sample size is determined from available records to be 30 trials of 100,000 man-hours each.
6. To disprove H_0, p must be significantly less than .511. Hence a lower limit value must be obtained, so Equation (6.23) can be used as follows:

$$\text{L.L.} = .511 - 2.327\sqrt{\frac{.511(.489)}{30}}$$

$$= .299$$

7. Since .433 is not below the lower acceptance limit of .299, accept H_0 and conclude that the sample value is not significantly different from the population value at the .01 level. It is interesting to note that the same results would be obtained if the .05 level of significance were chosen. The reader may wish to verify this statement.

---

*If the problem cannot be made to conform to this pattern, other tests should be applied. This is covered in most advanced statistical textbooks.

Example 6.8 Test of Two Sample Proportions

Suppose that a company has an eastern and a western division, both performing essentially the same type of function. The following proportions of lost-time accidents have been recorded:

	Eastern	Western
Number of accidents	28	40
Production (units)	20,000	25,000

Can the conclusion be drawn at the .05 level of significance that the western division has a higher relative frequency of accidents?

Solution: In this example two proportions are being compared. Therefore, the theory surrounding Equations (6.26) and (6.27) applies. Assume a common base of 100 units of production. Hence, $P(\text{eastern}) = p_1 = \frac{28}{200} = .14$; and $P(\text{western}) = p_2 = \frac{40}{250} = .16$, where $n_1 = 200$ and $n_2 = 250$. The standard procedure for testing hypothesis follows:

1. $H_0: p_1 = p_2$.
2. $H_a: p_1 < p_2$.
3. In the problem statement, α was chosen as .05.
4. Again the values of p_1 and p_2 are large enough, as are the values of n_1 and n_2 such that the normal approximation to the binomial, as exemplified by Equation (6.25), holds.
5. The sample sizes are determined by the available data and the unit chosen. In this case $n_1 = 200$ and $n_2 = 250$.
6. To disprove H_0, p_1 has to be shown to be significantly smaller than p_2. In other words, $p_1 - p_2$ has to be sufficiently small. This will occur when $p_1 - p_2$ is below the lower limit given by Equation (6.26). Now

$$\bar{p} = \frac{200(.14) + 250(.16)}{450} = .1511$$

so, according to Equation (6.26),

$$\text{L.L.} = -1.645\sqrt{.1511(.8489)\left(\frac{1}{200} + \frac{1}{250}\right)}$$

$$= -.056$$

Now $p_1 - p_2 = .14 - .16 = -.02$, which is not below the lower limit. Hence we conclude that the two samples are not significantly different.

7. Comment on the above example. If $p_1 - p_2$ had been less than the lower limit, it would have been concluded that a significant difference did exist and hence that $p_1 < p_2$. However, even if this could be stated as a statistically verifiable fact, the analyst would not be in a position to state the cause of this difference. Qualitative evaluation would still be necessary to determine whether the higher rate was due to differences in the workers, the management, safety techniques, age of facilities, or whatever. Care and good judgment in the use of statistics is imperative if the results are to be meaningful.

6.9 Control Charts

Because the concepts of control have been applied extensively in this book, the statistical presentation would not be complete without some discussion of the meaning of statistical control. The background for its understanding has been given in the preceding sections. A few definitions and examples should be sufficient to equip the reader with the basic tools necessary for the construction of control charts.

A control chart is a visual means by which an analyst judges whether a process is in control or not. The measurements plotted on the chart are those of any random variable. Hence the frequency and severity of accidents, as well as any other intermediate indicator of hazards, could be plotted. Judgments based upon these plots determine if the process is in control with respect to the random variable under consideration.

Figure 6.11 shows the typical layout of a control chart. The units of the random variable are given on the verticle scale, indicating that the height of the plotted point represents the value of the random variable for the indicated time period. The time scale, given horizontally, shows when the value occurred. The discussion of random variables above brought out that, although any one value cannot be predicted, measurements of central tendency and spread define the expected concentration and range of the variable. Thus, if the variable behaves in a nonrandom way, we can conclude that an outside influence is affecting the random variable. The most common way of identifying when this occurs is through the use of an upper and a lower control limit. These are generally placed at equal distances above and below the mean line.

The measured values as they are recorded in time are plotted as indicated in Figure 6.11. A point falling above or below the control limits, respectively, is indicative of an out-of-control situation, and assignable causes are generally sought. There are other indications of out-of-control situations, also. However, prior to discussing these, the means for obtaining the control limits will be given.

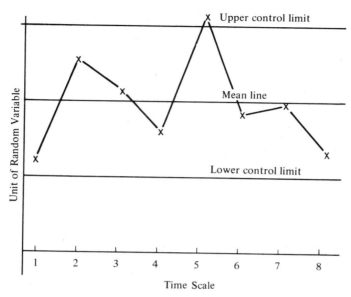

Figure 6.11. Example control chart

6.9.1 Determining Control Limits

The reader has probably already seen the analogy between testing a hypothesis and deeming a point "out of control." The procedures for setting control limits are essentially the same as those for setting the acceptance limits in a test of hypothesis.

The first step involves the establishment of α, that is, the probability of concluding that the process is out of control when in fact it is in control. If methods of identifying causes are expensive and the variable is not critical, a low probability can be tolerated. However, if an early indication of lack of control is necessary, then a high probability of this error should be specified.

Once the value of α is determined, the next question involves the definition of control. Quite often the state "out of control" occurs in one direction only. In sound-level readings, for example, rarely is the analyst concerned with the plant being too quiet. Here only an upper limit would be required, as it would in most cases of pollution measurements. Other monitoring of processes would require both an upper and a lower control limit.

In either case, the value of α chosen will represent the total area of probability in the out-of-control portion of the chart. The upper and lower control limits are obtained depending upon the random variable, its distribution, and the value of α chosen. This can best be illustrated by some examples.

Example 6.9 Accident-Frequency Control Chart

Assume that the frequency of accidents above a given severity for a plant has a normal distribution with a mean of 6 and a standard deviation of 1.5. Frequencies for the first 6 months have been 4, 7, 5, 12, 8, and 6. Set up a monthly control chart for frequency. Allow for a .05 probability of calling a point out of control when it is not.

In this example "out of control" is strictly in terms of an upper limit. However, the analyst chooses to set up a lower limit to provide possible evidence of a lowering of the accident frequency. Thus the .05 probability will be divided, .025 above the upper limit and .025 below the lower limit. The upper limit becomes [see Equation (6.8)]

$$U.L. = \bar{x} + z_{.025}\sigma_x$$
$$= 6 + 1.96(1.5) = 8.94$$

and the lower limit is

$$L.L. = \bar{x} - z_{.025}\sigma_x$$
$$= 6 - 1.96(1.5) = 3.06$$

The control chart is given in Figure 6.12. The fourth month was obviously out of control, and assignable causes should be sought. Any subsequent

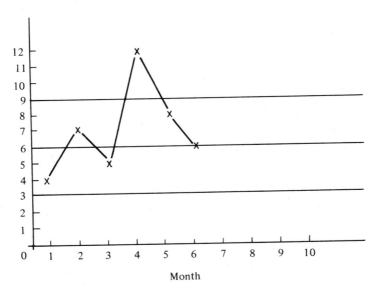

Figure 6.12. Control chart for Example 6.9

monthly reading that falls out of control should also prompt an investigation of the plant.

Example 6.10 Accident-Severity Control Chart

In Example 6.9 the normal distribution was used because frequencies were assumed to be normally distributed. This assumption should be tested, since it does not hold generally. Rather than charting individual random variables, whose distributions may be unknown, often sample means are plotted. As discussed in Section 6.5, the distribution of sample means can be closely approximated by the normal distribution.

The following weekly record of lost workdays per week has occurred for the first 24 weeks of the year: 2, 4, 3, 0, 0, 5, 2, 0, 1, 4, 6, 3, 0, 0, 0, 8, 10, 12, 2, 4, 3, 1, 0, and 0. Set up a control chart on the basis of a 6-week period.

Solution: Since the population variance is unknown, the t distribution must be used. The following results are required: $\bar{x} = 2.92$, $s = 3.32$. Using $\alpha = .05$, the $t_{.975}$ would be required if both limits are to be considered. Thus, using Equation (6.10),

$$\text{U.L.} = \bar{x} + t_{.975,6}\frac{s}{\sqrt{n}}$$

and

$$\text{L.L.} = \bar{x} - t_{.975,6}\frac{s}{\sqrt{n}}$$

yielding 6.40415 and zero, respectively (note that $n = 6$). Each group of six weekly readings should then be plotted as shown in Figure 6.13.

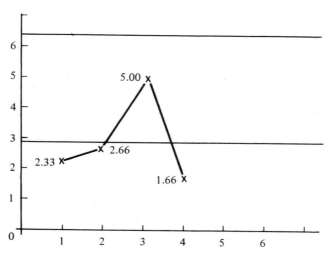

Figure 6.13. Control chart for Example 6.10

6.9.2 *Use of Control Charts*

The above examples indicate that any random variable can be plotted on a control chart. The construction of the chart is simply a matter of applying hypothesis testing on a continuous basis. The primary advantage is that continuous visual perception of the random variable is maintained.

This continuous picture enables the analyst to make judgments not otherwise discernible. This is not limited to the upper and lower control limits demonstrated above. Other factors that the analyst can use as indicators of abnormal operational behavior include:

1. Several points (four or more) in a row on one side of the mean line. The probability of four consecutive points on one side is approximately $.5^4$, or .0625.
2. Identifiable cycles. Here two or three years of history may be required to identify a given month or other period of time when the operation acts in an irregular manner.
3. Several points in a row, either monotonically increasing or decreasing away from the mean line. The probability of this type of trend is difficult to establish. However, since these points are all on one side of the mean line, the probability will be considerably less then $.5^n$, where n is the number of points exhibiting this characteristic.

In quality-control situations, 3σ control limits are generally used, based on the 1-in-1,000 value of α under the normal-distribution assumption. The 2σ and 1σ lines may also be set up, however, to help the analyst identify other out-of-control indicators. For example, two points in a row outside of 2σ limits would have an approximate probability of $(.025)^2 = .000625$, which is about the same as the probability of one point outside 3σ limits, assuming normality. Although control charts for safety applications should not be restricted to the $\alpha = .001$ value, the concept of intermediate lines to identify irregularities is a good one.

Example 6.11 *Use of Control Charts*

Control charts are used in some states for monitoring traffic accidents. Overall accident frequencies, as well as finer breakdowns by severity classifications, may be plotted. The following data, which were obtained from the State of Alabama,* demonstrate the application of control-chart theory to safety.

The following figures indicate the number of fatal accidents per month for 1973: 92, 70, 81, 87, 89, 93, 97, 79, 105, 86, 89, and 81. These figures are for "equivalent months." That is, the monthly total was prorated to a standard

---

*State of Alabama Highway Department, Accident Identification and Surveillance, *Monthly Accident Statistics*, 1973 and 1974.

30.44-day month and rounded appropriately. This distribution of fatal accidents has a mean of 87.42 and a standard deviation of 9.13.

The objective is to set up a control chart for individual values based upon the above data. Subsequent values are to be charted and evaluated based upon the results. Tests of the distribution of monthly fatality accidents indicated that the underlying distribution was very close to being normally distributed. Thus 95% and 99% upper and lower confidence intervals were obtained as follows:

$$UCL(95\%) = 87.42 + 1.96(9.13) = 105.31$$
$$LCL(95\%) = 87.42 - 1.96(9.13) = 69.52$$
$$UCL(99\%) = 87.42 + 2.5760(9.13) = 110.94$$
$$LCL(99\%) = 87.42 - 2.5760(9.13) = 63.90$$

Note the use of actual normal values rather than the use of 2σ and 3σ.

The control chart presented in Figure 6.14 includes the 1974 data for the first 10 months: 70, 57, 80, 60, 79, 65, 74, 62, 61, and 61. Each point should be posted as soon as its value is known.

Consider first the 1973 portion of the data. Except for the run of five successive increases from March through July, the points seem random. None are outside the 95% control limits. One out of 20 could be expected to be there, so it would not be upsetting if one point were between the 95% and

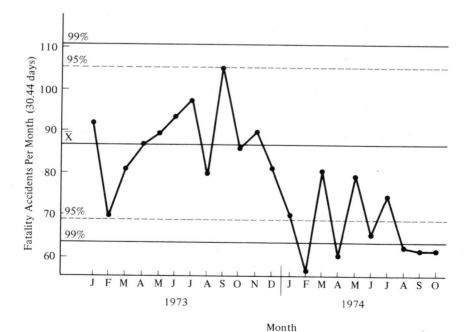

Figure 6.14. Control chart: Traffic fatalities per nonth

99% limits. The upward trend from February forward is interrupted in August and then again in October, November, and December. Thus, although suspicion of some underlying cause may have been created during the period of February through July, at the close of the year it appears that all has been operating in a random manner.

The first point in 1974 is also not unusual. It is within the 95% control limit. However, suspicion of a trend should start to form at this time in anticipation of the February point. Certainly, the February point, when posted, indicates a definite out-of-control situation. Definite causal factors are in effect. In this case, the fuel shortage and the call for voluntary reduction of speeds may have contributed heavily.

The March point would tend to negate the premise above. It may be surmised (without knowledge of future points) that the preceding causes had only a temporary effect. The mandatory 55-mph speed limit was signed into law in late February, but it was not fully implemented until late March. The speed differential between those who were voluntarily complying and others might be the reason for this point being closer to the original mean.

Another out-of-control point in April confirms that there has been a shift to a new mean value. The remaining points for 1974 further confirm the change. Since there appears to be an entirely new distribution in effect from January 1974 forward, new parameters should be set for future charting purposes.

Actual causes cannot be specified from the chart alone; only their effects can be determined. Causes might include fewer cars on the road, increased use of seat belts, safety improvements of the cars and the roadway, and the reduced speed limit. However, other data, such as traffic counts, can be used to focus on the specific cause or causes. The control chart tells, on the gross-data level, that there is something at work.

This section has just scratched the surface of both the theory and the application of control charts. The objective has been to expose the reader to the potential uses of control charts and to relate the techniques to basic statistical theory. The demonstration that some flexibility due to randomness is generally necessary in control relates this section back to the qualitative description of control in Chapter 1. If control-charting techniques are thought to have particular application in the reader's area of safety, he is greatly encouraged to consult the many basic statistics texts which discuss the subject more thoroughly.

6.10 Safety System Simulation

Although often shunned by the theoritician, simulation has been described by many practitioners as the most useful tool of operations research. This is especially valid in the area of safety, since one primary objective of simulation is to perform experiments on paper or within the computer, thus saving the expense of physical experimentation. In safety applications the expense might involve life or health. Thus, if at all possible, it would be extremely beneficial to test a countermeasure by simulation prior to implementing it.

In order to perform a simulation it is necessary to know the relationship between the random variables in the process and their effects upon the objective of the process. Further, the decisions (i.e., alternative counter-measures) will affect these relationships and/or the distributions of the random variables. Hence a distribution must be available for each random variable under each set of decisions. This will become clearer as the mechanics of simulation are discussed.

6.10.1 Monte Carlo Simulation

The techniques of simultation to be presented here will be restricted to the use of discrete distributions. The methodology used in sampling using random numbers is generally called *Monte Carlo*, whether the distributions are continuous or discrete. Manual methods will be discussed. These can easily be computerized, and many computer languages have been developed specifically for simulation.

The basic foundation of Monte Carlo simulation is the method for simulating a variable from its distribution. Consider the process of rolling a die. Suppose that instead of physically rolling the die and observing the out-comes, the process is to be simulated. Now since each of the outcomes is equally likely, they each have a probability of $\frac{1}{6}$. A line of length 1 can be divided into six equal lengths, as shown in Figure 6.15. Note that above the line are recorded the actual points of division of the line, rounded to the nearest .0001. Below the line are the numbered segments corresponding to the outcomes of the rolls of the die.

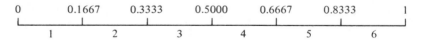

Figure 6.15. Unit line segmented for simulation

Figure 6.15 can be used to simulate the rolling of the die simply by selecting random numbers with a uniform distribution between zero and 1. Such a set of random numbers is given in Table 6.4. Using this list directly from the first column, the simulation results are given in Table 6.5. To summarize the results: 2 ones, 1 two, 1 three, 3 fours, 2 fives, and 1 six were "rolled." This is about what would be expected if the experiment were actually performed.

Before going into the usage of the sampling procedure, a step-by-step framework will be given and illustrated by an example. The following steps must be taken to simulate the outcomes of a random variable:

1. The distribution of the random variable is specified or calculated.
2. The real numbers between zero and 1 are divided into segments, as represented in Figure 6.15. The number of segments is equal to the number of possible outcomes, and the *length* of each segment is

Table 6.4 SAMPLE OF FOUR-DIGIT RANDOM NUMBERS UNIFORMLY DISTRIBUTED
BETWEEN 0 AND 1
(DECIMAL POINTS OMITTED)

1045	0207	3022	9546	1699	6865	3338	3158	1874	3972
6683	9168	9733	0982	2106	7735	3147	2342	2661	6680
1029	2603	1591	1169	3418	2354	1207	8388	2479	0627
9885	5331	2861	6540	4272	5524	1607	4118	9322	4809
5645	2623	8180	9957	2802	6263	7905	0805	2334	1017
1799	6764	6677	3223	6842	7373	8396	1640	8587	9694
3653	1112	8432	1343	6864	2221	2575	7092	2078	4577
8037	0269	7818	2093	7185	9793	0624	1228	1210	4851
5469	6591	0844	1315	3720	2591	6251	4303	9364	8842
6364	6153	2898	6456	0603	2867	1682	3091	1317	7493
5973	7121	1397	4748	4870	4876	1171	9009	8630	2836
4771	0196	9178	5650	0569	8688	5375	1397	0951	3402
8579	2277	9310	4628	5031	8085	7296	5169	7275	3131
7835	1079	3424	0848	5972	2671	7966	6388	7729	7118
6136	8253	9441	4977	2968	7853	7563	4922	9243	1680
1587	2297	0135	9306	7232	7791	9530	2434	8182	6335
8592	8678	6851	8421	3100	4734	3789	6938	0427	2933
6786	1286	2277	6291	7978	1589	1672	1114	1586	0419
5545	9143	2314	3897	6341	9694	6692	0549	2260	1179
7824	4679	7334	6574	4028	7894	9386	9482	0511	2136

Table 6.5 SIMULATION OF THE ROLLING OF A DIE

Trial	Random Number	Outcome from Figure 6.15
1	.1045	1
2	.6683	5
3	.1029	1
4	.9885	6
5	.5645	4
6	.1799	2
7	.3653	3
8	.8037	5
9	.5469	4
10	.6364	4

proportional to the probability of occurrence of each segment. Definite end points for the segments are specified.

3. From a table of random digits, a number is selected and appropriately scaled so that it is between zero and 1.
4. The random number selected must fall within one of the segments defined in step 2. The corresponding outcome is the simulated outcome of the experiment.
5. Repeat steps 3 and 4 an appropriate number of times as required by the experiment.

Example 6.12 Simulation of Two Dice

The example given above of the roll of one die was quite simple, and it may have caused the omission of some of the finer points of the simulation sampling technique. For example, each outcome was equally likely, which is generally not the case. In this example the problem is to simulate, in a roll of two dice, the total number of spots up. Note that this is not the same as simulating each possible outcome. There are 36 possible outcomes, six of which will yield 7 spots up. Here, all 6 are considered to be grouped as a single outcome.

Step 1. The distribution of the possible outcomes of the roll of two dice is given in Table 6.6. The probability of each occurrence is just the number of ways that it can occur, divided by the total number of outcomes. The third column gives the probability of occurrence, which specifies the distribution.

Step 2. The fourth column of Table 6.6 contains the segmental lengths. Note that since the probabilities add to 1, they are equal to the segmental lengths. This will always be the case; the concept of segmental length was

Table 6.6 Distribution and Segmentation for Example 6.11

Outcome (Spots Up)	Possible Combinations	Probability of Occurrence	Segmental Length	Segmental End Points
2	1	$1/36 = .0278$	.0278	0 to .0278
3	2	$2/36 = .0556$	.0556	to .0834
4	3	$3/36 = .0834$	.0834	to .1668
5	4	.1111	.1111	to .2779
6	5	.1389	.1389	to .4168
7	6	.1667	.1667	to .5835
8	5	.1389	.1389	to .7224
9	4	.1111	.1111	to .8335
10	3	.0834	.0834	to .9169
11	2	.0556	.0556	to .9725
12	1	.0278	.0278	to 1.0000
	36	1.000+		

merely for better visualization. The end points for the segments are given in the last column of Table 6.6. These are easily calculated by adding the segmental length to the previous end point.

Steps 3–5.   The results of 20 trials are given in Table 6.7. The random numbers came from the extreme right-hand column of Table 6.4. The outcome was determined by determining the segment according to Table 6.6.

Table 6.7 SIMULATION OF ROLL OF TWO DICE

Trial	Random Number	Outcome
1	.3972	6
2	.6680	8
3	.0627	3
4	.4809	7
5	.1017	4
6	.9694	11
7	.4577	7
8	.4851	7
9	.8842	10
10	.7493	9
11	.2836	6
12	.3402	6
13	.3131	6
14	.7118	8
15	.1680	5
16	.6335	8
17	.2933	6
18	.0419	3
19	.1177	4
20	.2136	5

6.10.2 Simulation Model

To this point the value of simulation might well be questioned, since the process of sampling from the distribution has in itself revealed little that was not already known. In fact, simulation is almost the reverse of statistical analysis, in that we are taking summary information and generating raw data. The process of sampling will not be useful until it is integrated into a model of a total system. However, the preceding examples do help to demonstrate the use of simulation to replace actual experimentation. In fact, the preceding techniques have virtually eliminated the need for rolling dice in games that require such exercises. Similarly, if the distribution of a random variable is fairly well known, its outcomes can be studied without costly experimentation.

To use these outcomes it is necessary to know their effect upon other parts of the operation. Thus some model of the total system must be available.

This need not be a sophisticated or complicated mathematical formulation. To describe this procedure, an example will be given, followed by some generalizations of the technique.

Example 6.13 Simulation of Maintenance Policies

Example 4.4 presented a comparison of three maintenance policies. Some assumptions in the approach used there will be relaxed here as the same problem is solved by simulation. Only policy IV will be simulated.

The system being simulated consists of the configuration given in Figure 4.13. The relaibility function, in terms of mutually exclusive events, was determined to be $R(T) = R(C) + R(A)R(B)R(\bar{C})$. Now this is the objective function, at least in the sense of the reliability aspect of the configuration. It forms the basic structure of the model, where the random variables to be simulated are $R(A)$, $R(B)$, and $R(C)$.

One other aspect, that of replacement, must also be considered as part of the model. Replacement will take place (1) when the policy so states and (2) if a failure occurs. The second type of replacement was ignored in Example 4.4. In simulation, any replacement rule can be applied. For example, we may specify a policy: "Replace all components that fail after three time periods with the exception of those causing system failure, which will be replaced immediately." For simplicity in this example the policy evaluated will be similar to that of policy IV of Example 4.4. However, in addition to replacing all components every five time periods, each component that fails will be replaced immediately.

Essentially the description above forms the model of the system. This would have to be put in mathematical terms and programmed if the computer were to be used. For manual calculations, however, this is sufficient.

The procedure for simulating the configuration $(T = AB + C)$ and the maintenance policy described above will be as follows:

1. Simulate the components operation for one time period, based upon their respective probabilities of success.
2. Evaluate if any component failed. If so, assume replacement for the next time period and modify component success probabilities accordingly.
3. Evaluate if the system failed. Keep a record of system success or failure for each time period.
4. If system life is five time periods since the last total replacement, make a total replacement by modifying the probabilities accordingly.
5. Repeat steps 1–4 for many time periods (at least 1,000). Calculate system reliability as the number of successful time periods divided by the total number of time periods. Also, based upon replacement counts, determine the system cost.

Notice that the reliability is not calculated at each time period. Simulation is used like an experiment, and the data obtained should be viewed as historical data. The simulation of policy IV in Example 4.4 is given in Table 6.8. Each component was simulated for its success by drawing a two-digit random number from Table 6.4. If the random number, scaled between 0 and 1, was greater than or equal to the reliability, the component was recorded to have failed. Otherwise, the component was not replaced but was continued in service. The random numbers are in parentheses after each component's probability of success.

Whenever a component failed, two things were done. First, the replacement was simulated by changing its probability for the next time period. Second, the effect upon the system was checked. If either A or B failed

Table 6.8 SIMULATION OF CONFIGURATION, $T = AB + C$

Trial	Reliabilities* and Random Numbers			S, Success; F, Failure	Comments
	A	B	C		
1	.99 (.10)	.90 (.66)	.99 (.10)	S	
2	.95 (.98)	.88 (.56)	.98 (.17)	S	A fails
3	.99 (.36)	.86 (.80)	.97 (.54)	S	A is replaced
4	.95 (.63)	.84 (.59)	.96 (.47)	S	
5	.90 (.85)	.80 (.78)	.95 (.61)	S	
6	.99 (.15)	.90 (.85)	.99 (.67)	S	All components replaced
7	.95 (.55)	.88 (.78)	.98 (.02)	S	
8	.90 (.91)	.86 (.26)	.97 (.53)	S	
9	.99 (.26)	.84 (.67)	.96 (.11)	S	A is replaced
10	.95 (.02)	.80 (.65)	.95 (.61)	S	
11	.99 (.71)	.90 (.01)	.99 (.22)	S	All components replaced
12	.95 (.10)	.88 (.82)	.98 (.22)	S	
13	.90 (.86)	.86 (.12)	.97 (.91)	S	
14	.84 (.46)	.84 (.30)	.96 (.97)	S	C fails
15	.76 (.15)	.80 (.28)	.99 (.81)	S	C is replaced
16	.99 (.66)	.90 (.84)	.99 (.78)	S	All components replaced
17	.95 (.08)	.88 (.28)	.98 (.13)	S	
18	.90 (.91)	.86 (.93)	.97 (.34)	S	A and B fail
19	.99 (.94)	.90 (.01)	.96 (.68)	S	A and B are replaced
20	.95 (.22)	.88 (.23)	.95 (.73)	S	

*From Table 4.2:

i	$R_i(A)$	$R_i(B)$	$R_i(C)$
0	.99	.90	.99
1	.95	.88	.98
2	.90	.86	.97
3	.84	.84	.96
4	.76	.80	.95

simultaneously with C, the simulation would indicate system failure. This did not occur in the 20 trials.

It should be clear that 20 trials are not a sufficient sample upon which conclusions regarding system reliability can be based. From Example 4.4 we know that the reliability of the system is at least .9804. Several thousand runs should be made in most cases to obtain accurate estimates of the reliability. Although this procedure is easily computerized, the simplicity of the calculations also makes manual simulation feasible.

6.10.3 General Applications

Simulation can be used for evaluating alternatives wherever random variables interact to produce a recordable outcome. The steps in applying simulation must include:

1. Determination of the random variables and the effect that they have upon the operation. In the example above, the random variables were the component reliabilities and the effect upon system performance was given by the expression $T = AB + C$.
2. Determination of the effect of the policy decision upon the random variables and the model structure. The policy applied changed the distributions of the random variables in the reliability example.
3. The formulation of a step-by-step procedure that emulates system operation. This was given in the example above.

Simulation provides a means for studying actual outcomes of an experiment without the necessity of performing the experiment. The simulation of fault trees should be well within the reader's grasp at this point. The advantage of fault-tree simulation is that actual outcomes can be produced in addition to averages. By including within the simulation model distributions of accident severities for given types of accidents, countermeasures can be evaluated in terms of actual outcomes as well as average or expected benefit. The development of such models is left to the reader as an exercise.

6.11 Conclusion

This chapter has presented methods whereby conclusions can be drawn regarding measurements subject to random variation. The need for such conclusions was established in Chapter 5, where one required input to the fault-tree cost/benefit technique was the amount by which a given alternative reduces the probability of a basic event. It is not recommended that statistical analysis provide direct input into the cost/benefit procedure. Rather, statistical studies should be conducted where applicable, to determine if the safety investments are, in fact, going to reduce basic event probabilities. Given this information, the analyst is in a better position to use the available data to estimate just what the effect might be.

The statistical techniques given here are but a small fraction of classical tests that have been developed. As with most disciplines, statistical analysis is continuing to develop. As a result, the presentation here is far from complete. However, the techniques presented are considered to be most useful to the safety engineer. It is hoped that by this "handbook" approach, statistical analyses will be adopted by many more safety engineers.

Statistical tests often have two distinct characters. Theoretically they are planned, sample sizes determined, procedures are established, experimentation is performed, the data are processed, and conclusions are drawn. In practice, the "experiment" is quite often long past before the analyst realizes that available data contain useful information that can be uncovered by the use of statistics. The above procedures have been presented in a general way to take both possibilities into account.

Finally, it should be emphasized that rarely, if ever, in practice do all the assumptions that underly any given statistical test hold up 100%. Hence statistical techniques must be viewed simply as another of the decision maker's tools used to aid him in the task of establishing the valid facts. When viewed in this perspective, they serve an essential function.

QUESTIONS AND PROBLEMS

1. Given $f(x) = \frac{1}{5}$ over $0 \leq x \leq 5$, and $f(x) = 0$ elsewhere, determine the probability that $x \leq 1$, $x \geq 2$, $2 \leq x \leq 4$.

2. Given $f(x) = \frac{1}{5}x$ over $0 \leq x \leq y$ and $f(x) = 0$ elsewhere, determine y. Find the probability that $x \leq 1$, $x \geq 2$, $2 \leq x \leq 4$.

3. Determine an estimate of the population mean, variance, and standard deviation from which the following sample was taken: 10, 12, 15, 13, 18, 22, 19, 13, 24, and 22.

4. Determine the estimate of the population mean, variance, and standard deviation from which the following sample was taken: 5, 10, 3, 15, 2, and 25.

5. Given a normally distributed random variable z, with a mean of 0 and a standard deviation of 1, determine the probability that (a) $z > .2$, (b) $z \leq .3$, (c) $z \leq -2.3$, (d) $-1 \leq z \leq 2$, (e) $-3 \leq z \leq 0$, (f) $-3 \leq z \leq 3$, (g) $-2.5 \leq z \leq 1$, (h) $-1.96 \leq z \leq 1.96$, (i) $-2.41 \leq z \leq 3.02$.

6. Given a normally distributed random variable x, with a mean of 25 and variance 9, what is the probability that (a) $x \leq 24$, (b) $x \leq 22$, (c) $x \geq 30$, (d) $x \geq 34$, (e) $19 \leq x \leq 31$?

7. Given a normally distributed random variable x, with a mean of 10 and a standard deviation of 2, what is the probability that (a) $x \leq 10$, (b) $x \geq 13$, (c) $5 \leq x \leq 15$, (d) $8 \leq x \leq 16$, (e) $12 \leq x \leq 13$?

8. Find the value of the random variable of x in Problem 6 that will be exceeded only 5% of the time.

9. Find the value of the random variable of x in Problem 7 that will not be exceeded 10% of the time.

10. Given a random variable (sample means) with a Student's t distribution of sample size $n = 9$, population mean of 20, and a population standard deviation estimated at 6, determine the following: (a) the probability of a mean value of 24 or greater, (b) the probability that $13.5 \leq \bar{x} \leq 24.612$, and (c) the value of the sample mean that will not be exceeded more than 25% of the time.

11. Find the average value of a sample of size 64 with a population standard deviation of 5 that will be exceeded only 10% of the time. Assume a population mean of 100.

12. Find the average value of a sample of size 9 with a sample standard deviation of 5 that will be exceeded only 10% of the time. Assume a population mean of 100.

13. A sample of 16 sound-level measurements are taken from a process with a known standard deviation of 2.6. If a mean of 81.2 dB(A) is obtained, test the hypothesis that the true population mean is 85.0 dB(A) at the .05 level of significance.

14. Determine the O.C. curve for Problem 13.

15. A sample of nine sound-level measurements were taken and a mean of 87.2 dB(A) with a standard deviation $s = 2.4$ was calculated. Test the hypothesis that the true mean is 85.0 at the .10 level of significance.

16. A sample of nine monthly accident frequencies from a plant yielded the following data: 2, 4, 3, 6, 5, 3, 1, 0, and 4. Test the hypothesis that the population mean is equal to 4.2 at the .05 level of significance.

17. The accident frequency of a plant is to be compared before and after the implementation of a new educational program. The monthly values of frequency before the installation were 2, 4, 8, 3, 1, 5, 4, 6, 2, 4, 0, and 1; and after installation the readings were 1, 3, 4, 6, 5, 2, 3, 5, 0, 3, 0, and 2. Test the hypothesis that the two means are equal at the .05 level of significance.

18. P-static dischargers are supposed to reduce the noise level in aircraft radios during periods of high static buildup on the wings. A certain manufacturer claims that his dischargers reduce the "static" level by 50% by means of continuous discharge. He guarantees that at least 18 in 20 will perform this way under rigid test conditions. A sample of 50 taken at random from 2,000 supplied by the manufacturer indicated that 43 would perform as specified. Can the manufacturer's claim be disputed based upon statistical tests at the .05 level?

19. A second manufacturer of P-static dischargers has supplied a sample of 50 at random from his production. When these were tested it was found that 41

would perform as specified. Is there a significant difference between these and those mentioned in Problem 18? Would these pass the test of Problem 18? Explain the results.

20. Assuming a mean of 4.2 and a standard deviation known to be 1.5, plot the data in Problem 16 on a control chart with 95% limits. Do the data exhibit statistical control?

21. Construct a control chart for daily sound-level samples of size 5. An estimate of the population standard deviation is 2.5, and the mean is 79.2 dB(A). The following readings were obtained for the first 10 days of the month: 76.2, 79.3, 84.2, 86.3, 81.6, 79.3, 84.2, 81.3, 85.2, and 79.3. Is the noise level in the plant under control at the .01 level of significance?

22. Simulate a simple parallel system of two components with the following time-dependent probabilities:

Time in Service (Periods)	Component 1	Component 2
0	.9	.6
1	.8	.5
2 or more	.7	.4

Assume that any component that fails is replaced at the end of that period.

SAFETY INFORMATION
SYSTEM DESIGN

7.1 Introduction

In Chapter 6 an introduction was given to some basic statistical techniques useful to the safety engineer. A dual problem often arises in obtaining quantitative data for statistical analyses as well as the other nonstatistical quantitative analyses. First, at the time that a given analysis is desired, rarely are the required data available. Therefore, data-collection procedures must be established and executed. This consumes valuable time and possibly renders the results obsolete when they do become available. The second half of the problem arises when countering the first. Rarely can the analyst know 1 or 2 years in advance all the data that he will need to fulfill the safety information requirements. Thus the safety information system cannot be expected to solve all the data problems. However, without a well-conceived safety information system, the quantitative techniques described in this book become little more than mental exercises.

The purpose of this chapter is to provide a basis for the design of safety information systems. The examples given are intended to provide a starting point from which an effective system can be tailored to the application. The basic concepts apply to any system, although, to have the broadest application, the examples are primarily directed at occupational safety.

7

The Occupational Safety and Health Act of 1970 mandates certain recordkeeping requirements. Similarly, government regulations affecting other areas of safety are often accompanied by recordkeeping standards. Because such recordkeeping is required by law, the recordkeeping function is certain to be established. The marginal costs involved in augmenting the recordkeeping procedures to make them responsive to information requirements is nominal when compared to the benefits that can be obtained.

This chapter begins by setting forth some basic concepts of safety information systems. The methodology is given for designing the system to fit operational requirements for statistical comparisons and data reduction. A general data-collection model is given in the next section. This is followed by elementary procedures for presenting and processing the data collected. The chapter concludes with a case study from forest harvesting.

7.2 Basic Concepts of Safety Information

The designer of the safety information system should constantly bear in mind the function that it will serve in the control mechanism. The measurement and evaluation functions are performed by it directly. The goal setting and correction functions are not performed, although they are strongly influenced by it. Hence the information system is a means to an end and not an end in itself. Managers often lose sight of this basic concept, especially in governmental organizations. This leads to the production and self-perpetuation of "statistical" reports that often fail to provide useful information for control.

Figure 7.1 provides a pictorial view of the boundary of the safety information system. Such a concept is necessary if the designer is to understand the limitations as well as the essential nature of this phase of the total system.

The first element within the safety information system is the function of measurement. Note that some measurements are to be made directly against the goals that are set. This is to determine if the safety system is accomplishing its objectives. However, another source of measurements is also necessary to lead to the proper corrections. These are such things as: (1) What caused the accident? (2) Where did the accident occur? (3) Upon what day of the week did it occur? And so on. The occurrence of the accident indicates a breakdown in the control process and hence the need for correction. These other questions are measurements which might pinpoint the type of correction that should be made.

There are two outputs from the measurement function. The first, the construction and maintenance of the data base, is still within the boundary of the safety information system. The other, however, provides a direct influence upon the correction element of control. This distinction is made to illustrate that the information system should not be a barrier to immediate and obvious action. Forcing all information through data-entry and processing procedures prior to action can slow the responsiveness of a good safety control system.

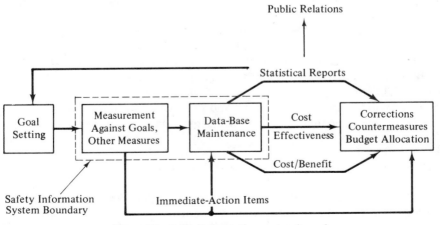

Figure 7.1. Safety information system boundary

Again, the purpose of the safety information system is to increase control; thus it should not decrease responsiveness to immediate needs.

Information in the data base provides a variety of functions. Some of these may not be for safety control purposes. For example, the use of statistical reports for public relations purposes would fall outside the realm of safety control. This is not harmful provided that the public relations aspect does not interfere with the primary objective of the data base and/or the statistical reports. The primary purpose of statistical reports should be for guiding the correction and goal-setting functions.

Figure 7.1 also shows that the data base should be designed to provide cost-effectiveness and cost/benefit information for the purpose of budget allocation. This will again have a direct effect upon the correction process.

At some point in the design of the safety information system a decision must be made as to the specific data that will be collected and stored. If the designer understands the function and boundaries of the information system, he will be in a better position to evaluate whether a given bit of information is worth recording and storing.

The trade-off that exists here is between the cost of recording and storing the information, and the value of its ultimate use. Given that a recordkeeping system is required by law, the basic cost of recordkeeping should not be considered to be part of the cost. The marginal cost of each additional bit of information recorded over that which is required by law must be compared with its value.

Certainly it seems a small thing to add one bit of information. For example, OSHA regulations* require a statement of the employee's function which includes the general department in which the employee was working, at the time of a recordable accident. The company may want to go beyond this and place the department number on the form. It takes very little effort for record-keepers to look up the department number while filling out the form. Also, if the same personnel fill out the majority of the forms, they may remember the numbers and not even have to look them up. Hence the cost of this phase of the work seems negligible.

However, other factors are involved, also. There are hundreds of pertinent bits of information that *might* be needed later. Each of these takes just a little time, but all of them together might take considerable time. Further, the cost of recording is not the only cost. The information must also be stored, and if it is going to be of any value at all, it should be processed. Costs here are not related linearly with the number of bits of information. That is,

---

*All references to the requirements under the Occupational Safety and Health Act of 1970 refer to federal regulations at the time of this publication. Since States are given flexibility within the Act, and all federal requirements are subject to change, the general statements made concerning the Act in this book should be verified with either state or federal officials if there is any question as to compliance.

after a certain point of adding information to a data base, the costs of retrieval can grow at a faster rate than that at which the number of bits of information grow. In a manual system this is obvious. If everything required can fit on one page, the cost of storage and retrieval will be only about half the cost of that when a second page is required. The situation is similar in computerized systems.

For these reasons it is important for the analyst to be selective in adding information requirements to the data base. He should consider the degree to which each will contribute to the overall control effort. On the other hand, information that is omitted from the data base is often of considerable value, both by itself and in conjunction with other available data.

For example, consider the specific day on which the accident happens. Now, OSHA requires the date to be specified both on the 100 Log and the 101 Supplementary Record; but they do not require the day of the week. Probably a summary of all accidents by day of the week would not be too revealing. There might be concentrations on given days, such as Mondays and Fridays, which would indicate the possibility of corrective action. However, if there were already another breakdown in the data base by type of accident, then each type of accident could be summarized by the day of the week. This might indicate, for example, that a high concentration of accidents involving machine contact occurred on Mondays during the weekly start-up procedures. Similarly, a breakdown by time of day could further enhance the value of the records.

The designer should ask the following questions when deciding whether to incorporate a given bit of information into the data base:

1. What is the source of the information?
2. What will it cost to generate and record the information?
3. What will it cost to store and retrieve the information so that it can be useful for decision making?
4. How will the presence of this information hinder the retrieval and processing capabilities of other, more useful information?
5. How will this information be processed, and what result will it bring in terms of increased control?
6. Can this information be reduced, coded, or in some way incorporated into another more important bit of information?

The decision to incorporate additional bits of information in the data base will depend largely upon the application and the facilities available to process the information. A company with unused computer time and storage capability will not react in the same way as a comparable company with no computer. Given that the questions have been thoroughly considered for each data element, the designer will be on a firm basis. He should include all data elements that he considers to be profitable from a cost/benefit point of view.

He should also include those questions which are of possible, although not necessarily proved, value and which have a negligible marginal cost. As time goes by, experience will dictate additional information bits to be added. Those which have not proved to be of use should similarly be dropped.

7.3 Information Sources

Figure 7.1 indicated that measurements were to be taken against the defined goals of the safety system as well as other measurements that might point the way toward more effective correction. The purpose of this section is to elaborate on the sources of these measurements. The discussion will be kept general until all sources are discussed. In the next section we shall give a specific method of constructing a usable data base, through coding. Since the keeping of records of highway accidents is well governed by law, the examples given will be from occupational safety. However, the principles expressed are pertinent to all areas of safety recordkeeping.

Generally all information about the physical plant will be obtained through some type of formal investigation procedure. There are basically two types of investigations: pre-accident and post-accident. The post-accident investigation tends to be more detailed and, because a specific chain of events has occurred, it can generally isolate the cause or causes of the accident for potential elimination. However, both types obtain the same general information and both have the same purpose: to eliminate future accidents. In effect, the safety staff attempts to turn the past accident into an asset. It would be better if it had not occurred, but now that it has, possibly a number of future similar occurrences can be prevented.

Figure 7.2 shows the two types of motivations for investigations. The

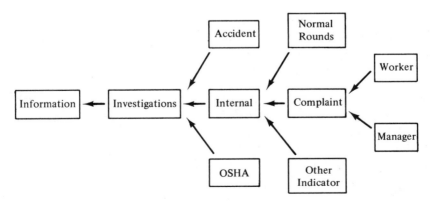

Figure 7.2. Sources of information outline

post-accident is called "Accident," and the pre-accident, which will be discussed below, is merely labeled "Internal." To these two has been added a third motivation, that of OSHA. When the OSHA inspector calls, he will investigate the physical plant to determine its compliance with the standards for the industry promulgated under the Occupational Safety and Health Act of 1970. He is bound by law to write citations for any standard violated. However, during the course of the investigation, he may provide valuable information as to potential hazards. All citations of violations of standards should certainly be abated or appropriately appealed. All trouble spots should also be investigated to provide information, as discussed below.

Continuing with the discussion with regard to Figure 7.2, internal pre-accident investigations may be motivated by the normal rounds of staff investigators. These should be at random intervals and should cover the entire plant periodically. Special investigations might be motivated by complaints from either workers or managers. Also other indicators, as discussed in Chapter 2, may lead to an investigation. For example, customer complaints on bad workmanship may pinpoint one department. If workers are negligent in workmanship, chances are good that they will also be negligent with regard to safety.

The information sources and measurements discussed above were within the plant. Information on employees, through psychological testing or interviews, might be pertinent also. Also, the most recent information on accident countermeasures is essential to effective control. The latest manuals, catalogues, and books on safety and safety countermeasures should be maintained and consulted frequently.

This section has briefly reviewed from Chapter 2 the sources of information for establishing the safety data base. The particular information as well as methods for coding and processing this information will be given below.

7.4 Coding Investigation Information

If safety information is to be processed effectively, it must be classified into predefined categories that can be used for sorting purposes. Qualitative information on investigation forms becomes unmanageable, for a variety of reasons. For example, two individuals may report the same injury using different terminology. Similarly, departments could be given various names by two different people at different times. A third individual may have the responsibility to sort and make sense out of the records. The problems that he would incur are obvious.

There is no inference here that the forms that a company may be using for investigations should be abandoned. In fact, some purely descriptive information is required by law, namely OSHA Form 101, Supplementary Record of Occupational Injuries and Illnesses. The law requires that this record be readable (i.e., not purely in codes) by the OSHA inspector. And, since it is rarely desirable to encode all the descriptive information, a "hard copy" basic reference document is desirable.

It is permissible to augment OSHA Form 101 with other information and also to *add* codes, provided that the codes are not substituted for information required on the form. Also, the form itself is not sacred. Another form can be used provided that it contains the minimum of information required by the 101 form and that it is as readable as the 101 form. Hence an inexpensive method for augmenting and encoding accident information would be to design a form suited to the company which could take the place of OSHA Form 101.

It should be remembered at this point, however, that recordable accidents are not the sole source of information for the data base. If a company depends only on OSHA recordable accidents, it will lose a major portion of the data base. Only *injury* accidents are recordable under OSHA regulations, and even some injury accidents are not recordable. For example, first-aid cases, one-time treatment that does not *require* medical care where the employee returns to work immediately, are generally not required to be recorded by OSHA. Certainly, this is valuable information for incorporation in the company's safety data base. Noninjury accidents also provide a strong indication of potential hazards, and they should be included.

Similarly, OSHA does not have any recordkeeping requirements for pre-accident investigations. The OSHA Form 101, when augmented and extended to include all accidents, can serve quite well for post-accident investigations. However, another hard-copy form must be designed for pre-accident investigations similar to that given in Figure 1.7. The discussion of specific form changes will be resumed in Section 7.5. At this point, information requirements and a coding scheme will be presented.

In designing the data-collection form, considerable thought should be given to the subsequent use of the data. Those items that will be processed further may need to be given special consideration. For example, keypunch instructions could be given on the form. This will facilitate handling and processing. To this end, a list of the information items that will be entered into the data base should be constructed prior to the design of the data-collection form. An example of such a list is given in Table 7.1. This table defines the items of information from each accident and investigation that will go into the data base. Note that with the exception of certain personal information about the injured individual, the same information requirements can serve for accident, injury, or hazard investigation. Placing these in the same format will enable the same programs to process all data, regardless of the source.

Table 7.1 EXAMPLE LIST OF DATA ELEMENTS TO BE CODED

Item Number	Spaces Required	Card Column	ANSI Z16.2	Description
1	4	1–4		Case or File Number
2	2	5, 6		Year
3	1	7		Hazard, Accident, Injury, Recordable Injury
4	9	8–16		Social Security Number
5	2	17, 18		Age
6	1	19		Sex
7	2	20, 21		Occupation Code
8	2	22, 23		Department Code of Employee
9	2	24, 25		Department Code Where Accident Occurred
10	2	26, 27		Location Code
11	1	28		Activity
12	3	29–31	A1.8	Unsafe Act
13	3	32–34	A1.1	Nature of Injury
14	3	35–37	A1.2	Part of Body
15	4	38–41	A1.3	Source of Injury
16	3	42–44	A1.4	Accident Type
17	3	45–47	A1.5	Hazardous Contition
18	2	48, 49		Month
19	2	50, 51		Date
20	1	52		Day of Week
21	4	53–56		Time of Day
22	3	57–59		Hours Worked Since Last Day Off Job
23	2	60, 61		Overtime Hours Worked Since Last Day Off Job
24	3	62–64		Severity: Lost Workdays
25	5	65–69		Severity: Total Cost in Dollars
26	2	70, 71		Investigator Code

The following is an example of the detailed descriptions that should accompany the data elements list (by item):

1. Case or File Number. This item is primarily for cross-reference purposes. Four digits are recommended so that up to 10,000 reports per year can be processed for each category of item 3. The case or file number may start over each year. Also, accident reports may index by 10, so that each injury in a given accident can be recorded with a unique case number.

2. Year. The last two digits of the year (fiscal or calendar depending on the organization) appears as the second item on the data card. Note that by sorting by the first six digits, cards can be arranged into case-number order within year. If sufficient information is to be added to the card to exceed 80 columns and a second data card is required, the first six digits should be repeated and a "card digit" should be added in a common position on both cards. By placing a 1 for first cards and a 2 for second cards in this position, a sort can be performed

with two cards per case number. If the designer is unfamiliar with procedures such as these, he should consult a data-processing analyst.

3. Hazard, Accident, Injury, Recordable Injury. The same basic format will be used for pre- and post-accident investigations. Thus a digit is required to designate the type of investigation being reported. The following provides a basic designation:

0. Hazard investigation, normal inspections.
1. Hazard investigation requested by management.
2. Hazard investigation requested by labor.
3. Hazard investigation prompted by hazard indicator (note indicator on form).
4. Accident investigation, no injuries.
5. Accident investigation, unrecordable injuries.
6. Accident investigation, recordable injuries.
7. Fatality.

There are three major divisions: (a) 0–3 are pre-accident investigations; (b) 4 and 5 are post-accident investigations, where the injuries, if any, were not recordable by law; and (c) 6 and 7 are recordable injury cases. The recordable cases are of particular interest, since extra copies may be required to show the OSHA inspector.

4. Social Security Number. This will completely identify the injured individual. If no one is injured, or for a hazard investigation, this should remain blank. Multiple entries should be made for multiple injuries; each will have a unique case number. An ordered list of names, addresses, and other pertinent personal information by social security number will make the storage of this information in the data base unnecessary. However, any pertinent information that could be sorted to pinpoint countermeasures should be listed: item 5 below, for example.

5. Age. The age of the injured individual, if available. Listings of accident frequency and severity by age, sex, occupation code, and so on, can help to pinpoint countermeasures.

6. Sex. Here the code will be 0 to indicate male, 1 to indicate female.

7. Occupation Code. A two-digit occupation code for the occupation of the injured party. There may be a different set of occupation codes for each department, since the department is identified in item 8.

8. Department Code of Employee. This is the department to which the employee is normally assigned, not necessarily the one in which the accident occurred. There may be a different set of department codes for each location, since the location is identified in item 10.

9. Department Code Where Accident Occurred. This will be different

from 8 if the employee is injured while outside his normal work area or during temporary assignment to another department. Note that for pre-accident investigations, items 8 and 9 should be both set to the department where the hazard occurs. In the case of a hazard at the interface of two departments, one may be listed under item 8 and the other under item 9.

10. Location Code. The location at which the accident or investigation took place. A coding scheme designed specifically for the company will be developed for items 7–10.

11. Activity. The following will generally pinpoint the activity that was being performed when the hazard was investigated or the accident took place:
    1. Normal activity in department.
    2. Setup for normal activity.
    3. Special activity not usually performed.

12. Unsafe Act. A good model for unsafe-act classification is given in Appendix A1.8 of ANSI Z16.2-1962(R1969). These codes are given in Appendix 1 of this book.

13, 14, 15, 16, and 17. Nature of Injury, Part of Body, Source of Injury, Accident Type, and Hazardous Condition. Good models for these classifications are given in Appendixes A1.1, 1.2, 1.3, 1.4, and 1.5 respectively, of ANSI Z16.2-1962(R1969). These codes are given in Appendix 1 of this book.

18. Month. The common numerical designation for the month in which the accident or investigation took place.

19. Date. The specific date on which the accident or investigation took place.

20. Day of Week. The code for the specific day of investigation or accident is as follows: 1, Sunday; 2, Monday; . . . ; 7, Saturday.

21. Time of Day. The 24-hour clock time during the day in which the accident took place or the hazard was observed.

22. Hours Worked Since Last Day Off Job. This could be a pertinent factor if fatigue could have contributed to the accident. It has been included, together with item 23, to illustrate some other possible data elements.

23. Overtime Hours Worked Since Last Day Off Job.

24. Severity: Lost Workdays. The number of lost workdays as defined by OSHA. This will provide a consistent measure of severity, as required by law.

25. Severity: Total Cost in Dollars. Total lost workdays do not reflect severity in the following: (1) first-aid cases, (2) permanent injury where the individual returns to the job quickly, and (3) situations where permanent transfer to another job or termination is required.

In cases such as these, it is beneficial to have an alternative method of measuring severity. An estimate of dollar cost is as good as any other measure of negative utility. It should reflect the personal loss to the injured individual, the medical and other compensation required, the time and lost experience by the absence of the worker to the company, the material goods, merchandise or machinery destroyed, or any other pertinent factors. In cases of hazard investigation, items 24 and 25 will be the estimated loss if an accident does occur.

26. Investigator Code. The number of the investigator who supplied the information.

The above list of data elements is presented as an example. Nine additional columns are available before encountering the inconvenience of two computer cards per investigation. In addition, elements that are thought to be of greater value for particular applications can be substituted for those of little value. If a large number of additional data elements are required, a second or third card might be justified.

As mentioned above, the same data elements, or very similar elements, are required in pre-accident (hazard) and post-accident investigations. Because of this, the same format is recommended so that hazard investigation can be summarized in the same way as accident investigations. This will make for ease in comparison as well as efficiency in programming. The processing procedures will be presented below. Prior to this, however, the forms for collecting the raw data will be discussed.

Before closing this section, some comment is in order on the necessity for keeping all the records discussed. Only a small fraction of the records listed are required by law (i.e., only the recordable injury post-accident investigation data). In addition, not all of the information presented in the example is required by OSHA. (Note, however, that some not mentioned is required, as will be discussed in the next section.)

Management may question the value of keeping all these records in the detail discussed, for two reasons: (1) the cost of maintaining the data base is not justified because it does not appreciably affect the number and severity of accidents, and (2) keeping such records in the detail exemplified lays the company bare to legal problems and possible additional citations by OSHA. To the first reason we can only reply that familiarity with the remainder of this chapter may dispel this belief. The value received from the basic reports to be demonstrated should more than offset the cost. Additional value from subsequent use in statistical evaluations will further enhance the value of the system.

It is the second objection, however, that we should like to address with the most strength. For, as discussed in Chapter 1, the primary role of management is in keeping the spirit of the law, not simply the letter of the law. There is no clearer demonstration that a company is keeping the spirit of the law than when they go beyond the required recordkeeping provisions to design an information system that will serve them and their industry in the reduction of accidents. The additional records (e.g., on first-aid cases and noninjury accidents) do not have to be shown to the OSHA investigator. However, evidence of an aggressive safety effort at all levels, particularly in record-keeping, can only help a company to receive the benefit of the doubt.

7.5 Source Documentation

Source documentation, "hard copy" as it is often called, will always contain more information than codes. However, the coded data elements should be defined first, as in the example above, so that the investigation forms can be designed properly.

As was probably noted in selecting the particular data elements of Table 7.1, most of the requirements to be coded are also requirements on OSHA Form 101, Supplementary Record of Occupational Injuries and Illnesses. Further, the order listed in Table 7.1 is roughly the same order as required by OSHA Form 101. This was to allow a minimum of changes to the form. As the augmented form is presented, it will be noted that none of the information required by Form 101 has been deleted. Therefore, the form for recording post-accident investigation information is acceptable as a substitute for Form 101, according to current legal interpretations.*

A second form will also be presented for recording hazard investigations. Although the questions are phrased differently on this form, the responses fit the format of Table 7.1. For the reasons mentioned above, this enables the same programs to be used to process both types of investigations. It is recommended that the hard copies of these two different forms be filed separately. In addition, another filing separation in the post-accident investigations might be required for OSHA recordable accidents.

7.5.1 Post-Accident Investigation Form

Figure 7.3 demonstrates a post-accident investigation form that fits the requirements of both the OSHA and the additional data elements required by Table 7.1. The data coding is performed right on the form, and it is in the format given in Table 7.1. Thus this form can be sent directly to keypunching for data entry. For the most part, coded information has been placed in the

---

*See "Recordkeeping Requirements Under the Williams–Steiger Occupational Safety and Health Act of 1970," U.S. Department of Labor, Occupational Safety and Health Administration, Bureau of Labor Statistics, Washington, D.C. 20212.

COMPANY XYZ ACCIDENT INVESTIGATION FORM

Case or File Number $\frac{}{1}$ – – $\frac{}{4}$ Year $\frac{}{5}$ $\frac{}{6}$

Accident Classification: $\underline{4}$ ◯ No Injury; $\underline{5}$ ◯ Unrecordable Injury
(check one; punch
col. 7) $\underline{6}$ ◯ Recordable Injury ; $\underline{7}$ ◯ Fatality

EMPLOYER

1. Name ....

2. Mail address ....

3. Location, if different from mail address ....

INJURED OR ILL EMPLOYEE

4. Name .... Social Security No. $\frac{}{8}$ – – $\overline{10\ 11}$ – $\overline{13}$ – – $\overline{16}$

5. Home address ....|

6. Age $\overline{17\ 18}$ 7.   Sex (check): $\frac{0}{19}$ ◯ Male; $\frac{1}{19}$ ◯ Female

8. Occupation .... Code $\overline{20}$ $\overline{21}$

9. Department of employee .... Code $\overline{22}$ $\overline{23}$

    Department of accident .... Code $\overline{24}$ $\overline{25}$

10. Place of accident or exposure .... Code $\overline{26}$ $\overline{27}$

11. Was place of accident on employer's premises? .... (yes or no)

12. What was the employee doing when injured? ....
    Activity Code (check one): $\frac{1}{28}$ ◯ Normal; $\frac{2}{28}$ ◯ Setup; $\frac{3}{28}$ ◯ Special

13. How did the accident occur ....
    .... Unsafe Act (ANSI A1.8) $\overline{29}$ – $\overline{31}$

OCCUPATIONAL INJURY OR OCCUPATIONAL ILLNESS

14. Describe the injury or illness in detail and indicate the part
    of the body affected ....

                    Nature of Injury (ANSI A1.1) $\overline{32}$ – $\overline{34}$

                    Part of Body (ANSI A1.2) $\overline{35}$ – $\overline{37}$

15. Name the object or substance that directly injured the employee
    ....

                    Source of Injury (ANSI A1.3) $\overline{38}$ – – $\overline{41}$

Figure 7.3. Example of post-accident investigation form

Accident Type (ANSI A1.4) $\overline{42} - \overline{44}$

Hazardous Condition (ANSI A1.5) $\overline{45} - \overline{47}$

16. Date of injury, accident, or initial diagnosis of occupational illness

.... Month: $\overline{48}$ —; Date: $\overline{50}$ —

17. Did employee die? .... (yes or no)

OTHER

18. Name and address of physician ....

19. If hospitalized, name and address of hospital ....

20. Day of the week (check one; 1̲ ◯ Sun; 2̲ ◯ Mon; 3̲ ◯ Tue; punch col. 52)

   4̲ ◯ Wed; 5̲ ◯ Thu; 6̲ ◯ Fri; 7̲ ◯ Sat

21. Time of day (24 hour clock): $\overline{53}$ —: — $\overline{56}$

22. Hours worked since last day off: $\overline{57} - \overline{59}$

23. Overtime hours worked since last day off: — — $\overline{60\ 61}$

24. Severity: lost workdays: $\overline{62} - \overline{64}$

25. Severity: total cost in dollars: $\overline{65}$ — — — — $\overline{69}$

26. Date of report .... Prepared by ....

   Official position .... Investigator code $\overline{70\ 71}$

Figure 7.3. (Continued)

right-hand margin adjacent to the pertinent descriptive information. This eliminates confusion to data-entry operators as well as to OSHA inspectors. Information not required by OSHA is generally placed at the bottom of the form.

Rules for filling in the descriptive portion of the form should be consistent with OSHA requirements. Much of this is explained on the official OSHA Form 101, Supplementary Record of Occupational Injuries and Illnesses. All companies subject to OSHA regulations should keep copies of this form available and refer to it when required. Figure 7.3 corresponds in item number to the OSHA form. In addition, each company should have the booklet "Recordkeeping Requirements Under the Williams–Steiger Occupational Safety and Health Act of 1970," which is available through the state or federal agency responsible for administering the Act in the particular state.

Methods for encoding the descriptive information were given above. The use of the ANSI model given in Appendix 1 is quite straightforward. For more information, the entire standard should be reviewed. This standard has been presented as a model. It should be modified by adding categories that fit the particular needs of the company. The standard itself, however, provides excellent checklists for all types of safety investigations.

The form given in Figure 7.3 is intended for all post-accident investigations. In noninjury accidents, data pertaining to the extent of injury, part of body injured, and so on, should be omitted. The person who was involved, or who caused the noninjury accident, should be recorded, however. This is for research and not for punitive purposes. For investigations that are not prompted by accidents, the form discussed in the next section is recommended.

7.5.2 Pre-Accident Hazard Investigation Form

Since much of the same information is required in a hazard investigation, the format will be much the same as the post-accident form. Certain modifications should be made, since no individual is involved. Also, since there are no OSHA regulations concerning general investigation recordkeeping, legal aspects need not be a consideration.

Figure 7.4 presents one example of a hazard investigation form. It is clearly patterned after Figure 7.3. However, certain items, such as social security number, and name and address of the injured party, have been omitted. The space for coding omitted items has been used either for other data or marked on the form as "Reserved." These spaces can be used for other bits of information if they should be deemed necessary. For example, hazard indicators that prompt investigations could be coded. In Figure 7.4, item 3, a check is given in space number 3 if the hazard investigation came about by means of a hazard indicator, such as damaged goods, customer complaints, or the like. It might be beneficial to set up a coding scheme for hazard indicators so that they can be more fully studied.

The benefits of having a hazard investigation form that is almost a mirror image of the post-accident investigation form should be obvious. It allows the same program to process both, and this facilitates comparisons. As one form is modified and improved, an attempt should be made to similarly modify the other. The forms given make no pretense of being universal. Each company should tailor the form to its own application.

Figure 7.4 presented a model hazard investigation form. It is recommended, if possible, that a form such as this be completed and entered into the data base for every potential hazard in the plant. This should be integrated into a cost-effectiveness procedure as presented in Chapter 2, or a more sophisticated budget allocation model, as will be presented in Chapter 8. At this point, however, concentration will be given to the use of the safety information system in producing more general reports.

COMPANY XYZ HAZARD INVESTIGATION FORM

1. Case or File Number $\overline{1}$ – – $\overline{4}$ 2.   Year $\overline{5}$ $\overline{6}$

3. Investigation Classification: $\frac{0}{7}$ ◯ Normal Inspection;
    (check one)

$\frac{1}{7}$ ◯ Management Requested; $\frac{2}{7}$ ◯ Labor Requested; $\frac{3}{7}$ ◯ Hazard Indicator

4. Reserved: $\overline{8}$ — — — — — — — $\overline{16}$

5. Mean age of employees involved $\overline{17}$ $\overline{18}$

6. Sex of majority of employees involved: $\frac{0}{19}$ ◯ Male; $\frac{1}{19}$ ◯ Female

7. Occupation of employees involved .... $\overline{20}$ $\overline{21}$

8. Department of employees involved .... $\overline{22}$ $\overline{23}$

9. Department where hazard exists .... $\overline{24}$ $\overline{25}$

10. Location or place of hazard .... $\overline{26}$ $\overline{27}$

11. Activity in which hazard occurs ....
    Activity Code (check one): $\frac{1}{28}$ ◯ Normal; $\frac{2}{28}$ ◯ Setup; $\frac{3}{28}$ ◯ Special

12. How could an accident come about? ....
    .... Unsafe Act (ANSI A1.8) $\overline{29}$ — $\overline{31}$

13. What type of injury could the hazard cause? ....
    .... Nature of Injury (ANSI A1.1) $\overline{32}$ — $\overline{34}$

14. What part(s) of the body would be affected? ....
    .... Part of Body (ANSI A1.2) $\overline{35}$ — $\overline{37}$

15. Name the object or substances that could cause injury ....
    .... Source of Injury (ANSI A1.3) $\overline{38}$ — — $\overline{41}$

16. Name the type of accident that could occur ....
    .... Accident Type (ANSI A1.4) $\overline{42}$ — $\overline{44}$

17. Name the hazardous condition that is present ....
    .... Hazardous Condition (ANSI A1.5) $\overline{45}$ — $\overline{47}$

18. Month that hazard was investigated: $\overline{48}$ —

19. Day of the month that hazard was investigated: $\overline{50}$ —

Figure 7.4. Example of hazard investigation form

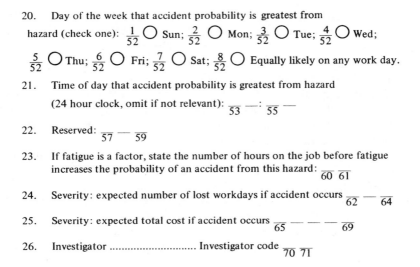

20. Day of the week that accident probability is greatest from hazard (check one): $\frac{1}{52}$ ◯ Sun; $\frac{2}{52}$ ◯ Mon; $\frac{3}{52}$ ◯ Tue; $\frac{4}{52}$ ◯ Wed; $\frac{5}{52}$ ◯ Thu; $\frac{6}{52}$ ◯ Fri; $\frac{7}{52}$ ◯ Sat; $\frac{8}{52}$ ◯ Equally likely on any work day.

21. Time of day that accident probability is greatest from hazard (24 hour clock, omit if not relevant): $\overline{\;\;53\;\;}$ —: $\overline{\;\;55\;\;}$ —

22. Reserved: $\overline{\;\;57\;\;}$ — $\overline{\;\;59\;\;}$

23. If fatigue is a factor, state the number of hours on the job before fatigue increases the probability of an accident from this hazard: $\overline{60\;\;61}$

24. Severity: expected number of lost workdays if accident occurs $\overline{62}$ — $\overline{64}$

25. Severity: expected total cost if accident occurs $\overline{65}$ — — — $\overline{69}$

26. Investigator .... Investigator code $\overline{70\;\;71}$

Figure 7.4. (Continued)

7.6 Processing the Safety Information

The entire purpose of the safety information system is to increase the effectiveness of the measurement and evaluation functions of control. Unless data are properly processed and presented, they will not serve this purpose. The objective of this section is to present some formats for information presentation.

Chapter 6 has provided a basic background for statistical analysis which will be utilized here. As indicated there, the processing techniques employed do not "create" information. Generally, they reduce the information in the data base to a form in which it can be understood and applied. There is never more information than is contained on the hard records. On the contrary, in data reduction, detail is sacrificed for inference. This is an important point, and it brings out the necessity to store the original uncoded information in an easily referenced manner. The specifics of the reports will be required frequently to verify and exemplify inferences that are drawn.

Prior to citing examples of potential statistical analyses, the basic concept of sorting will be illustrated.

7.6.1 Basic Data Presentations

Probably one of the most used, if not the most useful, aspect of computer technology is the ability to sort large quantities of data in a very short time. This section will present this capability in an elementary form so that the concept can be understood and applied. The techniques are applicable even if the company has access to only the most primitive card sorter. Of course, if the company has a computer center, it can draw upon the staff there to provide the expertise to actually perform the sorts and present the information. This section will aid safety personnel in communicating their needs to computer experts. Also, many of the summaries to be discussed can be performed manually. The presentation here will hold the middle road by discussing the techniques in terms of card sorting.

For a first example, consider the problem of basic data organization. Suppose that over the course of a year the data coding system discussed above has been in effect. Forms have been sent to keypunch at random intervals and the resulting data cards, including both pre- and post-accident cards, are stored together. Obviously for presentation purposes it would be good to separate the pre-accident investigations from the accident reports. This can easily be done by sorting on column 7. This will cause the different categories of investigations to fall in different bins so that the desired separation can be made. If coded as given above, 0–3 will be hazard investigations, and 4–7 will be accident investigations. It may also be desirable to isolate category 4, which was the noninjury accident classification.

Once this sort is performed, a simple count of each category will reveal the investigation history for the past year. Table 7.2 gives an example of the

Table 7.2 EXAMPLE OF INVESTIGATION SUMMARY

Code	Explanation	Investigation Frequency
0	Normal Inspection Investigation	85
1	Investigation Requested by Management	12
2	Investigation Requested by Labor	23
3	Hazard Indicator Investigation	18
4	Accident Investigation Noninjury	5
5	Accident Investigation Unrecordable	35
6	Accident Investigation Recordable	13
7	Fatality Accident	0

display of this information. This called a *univariate distribution* of the investigation history.

Now within each of these classifications, or within combinations of these classifications, additional sorts can be made and univariate distributions established. A good method of proceeding might be to combine 0–3 and 4–7 into two classes, called hazard investigations and accident investigations, respectively. Then sorts can be performed to determine and compare various

Table 7.3 EXAMPLE OF UNIVARIATE DISTRIBUTION BY
DEPARTMENT

Department Code	Hazard Investigations	Accident Investigations
01	5	0
02	2	1
03	10	2
04	4	0
.	.	.
.	.	.
.	.	.

subclassifications within each. For example, a sort on Department Code where accident/hazard occurred, variable 9 in Table 7.1, might produce the two univariate distributions given in Table 7.3. If it would be more informative, the department code might be augmented by the name of the department. This could be done manually or it could be computerized. Note the ease of comparing pre-accident and post-accident actions.

One procedure that is generally employed when the data base is computerized is to produce all the meaningful univariate distributions and arrange them in a report for ready reference by any user. This greatly increases the value of the data base, for two reasons: (1) every potential user knows the contents of the data base, and (2) there is no "computer barrier" to getting the information required. In addition, this report serves as an index to potential bivariate distributions, which will be discussed below.

One essential element of an effective presentation of univariate distributions is a usable table of contents. Quite often computer reports are produced without an easy guide to their uses, thus detracting greatly from their value. Table 7.4 provides an example reference list for the variables presented in Table 7.1. For each of these variables, a univariate distribution can be printed out such as exemplified for variable 8 in Table 7.3. Exceptions might occur in certain of the variables, such as "case number," which would give no meaningful information.

In certain variables, all the possible values of the variable should be broken into discrete intervals, for a more concise presentation. For example, variable 5 (age) could be divided into intervals of 15–20, 21–25, 26–30, . . . , instead of listing the frequency at each age. This would be useful in determining if accidents were correlated with a given age group. Another method of increasing the conciseness of univariate reports is to omit classifications that have zero frequency. This is especially useful in reporting variables coded by the ANSI Z16.2 codes, since many of the codes may not be used.

A listing of each univariate distribution will provide great insight into the causes and types of accidents that are occurring. This can guide hazard

Table 7.4 EXAMPLE OF REFERENCE LIST FOR CODED VARIABLES

Variable No.	Variable Description	Page
1	Case or File Number	—
2	Year	1
3	Hazard, Accident, Injury, Recordable Injury	2
4	Social Security Number	3
5	Age	4
6	Sex	4
7	Occupation Code of Employee(s) Affected	4
8	Department Code of Employee(s) Affected	6
9	Department Code Where Accident or Hazard Occurred	8
10	Location Code	10
11	Activity Code	11
12	Unsafe Act	13
13	Nature of Injury	14
14	Part of Body	15
15	Source of Injury	17
16	Accident Type	20
17	Hazardous Condition	22
18	Month	23
19	Date	24
20	Day of Week	24
21	Time of Day	25
22	Hours Worked Since Last Day Off Job	26
23	Overtime Hours Worked Since Last Day Off Job	27
24	Severity: Lost Workdays	28
25	Severity: Total Cost in Dollars	29
26	Investigator Code	30

analyses and other types of safety investigations. As important, however, is the ability to use these basic data to determine the types of subsequent analyses within the capability of the established safety information system. A first step in this direction is the establishment of bivariate distributions.

7.6.2 Bivariate-Distribution Analysis

The principle used for the construction of a univariate distribution (i.e., that of sorting) can be applied to construct a bivariate distribution. This tabularizes two variables simultaneously, enabling the analyst to recognize additional correlations. For example, consider the two univariate distributions, age and sex, variables 5 and 6 in Table 7.4. An example of these two univariate distributions is given in Tables 7.5 and 7.6 for accident frequency only. Now the first thing that should be done in evaluating these distributions is to compare them with the overall distributions of all workers. If the general age distribution does not vary from the age distribution of those involved in accidents, it might be concluded that age is not a determining

Table 7.5 Univariate Distribution, Variable 5: Age

Age	Frequency of Accidents
15–20	1
21–25	4
26–30	3
31–35	5
36–40	6
41–45	4
46–50	3
51–55	7
56–60	4
61–65	1
Over 65	0

Table 7.6 Univariate Distribution, Variable 6: Sex

Sex	Frequency of Accidents
Male	25
Female	13

factor. Similarly, with sex; if the number of male workers is roughly twice the number of female workers, then the results in Table 7.6 are expected.

However, suppose that the analyst has reason to believe that these two variables are correlated with accident history. Further information can be obtained from a bivariate distribution as given in Table 7.7. This indicates that the major portion of accidents are affecting the older women, as opposed to the younger men. Only when the variable of sex is broken out can this relationship be seen. Again the distribution of the general population of workers should be compared against these to determine if a true causative factor exists.

The method for obtaining a bivariate distribution is quite simple. Two sorts would be required, one for each variable. For example, the cards could be sorted first by variable 6, sex, and then each of the two subsets of the data base could be sorted by variable 5, age. This will produce two univariate distributions by age, one for each sex. By appropriately counting the accidents in each classification and tabularizing as in Table 7.7, the bivariate distribution could be produced.

Any two variables given in Table 7.4, or within the data base, can be used to produce a bivariate distribution. For example, if there is a major difference in the type of work that men and women do, it might be interesting to run a

Table 7.7 BIVARIATE DISTRIBUTION: VARIABLES 5 AND 6*

| | Sex | | |
Age	Male	Female	Total
15–20	1	0	1
21–25	3	1	4
26–30	3	0	3
31–35	5	0	5
36–40	6	0	6
41–45	2	2	4
46–50	1	2	3
51–55	3	4	7
56–60	1	3	4
61–65	0	1	1
Over 65	0	0	0
	25	13	38

*These are example data only. No general conclusions should be drawn. The data were tailored specifically to demonstrate the use of the bivariate distribution.

bivariate analysis of sex and source of injury (variable 15). Similarly, department (variable 8) and source of injury might provide insights into the hazards peculiar to given departments. Since there are so many possible combinations within Table 7.4, it is not recommended that all bivariate distributions be produced. Those which the safety staff feels to be most beneficial should be obtained. A thorough review of the univariate distributions will demonstrate those of greatest potential value.

7.6.3 Multivariate Distributions and Extensions

The principle of sorting can be extended further to produce a trivariate or a multivariate distribution. Since the increased dimensionality tends to obscure the usefulness of complete reports, ingenuity must be used in designing the output. For example, suppose that the analyst were interested in making certain distinctions, by age and sex, with regard to unsafe acts. According to the methods described above, he could first perform a bivariate analysis by sex and age. Then, by sorting the subsets within each classification by unsafe act (variable 12), the desired trivariate distribution could be produced. One simple method of presenting this distribution would be to list the bivariate age versus unsafe act for men in a different table from that for women. Placing the two tables side by side would facilitate comparison.

Variables with more than two possible outcomes would present a greater problem. For example, age versus department versus unsafe act might be an interesting trivariate distribution to determine the location and age of those performing in a given unsafe manner. This breakdown might be so fine, however, as to preclude the results from being meaningful. Unless particular patterns and densities within subclassifications can be obtained, the results might only be attributed to chance. Much data are usually required to form recognizable patterns in multivariate distributions.

The multivariate concept is useful from a special point of view, however. For example, suppose that it were necessary to determine the number of women that were injured during July in a particular department in the last 2 hours of the normal workday. A question such as this might come up when determining whether to curtail summer production by reducing the number of hours per day for these women employees. The procedure to obtain this information is merely one of sorting as follows: (1) by sex, eliminating men; (2) by month, eliminating other than July; (3) by department, selecting the one of concern; and (4) by hour, selecting those which occurred in the last 2 hours. Note that the sorts should be ordered to eliminate as many as possible first.

The above discussions of summary presentations have been limited to the frequency of accidents. The identical sorting procedures would be used in all cases (uni, bi-, or multivariate) if severity were of concern. However, instead of counting the frequency of accidents (or accident investigations) in each subset, the severity would be added. For example, in Table 7.7, age 21–25 under "Male," there is a frequency of 3. The severities of these three accidents in lost workdays might be added and placed alongside the 3 to reflect the severity of those three accidents. This can be done easily once the sorts have been made.

Severities, either in lost workdays (variable 24) or in total cost (variable 25), are very useful in establishing priorities. Thus it is recommended that both frequency and severity be given on reports. In this way the results will not be misleading. For simplicity, only frequencies were presented above.

Those who are familiar with computer programming techniques will recognize the ease with which the preceding reports can be prepared. In fact, many systems have package programs available that will perform all the above sorting and presentations. Even if these are not available, however, the simple sorting procedures presented are very powerful and easy to perform.

7.6.4 Statistical Processing

The safety information system shown above provides many of the data required to perform the statistical tests discussed in Chapter 6. The number of possible comparisons are practically infinite, even with the basic lists of variables presented in Table 7.4. Therefore, judgment must be exercised to prevent needless processing. Any variable may be compared statistically against itself for different ages, sex, occupations, locations, times, time periods (e.g., before and after), and so on.

In order to compare information statistically according to the methods of Chapter 6, generally a measure of the standard deviation is required. This may be obtained by subdividing the data base, by sorting, into months or weeks. The frequency or severity of the accidents within a given classification for that time period can be viewed as a random variable. Thus the random variables within one subclassification can be compared to the same measurements within another. A real example of such a comparison will be given below.

When the same variable is subclassified in a large variety of ways, a technique called *analysis of variance* (ANOVA) is often more applicable than a simple test of means. The ANOVA procedures are beyond the scope of this text but are given in most basic statistical texts.

Rather than going into the general use of statistical methods at this point, a case-study example will be presented that will demonstrate some of the applicable tecnniques. It will also demonstrate the type of tailoring required to fit a safety information system to a particular industry or company.

7.7 Case Study: Forest Harvesting

The following example of a safety information system is a subset of a total safety system designed for forest harvesting by the author. The project was partially supported by a grant from the US Forest Service (4730, FS-SO-3701-1.3) completed in July, 1972. Statistics given are obtained from actual field data.

Forest harvesting consists of the process of (1) cutting the trees down, (2) removing the limbs, (3) cutting the trunks into logs of manageable size, (4) transporting the logs to a truck, (5) loading the logs on the truck, and (6) transporting them to their destination. The purpose of this case study is to present an example of a safety information system designed for forest harvesting. In addition, some examples of the use of this system in processing 369 accidents over a $5\frac{1}{2}$-year period will be demonstrated.

The safety information system was designed for the purpose of processing information already on record. Therefore, many of the details given in Table 7.4 were not available from past history. The abbreviated design given below, augmented by the pertinent items in Table 7.4, was established for an ongoing system of data collection. To keep this example within its proper scope, only a subset of the total system will be presented.

Table 7.8 presents the variables under consideration. Variables 8 and 9 were specifically tailored to the industry. These are given in Tables 7.9 and 7.10, respectively. The codes are self-explanatory, with the possible exception

Table 7.8 REFERENCE LIST FOR FOREST HARVESTING

Variable Number	Description	Source of Coding
1	Month	
2	Year	
3	Code Number	Sequentially, restart each year
4	Producer	Code list of producers
5	Part of Body	ANSI Z16.2 A1.2
6	Nature of Injury	ANSI Z16.2 A1.1
7	Days Lost	OSHA interpretation
8	Accident Class	See Table 7.9
9	Operation	See Table 7.10

Table 7.9 ACCIDENT CLASSIFICATION CODES, VARIABLE 8

Code	Accident Classification
5	Automobile-Related
10	Ax Contact
15	Cable-, Chain-, or Binder-Related
20	Falls (Man Off-Balance)
25	Free Limb Contact
30	Tree Contact
35	Miscellaneous (Not-Equipment-Related)
40	Miscellaneous (Equipment-Related)
45	Overexertion
50	Saw Contact
55	Skidding-Machine-Related

Table 7.10 OPERATION CODES, VARIABLE 9

Code	Operation
1	Logmaking
2	Intermediate Transportation
3	Loading and Unloading
4	Final Transportation
5	Maintenance
6	Other

Table 7.11 RAW DATA COLLECTED (SUBSET)

Variable: 1	2	3	4	5 Part of Body	6 Nature of Injury	7 Days Lost	8 Accident Class	9 Operation
Month	Year	Number	Producer					
7	66	1	300	520	170	6	50	1
7	66	2	301	513	170	9	50	1
7	66	3	300	150	170	7	40	3
7	66	4	267	511	170	12	50	1
7	66	5	305	340	170	4	50	1
7	66	6	301	330	170	5	15	3
7	66	7	307	420	310	4	45	3
7	66	8	300	340	170	4	50	1
7	66	9	302	320	310	9	45	3
8	66	10	304	540	300	5	30	3
8	66	11	301	315	170	19	50	1
8	66	12	306	510	170	7	35	3
8	66	13	305	530	300	11	30	3
8	66	14	302	330	170	4	50	1
8	66	15	300	511	170	12	50	1
8	66	16	301	340	210	10	30	1
9	66	17	302	144	210	24	15	3
9	66	18	305	520	160	0	20	3
9	66	19	306	430	160	0	30	1
9	66	20	267	520	210	24	30	1
9	66	21	302	330	310	12	20	3
9	66	22	306	146	170	1	25	1
9	66	23	304	530	160	3	30	3
9	66	24	300	330	170	13	10	1
10	66	25	301	511	170	21	50	1
10	66	26	299	141	170	8	50	1
10	66	27	304	700	300	24	30	1
11	66	28	298	511	300	7	35	1
11	66	29	302	315	170	3	50	1
11	66	30	302	440	160	14	30	3
11	66	31	301	340	160	4	30	3
11	66	32	305	430	210	28	30	3
11	66	33	301	511	170	21	10	1
11	66	34	302	410	310	3	45	3
12	66	35	300	330	170	8	50	1
12	66	36	306	530	160	7	30	3
12	66	37	305	700	400	18	5	6
12	66	38	307	330	160	58	55	2
12	66	39	300	513	170	7	50	1
1	67	1	304	149	170	5	25	1
1	67	2	302	330	170	7	50	1
.	.	.	.	.	.	.	.	.
.	.	.	.	.	.	.	.	.
.	.	.	.	.	.	.	.	.

of local terminology. For example, "intermediate transportation" and "skidding" both represent the process of moving the cut logs from the woods to an appropriate place designated for a loading area.

An example of the coded raw data is given in Table 7.11. Notice that there is one line per accident. Each accident has a different accident number (variable 3), and this number is reinitialized each year. Thus in sorting to the order given, the year would be sorted first, followed by the case number. Each line of Table 7.11 represents one computer card.

Table 7.12 UNIVARIATE DISTRIBUTION BY VARIABLE 8

Summary of Accidents by Accident Classification

Accident Classification	Frequency	Days Lost
Automobile-Related	13	339
Ax Contact	6	75
Cable-, Chain-, or Binder-Related	27	250
Falls (Man Off-Balance)	21	550
Free Limb Contact	40	1,854
Tree, Log Contact	78	4,428
Miscellaneous, Not-Equipment-Related	21	55
Miscellaneous, Equipment-Related	63	737
Overexertion	10	63
Saw Contact	88	1,203
Skidding-Machine-Related	2	62
	369	9,616

Univariate distributions were produced for each variable, with the exception of variable 3. Variable 7 was displayed in 10-day intervals for all days lost above 10 days. For accidents of 1 to 10 days duration, the actual frequency was given. Both frequencies and severities in days lost were given for all classifications. Tables 7.12 and 7.13 present the univariate distributions for variables 8 and 9.

Table 7.13 UNIVARIATE DISTRIBUTION BY VARIABLE 9

Summary of Accidents by Operation

Operation	Frequency	Days Lost
Logmaking	173	6,400
Intermediate Transportation	24	624
Loading and Unloading	105	1,783
Final Transportation	14	282
Maintenance	34	414
Other	19	113
	369	9,616

Bivariate distributions of variables 8 and 9 were also produced, the first two of which are given in Table 7.14. From these the specific types of accidents within each operation could be identified and studied with respect to frequency and severity.

Table 7.14 FIRST TWO BIVARIATE DISTRIBUTIONS
OF VARIABLES 8 AND 9

Data Summary, Accident Classification Within Operation

Accident Classification	Frequency	Days Lost
Logmaking Operations		
Automobile-Related	0	0
Ax Contact	4	58
Cable-, Chain-, or Binder-Related	0	0
Falls (Man Off-Balance)	4	43
Free Limb Contact	21	1,734
Tree, Log Contact	39	3,301
Miscellaneous, Not-Equipment-Related	11	17
Miscellaneous, Equipment-Related	2	7
Overexertion	4	37
Saw Contact	88	1,203
Skidding-Machine-Related	0	0
	173	6,400
Intermediate Transportation Operations		
Automobile-Related	0	0
Ax Contact	0	0
Cable-, Chain-, or Binder-Related	1	57
Falls (Man Off-Balance)	2	0
Free Limb Contact	5	94
Tree, Log Contact	5	232
Miscellaneous, Not-Equipment-Related	4	22
Miscellaneous, Equipment-Related	4	157
Overexertion	1	0
Saw Contact	0	0
Skidding-Machine-Related	2	62
	24	624

In order to perform statistical tests, the total accident frequencies and severities within each classification were divided into 3-month periods. Hence the accident frequency or the accident severity during any 3-month period could be considered to be a random variable. Tests were run over several of the variables, including variables 8 and 9. For variable 9 (operation), the results are presented in Tables 7.15 and 7.16. Separate analyses are required for frequency and severity.

The statistical techniques employed included analysis of variance, the procedures for which are given in most elementary statistical texts. The basic tests given in Chapter 6 could also be employed, however, by recognizing that the average days lost (or average frequency) over a 3-month interval is a random variable. Using the standard deviation and the number of samples, the formulas of Chapter 6 could be employed.

Table 7.15 RESULTS OF ANOVA FOR OPERATION BY FREQUENCY

	Operation	Average Frequency per 3 Months	Standard Deviation of Frequency	.05 Significance*
1.	Logmaking	6.77	3.41	
3.	Loading and Unloading	4.58	3.36	
5.	Maintenance	1.77	1.56	
2.	Intermediate Transportation	1.15	1.41	
6.	Other	.81	.98	
4.	Final Transportation	.58	.70	

Calculated value of F 34.02
Table value of F at .05 significance   2.27
Conclusion: Significant differences in accident frequency exist between operations.

*A common line across from two operations indicates that they are not significantly different.

Table 7.16 RESULTS OF ANOVA FOR OPERATION BY DAYS LOST

	Operation	Average Days Lost per 3 Months	Standard Deviation of Days Lost	.05 Significance*
1.	Logmaking	246.15	525.27	
3.	Loading and Unloading	68.58	95.73	
2.	Intermediate Transportation	24.00	47.85	
5.	Maintenance	15.92	34.16	
4.	Final Transportation	10.85	27.45	
6.	Other	4.35	8.33	

Calculated Value of F 4.69
Table Value of F at .05 Significance   2.27
Conclusion: Significant differences in days lost exist between operations.

*A common line across from two operations indicates that they are not significantly different.

The purpose of this case study has been to present the general structure and the potential uses of a safety information system as designed for a specific application. No attempt has been made to interpret the results of the examples given. Specific questions regarding interpretation, however, are posed at the end of the chapter.

7.8 Closure

This chapter has presented some basic concepts of safety information systems. The philosophy has been to produce an efficient system by (1) effectively augmenting legal recordkeeping requirements, and (2) restricting coded information to that which can be profitably stored and processed. Examples were presented to demonstrate the concepts and the philosophy.

As in many other places throughout this book, we are dealing with new applications. Hence the burden in making the system work will fall upon the practicioner. It is essential that the material presented be modified to fit the application. The primary hope is that needless effort will not be wasted in re-establishing the basic system presented here.

QUESTIONS AND PROBLEMS

1. Consider a small manufacturing company with a single safety staff specialist. This company has a small computer with very little excess capacity. Present the positive and negative arguments in computerizing the safety information system.

2. Discuss the advantages and disadvantages of coding historical and investigation information.

3. Discuss the reasons that the OSHA recordkeeping regulations do not provide adequate information for the safety data base.

4. Why should the legal ramifications of keeping first-aid and noninjury accident records be discounted?

5. What are the advantages of having the pre- and post-accident forms as similar as possible?

6. Discuss the reasons that a bivariate distribution might yield more information than the two univariate distributions that make it up.

7. Discuss methods for establishing an estimate of varinace for the frequency and severity of accidents. Would this be valid if certain months (or time periods) always had higher readings than others?

8. In Table 7.12, which accident classification should be considered the most important? Why?

9. Which operation in Table 7.13 should be considered most important? How can you tell in which operation most of the "Tree, Log Contact" accidents occurred?

10. Set up the format of a bivariate analysis of Operation versus Accident classification, where the frequency is of primary concern. Fill in the first row under Logmaking, using Table 7.14.

11. According to Table 7.15, which operation has a significantly higher average frequency in a 3-month period? Why is there no significant difference between "Loading and Unloading" and "Maintenance" even though the difference in means is $4.58 - 1.77 = 2.81$ (answer in general terms).

ALLOCATION OF THE SAFETY BUDGET

8.1 Introduction

In Chapter 5 methods for taking available data from an industrial system and translating them into cost/benefits were given. These cost/benefits stated the expected dollars per workday saved, or a similar measure of benefit, for each alternative proposed. This, in turn, pinpoints those areas of the work environment that could return a maximum savings of lost workdays per dollar spent.

The primary use of fault-tree analysis in the past has been for analytical investigation. The methods given in Chapter 5 are the first quantitative procedures for using fault-tree analysis to obtain cost/benefits. The practicing safety professional probably recognizes the reason for the delay in integrating fault-tree and cost/benefit analyses. For, although the methods presented in Chapter 5 are probably the most precise techniques for determining benefits, the absence of available data may render these techniques impracticable to apply.

Before continuing, a word in defense of fault-tree analysis is required. At some point in any endeavor, the decision maker must decide whether he is going to commit himself to a given action or wait for more information. Similarly, in constructing a model of reality, it is necessary to determine the

8

level of detail at which further analysis is no longer justified (see Chapter 1). In order to make such a decision, a knowledge of what further *could* be done is essential. Otherwise, the decision to proceed no further with the analysis is based upon ignorance. The information in Chapters 3–6 was presented to demonstrate the techniques that could be applied. No apology is made for the existence of situations to which they do not or should not be applied. Rather, there was a warning against the application of these techniques when other, less sophisticated techniques would suffice.

The techniques presented in Chapters 3–6 were mathematically sound, and they presented an ideal level of sophistication which the safety systems engineer should master and be capable of applying should the opportunity present itself. However, as was discussed in Chapter 1, the safety systems engineer must be capable of not just applying techniques but also modifying and possibly inventing new techniques that will be suited to the problem at hand.

This chapter will present a compromise of the fault-tree technique as an operational tool for determining benefits of safety investments. This is presented as a "first-cut" method which safety personnel can use for estimating. In addition, methods by which this initial system could be sophisticated are given in the latter portions of the chapter. A new method of budget allocation, dynamic programming is introduced for taking the cost/benefit information and arriving at a definite policy. Equipped with this policy, the systems safety engineer is guided to the particular alternative to implement at each critical accident area.

8.2 Hazard Location Identification

The first step in determining an optimal allocation in the safety budget is to construct a list of "potential" areas of accident reduction. The procedure for constructing such a list might range in sophistication from pure management judgment to computerized sorting of accident records. In either event, the detailed record that is required by law (OSHA Form 101, Supplementary Record of Occupational Injuries and Illnesses) will provide an initial data base from which a list of high-hazard locations can be compiled. Many large companies have more detailed information than OSHA Form 101. In addition, noninjury accidents and non-lost-workday accidents, which in many cases are not recordable under OSHA regulations, provide excellent information from which a hazard location list can be compiled. This was thoroughly discussed in Chapter 7.

In the above paragraph the word "location" was used in a generic sense. Accident hazards accrue due to the environment, whether man-made or natural. The particular area, say in an industrial plant or on a roadway, where there has been a high concentration of accidents would be considered a high-hazard location. However, many hazards might not be tied to a physical location. A construction site may have a falling-obstacle hazard over the entire site. In situations like these, the word "location" should be interpreted to refer to the general hazard. Similarly with vehicles that cause hazards, the "location" would be interpreted to mean the location of the vehicle despite the fact that it moves from place to place. For simplicity, the word "location" will be used in the subsequent discussion.

Many large companies keep accurate records of all accidents even down to the "property damage only" level. Such a company is in a good position to sort these records to determine high accident locations. If the records are computerized, possibly a location code could be established to facilitate this sort. Note that the location code should be distinguished from the cause in that (1) there may be several contributing causes at one location, and (2) the cause at one location may also be a cause at another location.

In the absence of computerized accident records, the sorts by location will have to be done manually. Where there are a great number of locations and accidents involved, it may help to sort first by major department, then by divisions within departments, and finally to the individual machine or other article involved. Further, as will be illustrated in Chapter 9, when a very large number of accidents have occurred over numerous locations, it may be impossible to investigate and propose solutions for all locations. In these cases the sorting technique might be augmented by a selection procedure to generate those locations with the greatest accident reduction potential. For example, all single accident locations might be ignored in favor of multiple accident locations.

The larger the accident report data base, the more sophisticated must be the procedure for narrowing down the total number of locations to a subset of the right size for investigation. The size of this subset will depend heavily upon the investigation time available. Cases can be made for the use of both frequency and severity in determining most critical locations. Other factors that indicate a high potential savings may also be used to set investigation priorities. For example, a high frequency of the same *type* of accident in a given location may be more indicative of a potential for savings than a larger frequency of random accidents.

In the absence of other methods for determining high-potential-accident-reduction areas, a total loss-priority system should be used. This assumes that the potential accident reduction is directly related to the total loss that has occurred at any given location. A cost factor or *loss* may be assigned to each severity of accident, and the total loss from each location may be calculated. Or, actual workdays lost or dollars lost could be used. This total loss method is recommended to take the chance effects out of estimating. Since a cost/benefit is to be generated in terms of reduced accidental loss, the same terminology should be maintained throughout this procedure and that described below.

Whatever quantitative procedure is adopted to determine investigation priorities, good judgment and flexibility must be maintained. The "red tape" of determining which location to investigate cannot be allowed to stand in the way of the correction of an imminent hazard. While the investigation and evaluation procedures to be explained were designed with expediency in mind, any procedure is a potential deterrent to immediate action. But neither should the few cases in which obvious immediate action is required be allowed to negate a systematic procedure that eliminates arbitrary decisions.

The primary goal of hazard location identification is to find those locations and eventually those alternatives which will bring the greatest return in terms of accident reduction. Therefore, purely random occurrences which are not a function of the working environment need not influence the selection. If possible, *all* hazardous areas should be investigated, and alternative solutions should be recommended using the techniques given below. One place that this is impossible is in the area of traffic safety. An example of hazard location identification for that application will be given in Chapter 9.

8.3 Investigation Procedures

Although no procedure was given, the above section implied that a method of setting priorities on locations should be available to guide the investigators. This could range from a sophisticated computerized methodology to pure human judgment, or perhaps 100% inspection of all accident locations. In large systems, however, many of the locations will be recognized as noncontributing, and the concentration will go to the most critical locations first.

Given that current accident records have led to the establishment of investigation priorities, investigations should be performed in a systematic and consistent manner. This is of primary importance in order that comparisons between locations be meaningful. One of the best ways to bring about consistency is through a well-documented investigation form. Figure 8.1 presents an example of a form that could be used to obtain information for a cost/benefit analysis for alternative safety improvements. Further tailoring of this form to a particular industry is recommended. The general form and the reasons for this analytic approach will be discussed here. For a specific example of the general form applied to traffic and highway safety, see Chapter 9.

The general form in Figure 8.1 has spaces provided in the upper right-hand corner for an identification of the number of pages. Also, in the left-hand corner there are spaces for a location number, which may be completed centrally. Item 1 will contain the technical description of the department and item, possibly a machine, which is under consideration. Item 2 will be used to scale the number of accidents into a yearly rate, and therefore it should be expressed in years. Items 3 and 4 are self-explanatory.

The analysis required to fill out the form is primarily concentrated in item 5. Similar in most respects to the analysis portion of fault-tree analysis, the investigator will break down all the accidents that occurred and organize them into "Workplace Environment Causes." This is necessary since there may be more than one cause for the accidents at one location. The removal of one cause will not affect the accidents due to other causes. This would be especially pertinent if investigation locations were chosen based upon the frequency of accidents at given locations independent of cause. Additionally, if one remedy, such as an entirely new machine, affects some causes but not others, this needs to be noted. This will be done in item 14.

Many investigators tend to think immediately in terms of solutions instead of analyzing and thoroughly defining the problem. In essence, the solution may get into the way of a thoroughly objective look at the problem. The item 5 step tends to eliminate this problem by focusing attention on causes rather than conclusions.

CRITICAL LOCATION INVESTIGATION FORM

___ ___ ___
Location Number

(1) Location: _____

(2) Time Period of Accident History (Years): _____

(3) Date: _____   (4) Investigators: _____

Number of Causes _____

(5) Workplace Environment Causes

Severity During Accident History

(6) Number of Accidents	(7) Number of Fatalities	(8) Number of Permanent Partial	(9) Number of Temporary Total

1. _____

2. _____

3. _____

4. _____

Number of Alternatives _____

(10) Alternatives (Description)

(11) Cost	(12) Life	(13) Maintenance Cost/Year	(14) Effect on Cause			
			1	2	3	4

1. _____

2. _____

3. _____

Figure 8.1. Critical location investigation form

283

The breakdown within item 5 need not depend upon speculation. Rather, accident records, interviews, and past investigations of the location should serve as input. Causes that are beyond the control of management need not be included. Accidents that were only secondarily caused by the workplace environment (contributing causes) should be included. Allowances will be made for this problem in item 14. Finally, an attempt should be made to isolate the categories such that there are no overlaps in causes. Since this cannot always be done, it is recommended that accidents be assigned to their primary cause in items 6–9.

Item 6 contains the number of accidents assigned to each cause regardless of severity. This should be determined from the accident reports; each accident being assigned to but one cause. If an accident were caused by something completely out of the control of management, then it may be eliminated or placed in a special category. Each accident in item 6 will also be assigned to one of item 7, 8, or 9. If it were a fatality accident or permanent total,* it will be counted in 7; a permanent partial injury will go in 8, and a temporary total accident will go in 9. Thus the sum of items 7 + 8 + 9 for any one cause will equal item 6. Another category for "property damage only" accidents may also be set up.

Item 10 calls for a specification of various alternatives as recommended by the investigator. The investigators should be encouraged to specify many alternatives, since this will generally increase the total benefits obtained throughout the system. Bear in mind that maximum effectiveness at some locations will have to be sacrificed for greater overall total benefits. The system can only be optimized over the alternatives given. Therefore, if an alternative is excluded, it cannot improve the final solution.

For each alternative the cost, life, and maintenance cost per year is required in items 11, 12, and 13, respectively. This information can either be obtained from salesmen or estimated by the safety professional. The cost in item 11 should reflect purely the cost to be charged to this year's budget. If alternative financing plans are available, they should be considered as additional alternatives. As the development of the optimization routine continues, it will become clear that the objective is to spend the currently available funds (the current budget) such that the maximum in safety is obtained. Hence the item 11 cost figure should reflect this objective.

Item 12 is the life of the investment cost given in item 11. That is, the period of time that the benefits of the alternative will be in effect before the cost in item 11 will have to be repeated. The view here is toward a realistic simulation of what will happen if the alternative is adopted. A discrete entity will be purchased that will last for a given period of time. The benefits of that entity

---

*See National Safety Council, *Accident Facts*, 1972.

will accrue for that life and can be assigned to benefits obtained from a discrete outlay of funds. The discrete nature of both the cost (item 11) and the benefit, which will be based upon item 12, must be considered. Analyses that assume a continuum of both expenditures and benefits will yield erroneous results.

An example may be in order at this point. Assume that an alternative is to cost $1,000 and that this $1,000 will have to be repeated each year. Further, assume that the budget limitation is over a 1-year period. Then item 11 would be $1,000 and item 12 would be 1 year. On the other hand, if an alternative costs $1,000 and would be in effect for 5 years, then $1,000 would go in item 11 and a 5 would be placed in item 12. No decision will be made as to whether to replace the alternative or not until the 5 years of its life is used up, assuming it is funded.

In order to take into consideration the maintenance requirement of most physical assets, item 13 has been provided. Some alternatives weigh very heavily in this area, and the decision may be greatly influenced by maintenance requirements. Techniques for making these decisions will be discussed below.

The final item of the critical-location-information form is an estimate of the effect that each alternative has upon the causes as listed under item 5. Corresponding to each alternative are slots for four causes, one for each given in item 5. Here the percentage of accident reduction on each cause should be placed. For example, if alternative 1 will reduce 20% of accidents caused by cause 1 and 30% of accidents caused by cause 2, then a 20 and a 30 would be placed under cause 1 and cause 2, respectively, under item 14 and on the line corresponding to alternative 1. This "effect on cause" may be in reduction of severity or frequency of accidents. Those who are most experienced and most capable of making these estimates should do so.

Item 14 is clearly the weakest point in the input information. While it should be realized that any estimate of future occurrences will be subject to error, effort should continue to improve the ability to estimate item 14. Section 8.6 will discuss before–after studies that should be useful in this endeavor.

The form given in Figure 8.1 is intended to serve as a guide to more specific developments. A form such as this is necessary in that it provides a channel for the thought processes of the investigators. It provides a procedure for progressing from individual accident reports to cost/benefit information. The basic information required by the form is not unreasonable. It is information that investigators should already be generating in their minds in coming up with recommendations. Although it is true that now further quantification is required, this should result in an improvement in the precision of the estimates. Previous investigations that failed to consider all these aspects would only result in less precise decisions.

8.4 Cost Benefit Analysis

Assuming that the investigation techniques described above are in effect, the capability now exists to perform a cost/benefit analysis. This is nothing more than the calculation of a cost and a benefit for each alternative. The cost/benefit will be defined here as the cost of an alternative divided by its benefit. If the benefit is in some physical unit, then cost/benefit represents the dollar figure required to purchase one of these units of benefit. Although each alternative's cost/benefit is a measure of how good a "bargain" that counter-measure is, caution should be used in implementing alternatives based solely upon cost/benefit. The next section will show that cost/benefit is merely a means to an end, not an end in itself.

The first step in calculating the cost/benefit of a given alternative is to secure a good estimate of the cost. This has already been done if all alternatives have been investigated according to the procedures given in Section 8.3. Item 11 of Figure 8.1 will contain the cost requirement from the current year's budget. Further, the remainder of the information on the Critical Location Information Form will provide the required "benefit" estimate.

A comment concerning the calculation of "benefits" in general is required before detailing a methodology for their calculation. What will be given in the next few paragraphs is "a" method for calculating benefits, not necessarily "the" method. We lean toward the expression of benefits in terms of a dollar value. This reduces safety decisions to a pure economic decision to which some may object. However, we contend that if this "dollar" benefit is maximized, so will be the safety and health of the individual, since they are completely correlated. We hastily add that if another measure of benefit, such as some nonunitized utility function, is substituted for the dollar method given here, this will not negate the validity of the optimization technique that follows. The following methodology is set up for a guide with the hope that it can and will be improved for specific applications.

In general, the benefit of a safety investment is the amount by which that given investment will save life, limb, and health. In Figure 8.1 those most qualified were required to estimate a percentage reduction in frequency and severity of accidents to be brought about by each alternative. (Methods for improving this estimate are given below.) These percentage reductions will be applied to their respective causes to determine first an expected savings per year and finally a total expected savings over the life of the alternative.

In order to relate the different types of accidents, it is necessary to obtain a common unit. The National Safety Council* has published the following

---

Accident Facts, 1972, 425 North Michigan Avenue, Chicago, Illinois 60611.

compensation costs for the breakdown of accidents in Figure 8.1:

Fatal, Permanent Total	$21,000
Permanent Partial	2,500
Temporary Total	345

It is not the intent here to validify or negate these figures. Any set of figures would make the method work which is about to be presented. However, this is one common unit that is useful for considering severity. Other severity classifications will be illustrated below.

The first step in determining a benefit for the alternatives is to determine the total loss from the cases. This is accomplished by multiplying the number of accidents in each severity class by the corresponding loss. By dividing by the number of years of accident history, item 2, a yearly loss can be obtained for each cause of accidents. The sum of these will represent the total loss from the given cause.

By multiplying the percent reduction (expressed as a fraction) for each cause by the loss of that cause and summing, the yearly savings can be obtained. Finally, by multiplying by the life of the alternative, the total benefit can be obtained.

This can be seen easily by the use of an example.

Example 8.1 Calculation of Cost/Benefits

Assume that a Critical Location Investigation Form has been completed as given in Figure 8.2. This location was chosen for investigation because accident records indicated both a high frequency and high severity of accidents. The total of 11 accidents in the past 5 years were attributed to two causes: (1) the operator's exposure to the unguarded machine, and (2) visual distractions caused by a proximate operation. Three alternatives are suggested by the investigators as indicated. The third is a combination of the first two.

The cost figures in item 11 are the cost requirements from this year's budget. Hence for alternative 1, the cost is $1,000, which will have a life, due to anticipated remodeling, of 3 years.

To calculate the benefit from alternative 1, it is necessary to first calculate the expected loss if alternative 1 is not implemented. The total loss due to cause 1 is

$$(1 \text{ fatality}) \times (\$21,000/\text{fatality}) + (2 \text{ permanent partial})$$
$$\times (2,500/\text{permanent partial}) + (4 \text{ temporary total})$$
$$\times (345/\text{temporary total}) = \$27,380$$

0 0 9 9
Location Number

EXAMPLE OF CRITICAL LOCATION INVESTIGATION FORM

(1) Location: __Metal Cutter Department, Grinder #105__

(2) Time Period of Accident History (Years): __5__

(3) Date: __10/2/73__   (4) Investigators: __Brown, Walker, Loper, Rankin__

Number of Causes __2__

(5) Workplace Environment Causes	Severity During Accident History		
	(6) Number of Accidents	(7) Number of Fatalities	(9) Number of Temporary Total
1. Exposure of Operators	7	1	4
2. Distraction of Operators	4	1	3
3.			
4.			

Number of Alternatives __3__

(10) Alternatives (Description)	(11) Cost	(12) Life	(13) Maintenance Cost/Year	(14) Effect on Cause			
				1	2	3	4
1. Install screen to prevent distraction	$1,000	3	$100	0	50		
2. Install guards	$2,000	6	$200	30	20		
3. Combination (1) and (2)	$3,000	5	$300	30	70		

Page

Figure 8.2. Example of critical location investigation form

This is a 5-year total, so that the yearly expected loss from the location without improvements is $27,380/5 years = $5,476/year due to cause 1.

Similarly, for cause 2 the total loss is (0) (21,000) + 1(2,500) + 3(345) = $3,535/5 years or $707/year. Now alternative 1 is estimated to have no effect upon accidents caused by cause 1. On the other hand, it is estimated to reduce the accidents attributed to cause 2 by 50%. Hence the savings due to alternative 1 will be

$$(50\%) \times \text{(cost due to cause 2)} = .5 \ (\$707/\text{year}) = \$353.50/\text{year}$$

Now the life of alternative 1 is 3 years. That is, the $1,000 cost will alter the location's environment for a period of 3 years before additional investments are required. Hence the total benefit that will accure from the alternative is

$$(\$353.50/\text{year}) \times (3 \text{ years}) = \$1,060.50$$

The cost/benefit of alternative 1 is

$$\$1,000/1,060.50 = .943$$

That is, .943 dollars must be invested for every dollar return from this alternative. Note, however, that this alternative is discrete; that is, we cannot spend more nor less than $1,000 and get this return. That is significant in future decision making.

Now consider alternative 2. It will have an effect on the accidents attributed to both causes 1 and 2. The benefit per year for alternative 2 is

$$(.3)(\$5,476) + (.2)(\$707) = \$1,784.20/\text{year}$$

and its life is 6 years, so the total benefit over the life is 10,705.20. Since its cost is $2,000, the cost/benefit is .187 or a return of over 5 to 1. Clearly this is superior to alternative 1. However, realize that the decision is greater than the simple choice between two alternatives.

For one thing, there is alternative 3, which is the combination of 1 and 2. To evaluate this based upon the investigator's estimates, it is necessary to calculate the sum of the reductions as was done above.* That is, the expected savings from alternative 3 is

$$(.3)(5,476) + (.7)(707) = 2,137.70$$

with a prorated life (based upon cost) of 5 years. Thus the total benefit is

---

*Obviously the implementation of two alternatives simultaneously will not always result in a sum of resulting cost and benefits as is assumed here.

5 × (2,137.70), or 10,688.50. The cost/benefit is $3,000/$10,688.50 = .28, or nearly a 4-to-1 return.

The decision as to which is the correct alternative depends upon one other factor that manifests itself in two ways. That factor is the limitation of money. It manifests itself first in the overall yearly budget constraints. This, in turn, causes a competition between *locations* as well as the competition of alternatives at a location. Thus it is impossible to specify a best alternative without knowing (1) the total budget constraint, and (2) other locations' alternative cost/benefits.

To allow for this fact, one other alternative should be considered. This is the alternative "to do nothing" at this location. Hence all improvements might be sacrificed at one location for the sake of more total benefit at other locations. For the same reason, a more expensive alternative might be sacrificed and a less-expensive alternative implemented so that the dollar savings can be employed more profitably elsewhere.

The objective is maximum dollar benefit, which can be translated into reduced injuries and fatalities. Any time a savings at one location can be spent more profitably at another location, it should be. Determining the optimal policy generally requires a total enumeration of all combinations of alternatives, which would be virtually impossible by hand. However, with the use of an efficient algorithm, the optimum can be readily obtained by computer. Section 8.5 will discuss this optimization technique. First, however, an alternative method for determining benefits is considered. A computer program for calculating cost and benefits is documented in Appendix 2.

8.4.1 Aleternative Method for Calculating Benefits

The National Safety Council cites a figure for the cost of a fatal accident to be about 10 times the cost of a permanent partial. We have little reason to doubt these cost estimates and even less reason to question the *relative* value of one to the other. This being true, locations that have had fatalities are going to have, all other things equal, about 10 times the chance of being improved as other high accident locations.

This may not be bad, especially in a high-fatality-risk environment. However, in many occupational situations, fatalities are relatively rare and they are often highly subject to chance. That is, many permanent partials and temporary totals could have been fatalities if chance occurrences caused a variation of a few inches, a few seconds, and so on. Thus, in any location, the presence of a permanent partial or a temporary total might be the signal of a potential fatality some time in the future. For maximum fatality reduction, these too should be improved, not just those which had fatalities in the past.

The objective of this alternative method for calculating benefits is to eliminate the favoritism toward locations at which chance fatalities took place.

It will strengthen the position of locations where fatalities have not occurred just because of luck. The general approach will be to use the *ratio* of fatalities to the other severity categories to allocate fatalities rather than using the actual fatality figures themselves. In the event that the accident history does, in fact, reflect closely the true hazard situation, the use of the ratios can be overridden, and the method of the previous section can be used.

The first step is to calculate the ratios of each severity classification to total accidents. This is done by using company or industry figures. As an example, the National Safety Council figures for 1971 will be used. The frequency rates per million man-hours were as follows:

Fatal, Permanent Total	.05
Permanent Partial	.33
Temporary Total	8.99
	9.37

Thus the following ratios can be calculated:

Fatal/All Accidents = .05/9.37 = .0053

Permanent Partial/All Accidents = .33/9.37 = .0352

Temporary Total/All Accidents = 8.99/9.37 = .9594

Now suppose that the severities of the accidents reported were unknown. Suppose at a given location 10 accidents were reported. In the absence of other information, we could allocate these 10 accidents to the three categories by the ratio in which they occur. That is,

Number of Fatalities = (.0053)(10) = .053

Number of Permanent Partial = (.0352)(10) = .352

Number of Temporary Total = (.9594)(10) = 9.594

and the three categories add to approximately 10. Now, if the severity is purely a matter of chance, these estimates of future severity will be *better* estimates than the specific location's past history, assuming that the ratios generally represent the industry. This is because a larger sample is being used to determine the estimates. Note that the number of fatalities is now stated in terms of the expected number per year. Obviously, there cannot be .053 fatality in 1 year. However, over a long period of time, the .053 will yield better accuracy than the discrete integer obtained from past history. Finally, note that this technique reduces to "costing" accidents purely upon a frequency history basis.

The above technique only holds if the severity of accidents at a given location is purely a matter of chance, and the chance effects for the location reflect those for the industry. Quite often a slight modification will yield better results. This is based upon the knowledge that the location severity is not purely random. Since the most critical problem occurs in "costing" fatalities, it may be desirable to use ratios only on the two most severe categories. This would allow the severity history at a location to have some bearing upon the costing procedures. This can easily be done by calculating the following ratios:

Fatal/Permanent Partial or worse = .05/.38 = .1316

Permanent Partial/Permanent Partial or worse = .33/.38 = .8684

Now consider a location with 10 accidents, of which 7 were temporary totals. The procedure is to allocate the remaining three according to the above ratios. Hence

Number of Fatalities = (.1316)(3) = .395

Number of Permanent Partial = (.8684)(3) = 2.605

Number of Temporary Total from severity history = 7.000

This method is more sensitive to the location severity history, but it still takes the "chance effects" out of the fatalities. Hence this procedure is recommended for general application. Once the number of each severity class is calculated, the procedure for calculating costs is identical to that given in Section 8.4.

In closing this section, a generalization of the above techniques should be made apparent. The particular figures as well as the severity classes themselves were just *examples*. They were provided to make the explanation easier and for that reason alone. Safety personnel from each industry should determine severity classifications which fit their needs (possibly more than three classes). Ratios should be obtained from the particular accident history of that industry. The specific categories and numbers were used here because they were handy and familiar to all. In Chapter 9, an example will be given that will demonstrate this principle of using ratios applied to completely different categories than those above.

8.5 Allocating the Budget

It might seem that for optimal allocation of the safety budget (total amount of money to be invested in a given time or production period) that the decision maker need only select those minimum cost/benefit alternatives first and continue spending until the budget is exhausted. A crucial problem is encountered, however, in the straightforward application of this technique. This could lead to a very large expenditure first which would not allow room in the budget for other alternatives, the *sum of which* might have a larger return. Alternatively, a series of small investments might be made first, not allowing room for a larger expenditure that is next in line in terms of cost/ benefit. One or more of the smaller investments should be sacrificed to make room for the larger investment if the sum of the returns is larger.

For any large facility, alternative safety investments could easily number into the hundreds, thus precluding a "common sense" or trial-and-error approach. Hence some optimization technique should be applied to cost/ benefit information generated by the cost/benefit analysis in order to secure an optimal solution under the assumed conditions. One technique that accomplishes this task is dynamic programming. Although it is a relatively efficient technique, a large number of computations are still required, and it is helpful to set up a computer program for performing the calculations. Such a program has been established and it is documented in Appendix 2. The remainder of this section is dedicated to the general philosophy behind the use of this algorithm and the mathematical concepts involved.

At this point it should be emphasized that a thorough understanding of the mathematical techniques applied are not essential for an understanding of the use of the methods discussed here. A fitting analogy is the fact that most people very successfully operate automobiles without knowing the principles or the detailed operation of the internal combustion engine. Knowing enough about the capabilities and limitations of a technique is often enough to obtain its full use. These will be demonstrated presently, followed by an example. For those who care to modify or improve the technique itself, the mathematical framework is presented in reference 1.

It is convenient to depart from the concept of cost/benefit as presented in Chapter 5 for a short while and reconsider some of the concepts presented in Chapter 1. It is the objective here to synthesize a total safety system composed of a number of safety investments each of which has been selected from a large number of possible alternatives. The qualitative aspects of performing this synthesis were discussed in Chapter 1 and illustrated in Figures 1.1–1.4. The purpose here is to perform a parallel quantitative procedure to aid in obtaining the maximum reduction in accident loss. Figure 8.3 illustrates a concept useful in this endeavor.

In Figure 8.3, X_s can be viewed as the total budget, which is to be allocated to the several stages. The number of these stages is s, and numbering proceeds from right to left. A decision is made concerning stage s, which is depicted by D_s. This decision will require the expenditure of money and will result in an expected savings in life, health, workdays, and so on, depicted by R_s. Once an expenditure is made to stage s, the total available budget is

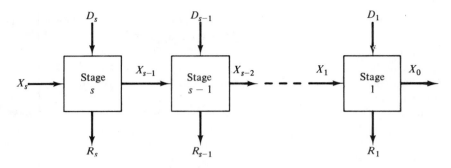

Figure 8.3. Depiction of the stages of the total safety system

reduced, and now only X_{s-1} remains. This procedure is followed for stage $s - 1, s - 2, \ldots, 1$, hopefully in such a way as to maximize the sum of the returns.

These stages are nothing more than the "locations" discussed in Section 8.3. Of course, some of the alternatives need not be limited to a specific location. An example might be education or safety personnel, both of which would bring benefits in a variety of areas. Hence it is good to preserve the concept of stages and not limit its interpretation to locations.

The general rule for organizing alternatives into stages is simple. Only one alternative will be chosen for each stage. Thus *if one countermeasure excludes another, they must be considered within the same stage.* On the other hand, if two countermeasures may be implemented simultaneously, they *may* be considered in separate stages. Note that it is also possible to structure two countermeasures that may be implemented simultaneously within the same stage. For example, if countermeasure A and countermeasure B can be implemented simultaneously, there are four possible alternatives which they can form which cannot be implemented simultaneously. These are: (1) neither A nor B, (2) A only, (3) B only, and (4) both A and B. Hence these four alternatives could be placed within one stage. Generally, the problem is simplified by setting up more stages, as opposed to more alternatives within a stage.

The reader may wonder why an organization of alternatives into stages is necessary. The solution algorithm presented below uses the concept of dynamic programming, which requires the framework discussed above. This framework will now be demonstrated by an example.

Example 8.2 *Organizing Alternatives into Stages*

Suppose that a current machining operation is producing dust which is hazardous to the eyes. This "location" will be assigned to stage 1. The following alternative solutions were proposed:

1. Machine change to dustless process.
2. Protective Glasses Style 1.
3. Protective Glasses Style 2.

Of necessity, investments 2 and 3 were assigned to the same stage because of the rule given in the preceding section. The dynamic programming characteristic of choosing only one of the alternatives can be used to advantage as far as 2 and 3 are concerned, since obviously it would be impossible to wear two sets of glasses at once. However, with investment 1 this is not the case. It is possible to implement simultaneously both 1 and 2 or both 1 and 3. Hence there are five possible alternatives as follows:

1. Item 1 only.
2. Item 2 only.
3. Item 3 only.
4. Items 1 and 2.
5. Items 1 and 3.

This covers all the possible affirmative alternatives. However, it is also necessary to allow for the fact that the money allocated for any one of these alternatives might be better spent in another stage. Hence it is necessary to include a sixth alternative: (6) None of the alternatives in this stage. The dynamic programming algorithm will select exactly one alternative from each stage. It will be an optimal selection in that the budget will be spent in such a way that the maximum total return will be derived.

To repeat: all possible feasible combinations of safety investments must be generated for input *for a given stage*. It is not necessary, however, to generate all possible combinations across stages. This is, in fact, the basic advantage that dynamic programming has over a total enumeration of all the alternatives. To reduce the number of combinations that must be generated by the user, it is recommended that the number of independent safety investments within a given stage not exceed four. However, the program documented has a total number of alternatives per stage for 30 combinations. The number of stages has been set at 50, allowing the number of independent alternatives at each stage to be kept small. This may be enlarged by altering the appropriate DIMENSION statement (see Appendix 2).

8.5.1 Input-Data Preparation

In order to use the computerized algorithm, the input data must be preprocessed so that it will fit the requirements of the program. The specific details as to key punching instructions are given with the program documentation in Appendix 2. At this point concentration will be on the preparation required to fit those input specifications. Even if the processing were to be performed by hand, these preparatory steps would have to be taken.

The cost/benefit analysis performed in Section 8.4, possibly augmented by the revised benefit calculations of Section 8.4.1, provide all the data required to perform the optimization. Some other parameters are required to run the program; but the discussion of these will be deferred to Appendix 2.

The following summary of the data-preparation steps should prove helpful:

1. All critical hazard locations, or as many as resources will permit, will be investigated and the resulting data will be compiled according to Section 8.3. The results will be in a standard format, such as that given in Figure 8.1. Each location will be considered one stage.
2. Alternatives that do not logically fit specific locations will also be investigated and data will be compiled on the same standard forms as in step 1. These alternatives will be organized into stages by the rules given in Section 8.5.
3. All information obtained in steps 1 and 2 above will be processed by the techniques of Section 8.4 so that a unique cost and benefit can be attributed to each alternative. This cost and benefit information will be maintained in subsets by stages.
4. The costs and benefits obtained in step 3 will be used as input to the optimization routine to determine the alternatives to implement.

Example 8.3 Complete Optimization Procedure

The four steps above will be applied to a two-stage problem to demonstrate the complete procedure.

Step 1.   The basic data given in Figure 8.2 will provide an example of a complete iteration of the cost/benefit procedure.

Step 2.   Assume that a second stage has been investigated that does not necessarily fit any given "location." Space will not be dedicated to performing the complete cost/benefit analysis since it will be identical to that given for the first stage. The assumed results will be given below.

Step 3.   Using the raw data in Figure 8.2, the cost/benefit analysis will be performed under the assumption that the fatality was a random occurrence. Hence the two higher-severity classifications will be pooled, and a ratio will be used to allocate the severity for costing. Assume the following parameters for this example:

Parameter	Item	Value
1	Average cost of accident, item 7*	$50,000
2	Average cost of accident, item 8	5,000
3	Average cost of accident, item 9	500
4	Ratio of item 7 accidents to both items 7 and 8	.1
5	Ratio of item 8 accidents to both items 7 and 8	.9

Now using the data in Figure 8.2, the following costs of accidents will be incurred due to cause 1:

$$(.1)(3)(\$50,000) + (.9)(3)(\$5,000) + 4(\$500) = \$30,500$$

Notice that the 1 accident in item 7, cause 1 of Figure 8.2, and the 2 accidents in item 8 were pooled and then prorated according to the .1 and .9 ratios for the maximum severity and middle severity, respectively. This uses the techniques discussed in Section 8.4.1.

The second cause has only one accident in the two upper severity classifications, but it will also be prorated as follows:

$$(.1)(1)(\$50,000) + (.9)(1)(\$5,000) + 3(\$500) = \$11,000$$

Since a 5-year history was used, the following can be obtained:

cost per year for cause 1 = $30,500/5 = $6,100/year

cost per year for cause 2 = $11,000/5 = $2,200/year

Now alternative 1 is estimated to reduce accidents due to cause 2 by 50%, and therefore its savings per year is

$$(.50 \text{ savings})(\$2,200/\text{year}) = \$1,100 \text{ savings/year}$$

Since alternative 1 has a 3-year life, the total savings to be anticipated by the implementation of alternative 1 is

$$(\$1,100/\text{year})(3 \text{ years}) = \$3,300$$

Alternative 2 is estimated to reduce both cause 1 and cause 2, so its

---

*One of the purposes of this example is to demonstrate that the severity classifications need not follow National Safety Council categories. These should fit the industry, as should the ratios and the costs.

expected savings per year is

$$(.30)(6,100) + (.20)(2,200) = \$2,270 \text{ savings/year}$$

which over 6 years comes out to $13,620.

Similarly, alternative 3 at this location will yield a total return of

$$5 \text{ years } [.30(\$6,100/\text{year}) + .70(\$2,200/\text{year})] = \$16,850$$

The following set of costs and benefits can now be compiled for this location, which will be called stage 1:

STAGE 1

Alternative	Cost	Benefit	Cost/Benefit
0	0	0	0
1	$1,000	3,300	.3030
2	$2,000	13,620	.1468
3	$3,000	16,850	.1780

Assume further that a second stage exists and that the cost/benefit analysis has been performed, yielding the following result for stage 2:

STAGE 2

Alternative	Cost	Benefit	Cost/Benefit
0	0	0	0
1	$2,000	9,175	.2180
2	$3,000	16,500	.1818

This completes step 3 for the example 2 stage system. All the information is now available to find the optimum allocation of funds.

Step 4.   One way to find the optimum policy is by total enumeration of all possibilities. This will be done here to demonstrate the need for a total system optimization as opposed to a minimum cost/benefit approach. The dynamic programming procedure given in Appendix 2 is more efficient for systems with more than two stages.

It should be evident by now that the optimum allocation of funds is almost completely dependent upon the total budget constraint. For example, if only $3,000 were available, the optimal allocation would be quite different than if $5,000 were available. Note that at least $5,000 must be available to implement the minimum cost/benefit alternatives for both stages. For illustrative purposes, all budgets are evaluated in Table 8.1.

There is an optimum policy for each maximum budget constraint. Generally, an alternative will not be implemented at every stage. The budgets of less than $5,000 illustrate this situation and they demonstrate that it is not always optimal to implement the lowest cost/benefit first. Also, when spending may exceed $5,000, it is, of course, more beneficial to go to alternatives that are not so good from a pure cost/benefit view.

Table 8.1 ALL OPTIMAL(*) POLICIES FOR EXAMPLE 7.3

	Feasible Alternatives			
Maximum-Budget	*Stage* 1	*Stage* 2	*Total Benefit*	*Optimal*
0	0	0	0	
$1,000	1	0	3,300	*
$2,000	1	0	3,300	
$2,000	2	0	13,620	*
$2,000	0	1	9,175	
$3,000	3	0	16,850	*
$3,000	1	1	12,475	
$3,000	0	2	16,500	
$4,000	1	2	19,800	
$4,000	2	1	22,795	*
$5,000	3	1	26,025	
$5,000	2	2	30,120	*
$6,000	3	2	33,350	*

With the list given in Table 8.1, all that remains is that the total budget be determined and the alternatives implemented. The determination of the total budget is a management decision, the details of which will be considered more fully in Chapter 9.

In this example, a total enumeration of all alternatives was easy to perform. It should be easy to see that each time another stage is added, the total number of alternatives is multiplied by the number of alternatives in the new stage. Hence 30 stages with one alternative each (two real alternatives) would lead to the evaluation of over 1 billion alternatives. Since dynamic programming can handle this problem in a matter of minutes, the use of this computerized algorithm is greatly recommended. Any company or industry that spends the thousands of dollars required to construct the data base for the types of analyses given above can afford the $10 or $20 of computer time required to ensure that their solution is the best they can obtain. The completely documented program is given in Appendix 2.

8.5.2 Budget-Allocation Manual

An individual would have to be quite naive to believe that the techniques for budget allocation could be fully implemented in most organizations in a month or even a year. Two things stand in the way of such an innovation. As already discussed, data requirements present a formidable barrier. Data-collection schemes, such as those described in Chapter 7, are not hard to implement, but their benefits are not obtained overnight. This contributes to the second barrier, which is usually more important, that of management rejection. Management might correctly reason that efforts to construct an effective budget-allocation system may not pay off for several years. When results are not forthcoming as quickly as anticipated, they might decide to terminate the entire effort.

In Chapter 9 the modular concept of system development will be introduced to counter this problem. It is sufficient to say at this point that the benefits of the budget-allocation system must be visualized by uper levels of management as quickly as possible. The benefits obtained by its evolutionary implementation far surpass the ultimate future result. The many evaluations and cost/benefit analyses provide immediate factual information for decision making. By designing such studies to provide input to the budget-allocation system, overall direction is given to this research-and-development effort.

The purpose of this section is to provide the vehicle for the evolutionary implementation of the budget allocation system. This consists of a flexible manual that can be updated frequently to reflect the status of the system. The manual should contain the following:

1. An introduction, providing the decision maker with a detailed explanation of its use.
2. Budget allocations [i.e., computer outputs from DPM (see Appendix 2)] for a range of possible budgets. Since these will be based upon the data currently available, they should be interpreted by the decision maker accordingly.
3. Budget-allocation forms, which provide a brief description of each countermeasure for each stage. The cost and the benefit for each alternative should also be presented for ready reference.
4. Budget-allocation worksheets, which present the economic analysis and the method by which each benefit was estimated. These will contain actual computations of costs and benefits condensed to one page. Longer detailed reports upon which the computations are based should be referenced.

Note that the above contents are in order from general to specific. It is tailored for the user on an exception basis. That is, if the decision maker has confidence in the results (i.e., the budget allocations), he may only use that section. Where questions arise as to the particular countermeasure or the input data, he would consult the "forms" section. In cases where the source of such data is in question, he might consult the actual calculations and fundamental assumptions as given in the "worksheets" section. Finally, for the most detailed information, he would consult the studies that are referenced in the worksheets, although they will not be included in the manual.

An attempt should be made at all times to improve upon the final policy. This will draw upon inferences obtained from studies conducted either in-house or by independent research groups. Whenever the cost or benefit of a countermeasure is to be changed by new information, revisions should start with the "worksheets." This will lead to changes in the "forms," and thus a new set of input data will be generated for the budget allocation. All pages should be dated, and pages should be modular such that updates can be made easily.

Before proceeding with an example, one additional value of this manual should be emphasized. During the initial establishment of the manual, information will be totally lacking for some stages. Others will rely upon pure estimates of cost and benefit. This will cause a lack of confidence in the results, and rightly so. However, there is nothing to prevent the input estimates, on the worksheets, from being revised. Calculations can easily be made to revise the input to DPM. In turn, the results of those changes on policy can be viewed and compared to the original. This procedure, called *sensitivity analysis*, enables the decision maker to determine his confidence in the final policy. More important, however, it enables the analyst to determine where future data collection and evaluation efforts should be directed.

Quite often studies are performed to determine the effectiveness of a given countermeasure to an unnecessary degree of accuracy. If the sensitivity analysis reveals that an error in an estimate of 100% will not significantly alter the policy, obviously it is not necessary to fine-tune that estimate to the nearest 10%. Cost/benefit analyses cost money; and that money should also be allocated to obtain accuracy on a broad front. Accuracy should be sacrificed where it is not warranted. The systems analysis principles of optimization apply here as they do elsewhere.

Example 8.4 Budget-Allocation Manual for Traffic Safety

Space will not permit the inclusion of an entire budget-allocation manual. However, excerpts from a manual that the author developed for the Alabama Office of Highway and Traffic Safety* should provide enough of a guide for a

---

*Alabama Office of Highway and Traffic Safety, *Budget Allocation Manual*, Montgomery, Alabama, Jan. 1975.

generalization to most safety applications. This example will not restrict input to the format of the Critical Location Investigation Form of Figure 8.1. The case study of Chapter 9 will present a further exemplification of the use of this form. At this point the flexibility in the types of studies that provide input to the optimization procedure should be noted.

Table 8.2 presents a description of the stages, which would be discussed in the introduction to the manual. Tables 8.3 and 8.4 present two of the budget allocations, one for $100,000 and the other for $500,000. In actuality, allocations were produced for budgets ranging from $50,000 to $1,500,000 in increments of $50,000. This enabled an easy comparison of policies for the various-sized budgets.

Using Table 8.2 together with the appropriate budget allocation (as exemplified by Tables 8.3 and 8.4), the decision maker can determine the allocation to areas that the model indicates to be optimal. In most cases, however, the specific countermeasure will be of concern. Ready reference is available in the budget-allocation forms, two of which are given in Tables 8.5 and 8.6 for stages 4 and 13, respectively. Note that some of the stages may not

Table 8.2 Description of Stages

Stage	Description
1	Traffic Records
2	Alcohol-Related
3	Seat Belts
4	Motorcycle Safety
5	School Buses
6	General Motor Vehicle Inspection
7	Beginning Driver Education
8	Pedestrian Safety (Bicycle)
9	Driver-History File
10	Selective Enforcement
11	Radar Equipment
12	Motorist Aid and Debris Removal
13	First-Aid Courses
14	Traffic-Court Relations
15	Experienced Driver Education
16	EMS Training and Equipment
17	Training Design Engineers
18	Inventories of Guardrails, Traffic-Control Devices
19	Skid Test and Asphalt Study
20	Consultant Services
21	Driver Education for EMR
22	Police Motorcycles
23	Administration and Surveys
24	Laws, Registrations (truck weights)
25	Death-Trap Program
26	Training
27	Maintenance

Table 8.3 ALLOCATION OF THE $500,000 BUDGET

Stage	Alternative Number	Cost	Return
21	1	$10,000	304,719,104
20	1	$10,000	302,719,232
19	0	0	301,424,384
18	0	0	301,424,384
17	0	0	301,424,384
16	5	$100,000	301,424,384
15	4	$40,000	172,534,768
14	1	$100,000	160,470,544
13	2	$25,000	147,593,056
12	0	0	135,113,056
11	1	$10,000	135,113,056
10	0	0	132,651,808
9	0	0	132,651,808
8	0	0	132,651,808
7	0	0	132,651,808
6	0	0	132,651,808
5	0	0	132,651,808
4	1	$15,000	132,651,808
3	3	$140,000	130,035,632
2	5	$50,000	28,343,136
1	0	0	0

Table 8.4 ALLOCATION OF THE $100,000 BUDGET

Stage	Alternative Number	Cost	Return
21	0	0	175,709,680
20	0	0	175,709,680
19	0	0	175,709,680
18	0	0	175,709,680
17	0	0	175,709,680
16	1	$20,000	175,709,680
15	0	0	89,409,552
14	0	0	89,409,552
13	1	$10,000	89,409,552
12	0	0	79,100,000
11	0	0	79,100,000
10	0	0	79,100,000
9	0	0	79,100,000
8	0	0	79,100,000
7	0	0	79,100,000
6	0	0	79,100,000
5	0	0	79,100,000
4	0	0	79,100,000
3	1	$60,000	79,100,000
2	1	$10,000	16,520,000
1	0	0	0

Table 8.5 BUDGET-ALLOCATION FORM (I)

STAGE 4 Motorcycle Safety

Countermeasure	Description	Cost	Benefit
1	Teacher training for public schools (10 teachers trained)	$15,000	2,616,183
2	(20 teachers trained)	$30,000	3,360,000
3	(30 teachers trained)	$45,000	3,924,275
4	(40 teachers trained)	$60,000	4,257,710
5	(50 teachers trained)	$75,000	4,488,550
6			
7			

Table 8.6 BUDGET-ALLOCATION FORM (II)

STAGE 13 First-Aid Courses

Countermeasure	Description	Cost	Benefit
1	Officer training: Department of Public Safety 120 officers trained	$10,000	10,309,565
2	300 officers trained	$25,000	12,480,000
3			
4			
5			
6			
7			

have proposed countermeasures, owing to lack of proposals or lack of evaluation data. This should be clearly indicated to the decision maker, and plans for future action (i.e., evaluation or removal of the stage) should be documented. There should be one form for each of the stages under consideration.

Corresponding to each form is a worksheet. The example stage 4 and 13 worksheets are presented in Tables 8.7 and 8.8, respectively. Please note that these are for example only, and they are not to be an endorsement of motor-

Table 8.7 BUDGET-ALLOCATION WORKSHEET (I)

STAGE ___4___   Motorcycle Safety

General Statistics

Year	Accidents	Fatal	Injury	Property Damage	Loss
1971	1841	57	1140	644	13,748,960
1972	2119	48	1463	708	13,826,389
1973	2668	54	1768	846	16,104,134
Average	2209	53	1457	733	14,559,828

The total number of registered motorcyclists was 65,560 in 1973.

Countermeasure 2
The probability of any motorcyclist being involved in an accident in a given year is $2,209/65,560 = .034$, yielding an expected loss of $(.034)(14,559,828/2,209) = \$224/year$ for each motorcyclist.
The cost per teacher trained is $1,500.
One teacher trains 120 motorcyclists/year or 600 in 5 years.
The reduction in accident probability averages 40% over a 2-year period for each motorcyclist so educated and 15% for the next 3 years. Thus the total benefit per teacher is $600(224)1.25 = 168,000$.
Countermeasure 2 will be the present system of $30,000 expenditure, training effectively 20 teachers. The total benefit is $(20)(\$168,000) = \$3,360,000$.
Other Countermeasures
The other countermeasure costs were prorated according to the number trained, as indicated. Benefits were obtained by estimated diminishing returns with alternative 2 as a base.

Table 8.8 BUDGET-ALLOCATION WORKSHEET (II)

STAGE _13___   First-Aid Courses

Countermeasures 1 and 2.   See Report 300-74-001-002-008. The benefit per officer per year was determined to be:

Severity-Reduction Effect		Value	Benefit
Critical to severe	.24	$150,000	$36,000
Severe to moderate	.94	$4,000*	$3,760
Severe to minor	.94	$2,000*	$1,880
			$41,640

*Estimated.

The cost of training per officer given in Report 300-74-001-002-008 is $83.
The benefit estimated from training 300 officers is approximately $(41,600)(300) = 12,480,000$, as given in countermeasure 2. The benefit of countermeasure 1 was estimated based upon a large anticipated immediate benefit of getting nearly half of the lacking officers trained.

cycle safety training or first-aid training over other types of safety expenditures. Benefits are high compared to cost because a large fatality cost was used ($150,000) and because the cost of the countermeasure was solely this governmental agency's contribution, which in some cases was only a small portion of the total cost.

Table 8.7 and 8.8 show how probabilistic inference, as discussed in Chapters 4, 5, and 6, are used to estimate benefits. At times estimates are made that will require future verification, as in the case of the per cent accident reduction in Table 8.7. In other cases, such as in Table 8.8, the estimates will be based upon documented reports.

The worksheets provide a focal point for criticism, and this is good. They provide a means of determining the parameters necessary for logical policy formulation. Simple sensitivity analysis can be performed to determine the degree of accuracy required.

Although this example is of necessity sketchy, the value of a budget-allocation manual should be clear. It serves first to influence policy, thus making studies performed and data collected more than meaningless exercises resulting in unread documentation. But it also provides a mechanism for directing the evaluation studies such that they will all be meaningful. The manual itself should never be considered completed. Rather, it should be continuously updated and improved until it becomes both reliable and worthy of higher management acceptance.

8.6 Before–After Studies

The purpose of this section is to construct a framework for control of the optimization technique. The job is not completed when the final policy is determined and the optimal alternatives are implemented. Measurements should be taken on these implementations to determine whether the optimization system is functioning properly. The "optimization system" includes (1) the hazard-location-identification subsystem, (2) the investigation subsystem, (3) the cost/benefit subsystem, and (4) the optimization subsystem.

The last two subsystems are theoretically sound. And, whereas improvements and modifications should be continuously made on these subsystems, a failure to improve safety conditions is more apt to come from the first two subsystems.

In Section 8.2 methods for identifying high-potential-accident-reduction areas were discussed. Methods for evaluating these methods will be discussed here. Since high-hazard-location identification is the initial step in the total procedure, it is crucial to the success of the system. If for some reason a location is excluded from investigatory consideration, there is no way that an improvement would be considered there.

The evaluation of the methods for selecting locations for investigation is quite difficult. It cannot be evaluated based upon past accident history, for that is exactly what is used as its determining criteria. It can only be evaluated on the overall effectiveness of the total system itself. Therefore, as time goes by after implementation of the system, the hazard-location-identification procedure should be repeated, and a comparison should be made of the results before and after.

Methods for performing such a before–after study will not be detailed here, since it is not seen to be a problem in regard to occupational safety. Obviously, for maximum accuracy, every potential hazardous location should be studied. The potential accident forecast would have to be estimated in those areas which do not have an accident history. Hence items 6–9 in Figure 8.1 would have to be estimated as to the accidents that will occur in, say, a 1-year period. If 100% of all potential hazards could be studied in this way, the evaluation of this subsystem would become unnecessary. Chapter 9 will present a traffic safety example in which 100% investigation is impossible. Some methods for evaluating the methods of hazardous location identification are given there.

Aside from the possibility of not selecting a location for investigation, another problem exists in making improper estimates during the investigation. There is only one critical part of Figure 8.1 which could cause a serious problem; this is item 14. All other items for determining the costs and benefits either come from the known accident history, or from fairly well known costs and life expectancy.

To be perfectly accurate in item 14, the estimated accident reduction, a knowledge of the future is required. Those who have the most experience and are most able to make such a prediction should do so. Note that this has little to do with position in the organizational chart. "Experience" is the accumulation of facts relative to the situation under consideration. A substitute for subjective experience is the hard, cold fact of what the implementation of the alternative has done in the past under similar circumstances.

Before–after studies can be performed to determine these facts, using nothing more than the ongoing accident records. Whenever a safety alternative was implemented at a given location, the accidents at that location can be studied over a period before and over a period after the implementation. Hence many accident data bases already contain a wealth of information useful for estimating effects of improvements.

For a new improvement, either because it was never tried before or because the improvement was tried in a different environment, an estimate will have to be made. However, the implementation point in time should be marked so that a before–after study can be made. This may not be useful until 1 or 2 years hence. Many safety people are too busy "fire-fighting" to think this far ahead. But consider this: if these studies were started a few years ago,

308 / *Allocation of the Safety Budget*

much of the information required for making estimates would now be available.

Before–after studies should be initiated, and captured from past data, as part of the total optimization effort. In this way the system will continually improve itself. Although estimates may have to be made initially, eventually they will be replaced by historical fact. Thus a transition will take place from "the best we can do under the circumstances" to a system in which total confidence can be placed. Chapter 9 gives an example of a quantitative before–after study.

SELECTED REFERENCES

1. R. BELLMAN,; *Dynamic Programming*, Princeton University Press, Princeton, N.J., 1957.

2. *BLS Report* 406, U.S. Department of Labor, Washington, D.C., 1972.

3. D. B. BROWN,; "Application of Fault Tree Analysis to Forest Harvesting Operations," Final Report, *Report* 4730-*FS-SO*-3701-1.6, U.S. Forest Service.

4. BROWN, D. B., "Cost/Benefit of Safety Investments Using Fault Tree Analysis," *J. Safety Res.*, Vol. 5, No. 2, June 1973.

5. RECHT, J. L., "System Safety Analysis: Error Rates and Costs," *National Safety News*, June 1966.

6. RECHT, J. L., "Systems Safety Analysis: Failure Mode and Effect," *National Safety News*, Dec. 1965 and Feb. 1966.

7. RECHT, J. L., "Systems Safety Analysis: The Fault Tree," *National Safety News*, Apr. 1966.

8. ROCKWELL, THOMAS H., "A Systems Approach to Maximizing Safety Effectiveness," *ASEE Monograph* 1, p. 14, June 1971 (System Safety Techniques . . .).

9. ROCKWELL, T. H.; and BURNER, L. R., "Information Seeking in Risk Acceptance," *ASSE Journal*, Feb. 1968.

10. RUBINSKY, STANLEY; and SMITH, N. F., "Process Simulators for Accidental Research," *ASSE Journal*, Nov. 1971.

QUESTIONS AND PROBLEMS

1. Discuss the reasons why the safety professional should be aware of sophisticated techniques even though he may not use them in common practice.

2. Given the following costs to be input for each type of accident, respectively:

Type	A	B	C	D	E
Cost	$300	$1,000	$5,000	$20,000	$50,000

calculate the expected benefit of a countermeasure that reduces 5 *A*, 3 *B*, 4 *C*, 2.5 *D*, and .3 *E* accidents to 7 *A*, 2 *B*, 3 *C*, .6 *D*, and .1 *E* accidents.

3. Perform Problem 2 assuming that the ratio of *D* to *E* accidents is 10:1, and pool the *D* and *E* types.

4. Accidents are divided into three categories in an industry as follows:

Type	A	B	C
Cost	$30,000	$3,000	$300

Two critical locations have had the following average yearly history:

Location	A	B	C
1	1	2	4
2	2	1	3

The ratio of type *B* to type *A* accidents historically has been 10 to 1. Compute the expected loss from each location using actual history and the ratio method.

5. Use the following data:

		Effect on Cause		
Alternative	Cost/Year	1	2	3
1	$1,000	20%	0	0
2	$4,000	0	30%	50%
3	$10,000	40%	60%	50%
Cost/Year Assigned to Cause:	$20,000	$30,000	$10,000

Determine a cost/benefit for the three alternatives.

6. Consider three causes of the problems at the two locations discussed in Problem

4. The accidents can be attributed to these causes, as follows:

	Location 1		
Cause	A	B	C
1	1	0	0
2	0	2	4
3	0	0	0

	Location 2		
Cause	A	B	C
1*	0	2	0
2*	1	0	0
3*	0	0	3

*Not necessarily the same as at location 1.

Now suppose that the following alternatives are proposed:

LOCATION 1

			Effect on Cause		
Alternative	Cost	Life	1	2	3
1	$5,000	5	10%	5%	0%
2	$10,000	7	5%	20%	10%
3	$15,000	8	0%	40%	30%

LOCATION 2

			Effect on Cause		
Alternative	Cost	Life	1	2	3
1	$2,000	4	10%	20%	30%
2	$8,000	15	0%	40%	10%

The life is in years, and the history is for a 1-year period. Calculate a cost, a benefit, and a cost/benefit for all alternatives. Use the ratio method of combining accident classes A and B.

7. Determine the optimal policy in Problem 6 for budgets of $10,000, $20,000, and $30,000.

8. Determine the optimal budget allocation given the following data:

Location	Alternative	Cost ($000)	Benefit ($000)
1	1	2	20
	2	8	28
	3	10	50
2	1	12	30
	2	16	25
3	1	6	18
	2	12	24
	3	18	28

Assume budgets of $10,000, $20,000, . . . , $50,000.

CASE STUDY: BUDGET ALLOCATION
APPLIED TO TRAFFIC SAFETY

9.1 Introduction

This chapter is intended to demonstrate a real application to the techniques presented, especially those in Chapter 8. Although this case study is based upon the experience of the author, it is not intended to be an accurate historical account. Certain facts have been altered to make the case more representative of common problems and practices. This may have sacrificed some local realism, but the intention was to enhance the general usefulness of the case study.

The particular application involves a governmental agency at the state level charged with the responsibility of increasing traffic safety by making spot roadway improvements. This task in the past had been performed by upper levels of management* on a nonquantitative basis. Recognizing the inadequacy of this arrangement, a small staff group had been formed to develop and implement a system by which this could be done quantitatively.

The basic problems of a group such as this are the same regardless of the

---

*It should be understood that the term "management" applies to governmental service organizations as well as to the private sector.

9

organization or the mission. They were threefold: (1) technical, (2) organizational, and (3) political. Concentration in Chapters 3 and 4 has been upon the technical requirements of the systems approach. Although the emphasis in the case study will be toward technical soundness, consideration will also be given to the other two problems. The discussion in Chapter 1 brought out that technical soundness, although essential, is generally not sufficient for success.

The organizational problems of a staff group in obtaining necessary information can be fatal to the system development. Ways must be devised to obtain information without invoking the strict lines of authority and responsibility. Staff groups technically do have the power, through the line manager to which they report, to force lower-level personnel to supply information. But power once used is power lost. Forced cooperation would have been the last cooperation that the group would obtain for some time.

A second problem involved the political aspects of "selling the product." The term "political" is used loosely, and it is not intended to be a slur to the mechanisms for getting things done in government. It is hoped that most industrial managers will freely admit what we candidly state to be a fact: the politics involved in decision making in most large companies would make many government workers shudder.

This being the case, generalizations of the factors that cut through the politics in a state government organization could be applied equally well to industrial or even labor organizations. Hence these factors will be presented in their historical context. It will be up to the reader in most cases to make the generalizations and the reapplications.

This chapter will progress by giving an historical background in which the case study will be couched. Then some of the overall "psychological" considerations will be discussed in terms of the organizational and political problems discussed above. The cost/benefit system itself will be presented in the next sections with an attempt not to repeat the basic principles given in Chapter 8. The concurrent development of other programs to determine optimum policies will be presented. Latter sections deal with methods by which the usefulness of the system is communicated to higher levels of management.

9.2 Historical Background

Two years prior to the design of a cost/benefit system, federal guidelines were applied in the state to initiate a standard traffic accident recordkeeping system. This formed a multipurpose computerized data base that could be used for enforcement, roadway improvements, and a variety of other applications. Shortly after the beginning of the creation of this data base, a small group was formed within the Highway Maintenance Department to develop programs and procedures by which the accident data base could be used for guiding roadway safety improvements. This group was called the Hazardous Location Identification and Correction Group (HLIC).

HLIC was initially composed of three persons. It was headed by a computer systems analyst, whose primary experience had been in computerized systems design. That is, his function was to determine what upper management required in terms of a computerized report, and then to translate this into instructions or documentation that could be programmed by specialists. Of course, in his new role he would do this as well as serve as an administrator and leader to the group. He was supported by a programmer and a traffic engineer.

The organizational position of HLIC is also of significance. Its head was essentially on a fourth level, having two levels of management between himself and the Highway Department Director. Most of the decisions regarding the allocation of safety funds in the past had been made at the highest levels of management. Therefore, the HLIC group was not in a position to *directly* influence those who made the decisions. This was not so much a flaw in the organizational design as it was the result of the evolutionary functional expansion necessitated by federal funding. In any event, the problems of communication and selling were inherent in the organizational structure.

This organization was not without its advantages, however. Communication with the state division offices were facilitated by the group's placement within the Maintenance Department. As the development continued, more and more information was required from the local field personnel. In addition, the initial computer reports produced were directed at aiding these local people in their formulation of proposals, which were sent to higher management at the central office.

Development of the computerized cost/benefit system began against this backdrop. There were, as usual, many who were greatly in favor of this system, which they considered long overdue. Others were afraid that it might upset the status quo. Still others were directly opposed to it because of their experience with "computerized failures" of the past.

Development began with a series of sort programs to take the chronologically ordered accidents and rearrange them according to location. Subsequent efforts were aimed at scanning these records and determining those locations which had high concentrations of accidents. For example, any area that had more than eight accidents over 1 year in .4 mile was flagged by the computer.

A second program was developed which took the high-accident areas and assigned tentative priorities to their improvement capability. The complete details as to property damage and injuries were available for each accident at a given location. Thus a total loss figure was obtained for each location by summing the property damage and the injury cost for all accidents there. Since the actual injury costs were unavailable in most cases, the National Safety Council* figures of $37,000/fatality accident, $2,200/injury accident, and $360/property damage accident were used.

Finally, it was recognized that investigators need historical information to determine the causes and estimate the remedies for accidents. A program was developed so that at any location the complete accident history could be obtained. This provided the investigators with the who, what, when, where, and why of each accident at each critical location. An investigative procedure was established such that the investigators could draw diagrams to graphically represent the accident from the computer printout. These "collision diagrams" provided investigators with the insight necessary to evaluate accident patterns.

The development of these programs was not without its problems. Aside from the normal personnel and organizational problems, there were the complications of getting a reliable data base. Interpretations in coding locations as well as a failure on the part of many localities to report accidents, resulted in some deficiencies. These were primarily problems of "getting

---

*U.S. Department of Transportation, PPM 21-16, Mar. 7, 1969, as given by the National Safety Council.

started"; and after a few months most of the bugs of the data acquisition and entry were smoothed out.

Nevertheless, because of the emotional impact of traffic accidents, the system in its infantile state was subjected to much scrutiny. This caused the minor "bugs" of development to be magnified to a point where the reports generated by HLIC were given little credence by upper levels of management. However, distributions of the collision diagram and cluster information were well received by field investigators. And, in a roundabout way, this was starting to have some effect upon implementation decisions.

Thus the historic review brings us to a point where some basic foundational development had taken place in the areas of (1) hazardous location identification, (2) priority arrangement for locations, and (3) providing investigators with needed details and facts. But the system was not having a great deal of influence on decisions regarding implementation of spot improvements. At this point, the case study history will go into more detail as to the cost/benefit system which was implemented to supply the critical missing link in the system. This will begin by demonstrating the planning and overall "level I" strategic decisions. Then the details of the cost/benefit system will be discussed.

9.3 Level I Development

Near the end of the second year of the existence of the HLIC group, a decision was made to augment the group with a specialist in cost/benefit analysis and optimization. This was partially in response to federal requirements for more quantitative analyses of safety expenditures. Primarily, however, middle management felt the need for further use of the valuable data base that was accruing from the accident records. This specialist was charged with the development and implementation of cost/benefit techniques to roadway spot safety improvements.

In addition, a good hard look was taken at past accomplishments and failures. In an early meeting of the newly augmented group with interested members of middle management, it was determined that the technical development was superb to that point. Efforts to continue to improve the system showed aggressive forethought. But the inability of the system to affect the decision-making process was disappointing and discouraging.

The demand for the type of information produced by HLIC was apparent from the requests for information by high-level personnel. Here a problem was isolated and identified in that the reports being generated actually contained *too much* information. It was common to send out computer printouts several inches thick. Obviously further data reduction was required, and it was hoped that the cost/benefit system would help. It was also felt that much of the information sent out was being used more for toys then for tools.

The interdepartmental meeting also concluded that the current data base was an insufficient foundation for determining specific spot improvements. All agreed that, although the accident data base could pinpoint problems, it was not sufficient to evaluate alternatives. Thus HLIC was charged with the task of obtaining the additional necessary information through investigations, and reducing this information to a useful tool for decision making regarding spot improvements.

The HLIC group took this mandate and plotted their course of action. Since they had been working closely with field investigators, the solicitation of their skills was not considered a problem. In fact, most of the projects submitted for proposed funding already required a site investigation by local personnel. However, there would probably be some problem in persuading them to make the quantitative estimates necessary for cost/benefit analyses.

A more crucial problem was seen in the other direction. The decision makers' confidence in the current system was waning. Some even felt that the total system development effort had been a waste of money. Planning sessions revealed that this was not completely the fault of upper management. In fact, up to this time very little effort had gone into selling the system to upper management.

An internal decision was made to publicize both the past developments and future plans. This would serve two purposes. First, it would demonstrate to the field investigators that the information they were supplying would be used. This would increase their support as well as the accuracy of their estimates. Second, publicity would announce to the decision makers that a tentative system was developed and a final system was on the way to aid them in their efforts. This would prevent HLIC from being "ignored" by upper management.

Problems had arisen from the inability of others to identify with the HLIC "products." Thus names were assigned to the programs already developed: (1) Location Identification Module (LIM), (2) Priority Selection Module (PSM), and (3) Collision Diagram Module (CDM). In addition, planning revealed the necessity for at least three more modules: (1) Cost/ Benefit Module (CBM), (2) Dynamic Programming Module (DPM), and (3) Before–After Module (BAM). Although a primary reason for the assignment of names was to aid publicity, they also served quite well as an aid to internal communications. The total system was called HALCS, the High Accident Location Correction System (pronounced "Hawk" system).

Publicity was accomplished in two ways. First, a meeting was called of all local investigators to announce the system and discuss the procedures to be implemented. Details of this are contained in Section 9.4. Also a brochure was printed and given wide distribution. This brochure discussed the past accomplishments and future plans of HLIC and set a definite timetable for their fulfillment.

The system was billed as a modular design such that each module was independent and yielded a unique benefit. Once all the modules were completely designed they would be integrated to form the HALCSystem, which would specify the optimal alternative to implement at each location. In the meantime, the benefits of the individual modules could be reaped. Of course, this was already true of the first three modules, although many did not realize it before reading the brochure.

The purpose for the modular approach was twofold: (1) it would get both the decision makers and the investigators involved and participating prior to the total system completion. More important, however, were the psychological effects of turning over a completed product. By successfully completing each module, a pattern or habit of success would be formed. Each completion could be publicized and this would establish a reputation of success for the group. Finally, if difficulties did arise and the total system integration were delayed, the independently operating modules would be able to bring their benefits, thus compensating for the temporary setback.

The concept of the system's "momentum" was considered to be in effect. That is, the system's momentum depends upon how large it is and how fast it is moving. At this point in the development, the size was increasing, but the speed (impact on upper levels of management) was questionable. The momentum began to increase with the announcement of the total system. However, the vulnerability of the system was also being increased.

Before leaving these psychological factors for the more technical history, it is notable to point out the willingness of the HLIC group to "lay it on the line." Developing a system in the dark is a hedge against failure and, as such, it makes failure easier to accept. The NASA program is an example of a development in the light of public scrutiny. While the group knew that they could be subject to much criticism, they felt this was a risk worth taking to ensure, not just the successful development, but the successful *application* of the system. They were resolved that the system was not worth developing if it would not be used. Developing it in the dark would have guaranteed failure in this respect.

9.4 Investigation Procedures Subsystem

Early in the development of HALCS it was recognized that a broad base of support would be required from local traffic and maintenance engineers to perform the required investigations. Therefore, the first subsystem to be developed involved procedures for investigation. Current practices in investigation were used to shape a system that would cause a minimum of change. However, one significant change was necessitated by the quantification of benefit estimates. These estimates would have to be structured so that they could be subjected to accuracy and consistency checks.

Once a standard form was assembled which contained all the information required for a cost/benefit analysis, a meeting was called of all interested persons to review the specifications. The use of the High Accident Location Investigation Form, or HALIForm, was demonstrated. Figure 9.1 shows the first part of the form, which contains all cost/benefit information for three alternatives. If additional details are required, Part 2, as given in Figure 9.2 would be used. This would allow detailed specification of alternatives without crowding Part 1, which was designed to be used directly in keypunching. Also, if more than three alternatives or four causes were required, the supplementary form could be used.

The following detailed description was given of each item on the form:

1. Location: description of the high accident location, which can be cross-referenced to the high-accident-location computer output or other supplementary records. On this line also check whether the location is in an urban or rural area.
2. Time Period of Accident History: period for the history of accidents available to the investigators, in years. Example: If 18 months of accident history is available, enter 1.5 (years).
3. Date: date of the investigation.
4. Investigators: list of persons on the investigating team.
   (*Note:* The "Number of Causes" space will be completed centrally.)
5. Roadway Environment Causes: only causes that can be remedied by roadway improvements should be included. Benefits are related to the elimination or partial abatement of causes. There may be several causes at a given location. Therefore, each cause and its severity must be listed.
6. Number of Accidents: enter the total number of accidents that occurred during the accident history period for each cause. Assign each accident to only *one* roadway cause. In the event of multiple causes, assign the accident to its primary cause. (*Note:* This is the sum of items $7 + 8 + 9$.)
7. Number of Fatalities: number of fatal accidents within the accident cause, as given by the accident records.
8. Number of Nonfatal Injuries: number of injury (nondeath injury) accidents, as given by the accident records for the given accident cause.
9. Number of Cases of Property Damage Only: number of accidents that resulted in property damage only (no injuries) for the accident cause as given by the accident records.

HIGH-ACCIDENT-LOCATION INVESTIGATION FORM

(1) Location: _____ Urban ( )   Rural ( )

(2) Time Period of Accident History (Years): _____

(3) Date: _____   (4) Investigators: _____

(5) Roadway Environment Causes   Number of Causes _____

		Severity During Accident History		
(6) Number of Accidents	(7) Number of Fatal Accidents	(8) Number of Non-fatal Accidents	(9) Number of Cases of Property-Damage Only Accidents
1.				
2.				
3.				
4.				

Number of Alternatives _____

(10) Alternatives (Description)

	(11) Cost	(12) Life	(13) Maintenance Cost/Year	(14) Effect on Cause			
				1	2	3	4
1.							
2.							
3.							

Figure 9.1. HALIF part 1

320

HIGH-ACCIDENT-LOCATION INVESTIGATION FORM

Supplementary Information (check one):

(   )   Roadway Environment Cause No. _____

(   )   Alternative No. _____

(   )   Other_____

Description:

Figure 9.2. HALIF part 2

(*Note:* The "Number of Alternatives" space will be completed centrally.)

10. Alternatives (Description): the "*best*" alternative at a given location will depend heavily upon availability of funds, and, in turn, the amount spent elsewhere. Therefore, this "best" alternative cannot be determined without considering all other alternatives at all other locations. The investigating team should report all alternatives that they wish considered. A suggested procedure is to start with the bare minimum to abate the hazard as the first alternative. A second alternative may be composed of this, plus other roadway improvements considered most critical. The third alternative can be the most expensive, possibly a combination of all feasible alternatives. Place the alternatives in the order of lowest cost first. If more than three alternatives are required, attach the additional information required, using another form.

    The description may be brief with reference to the more detailed description attached. Only one of the alternatives will be chosen for any given location. Therefore, if it is feasible to simultaneously implement several remedy items, this should be considered as a separate alternative and treated appropriately.

11. Cost: this will be an estimate of the dollar cost required to implement the entire alternative, as described in item 10. If this is a one-shot cost, nothing further is required. On the other hand, if a yearly cost will be incurred or if other special means of financing is employed, a special note should be made.

12. Life: this is the life of the alternative in years. Generally it represents the time at which the cost given in item 11 will have to be repeated to continue the implementation of the alternative. For an alternative that has several items of varying life, use a weighted average of the individual lives (weighting by cost of each).

13. Yearly Maintenance Cost: yearly costs that must be incurred throughout the life of the improvement.

14. Effect on Cause: in this space the effect of each alternative upon the particular cause listed in item 5 will be indicated. Certain alterna-

tives will affect some causes but not others. The *per cent reduction* estimated for each cause will be entered in the appropriate cell. This will be in per cent reduction, where 0% is no reduction and 100% is total elimination of the accident cause. For example, if it is estimated that alternative 2 will reduce 80% of the accidents caused by cause 3, then 80 would be entered in the alternative 2, cause 3 cell. More than one cause may be affected by each alternative, and, similarly, several alternatives might effect the same cause.

There is always a degree of uncertainty in estimating future results of current actions. Further investigation and development is continuing to establish guidelines for these estimates. In certain cases the hazard "cause" is completely eliminated, resulting in a simple case of 100%. In other cases the elimination of one hazard might introduce another (as with guardrails). The objective here is to get the best estimate in the reduction of either frequency and/or severity of accidents associated with each cause. Perfect answers are neither available nor required. Estimates will be checked for consistency among investigating teams.

Examples were also given of the use of the form. Questions were raised with respect to item 14. Some information based upon before–after studies conducted in California* was dispensed. However, investigators were advised to temper the averages given with their judgment. Mention was made of before–after studies planned in the state. Since money was to be allocated to safety improvements prior to the before–after studies, the evaluation would have to proceed based upon the judgment of the investigators.

Before the meeting of investigators, the Location Identification, Priority Selection, and Collision Diagram Modules had been run to provide each investigation team with a knowledge of the locations they would investigate. This also would enable them to fill out items 5–9 on the HALIForm. A pilot project of 80 locations was chosen to test the developing system.

9.5 Implementation the Cost/Benefit System

The theory behind the cost/benefit method developed was given in Section 8.4. Also, a detailed application to traffic safety is presented with the CBM program documentation in Appendix 2. Therefore, this section will deal with the general integration of the cost/benefit subsystem into the total HALCSystem.

A computer program, called the Cost/Benefit Module (CBM), was

---

*Roy Jorgensen and Associates, "Evaluation Criteria for Safety Improvements on the Highway," Parts II and III, Westat Research Analysts, Inc., Gaithersburg, Maryland, 1966.

developed to produce the required cost and benefit information. It was tested by invented examples and found to be reliable. When the results of the investigations began to trickle in from the field, they were immediately processed through CBM to obtain the type of output, as given in Table 9.1.

This output report was designed for use by management in determining which alternative to implement at each location. Of course, one of the alternatives is "no change," which should be considered at each location, although it is not specifically listed. The output design was directed toward a focus on the results of the cost/benefit analysis. However, it was felt that the raw data used in obtaining the results should not be withheld. Thus the portion of the report for each location above the subtitle "COST/BENEFIT ANALYSIS" contains the raw data in much the same form as it appears on the HALIForm.

Table 9.1 SAMPLE CBM OUTPUT

Ref No.							
136	(57-UI) McFarland Blvd at Rice Mine Road						
	Accident History 2.50 Years. Month 9, Year 73, 3 Causes.						
	Area Considered Urban						

Roadway Cause	*TACC*	*NFAT*	*NINJ*	*NPRO*			
1	32.	0.	11.	21.			
2	21.	0.	3.	18.			
3	7.	0.	3.	4:			
Totals	60.	0.	17.	43.			

					Effect on		
Alternative	*Cost*	*Life*	*Main Cost*	1	2	3	
1	1500.00	10.	100.00	0.0	0.0	0.50	
2	3000.00	5.	300.00	0.0	0.20	0.0	
3	17000.00	10.	500.00	0.50	0.25	0.0	

Cost/Benefit Analysis			
Alternative	*Cost*	*Benefit*	*Cost/Benefit*
1	1500.00	20174.11	0.0744
2	3000.00	6050.82	0.4958
3	17000.00	93658.75	0.1815

Cost/Benefit Analysis, Maintenance Included			
Alternative	*Maintenance*	*Total Cost*	*Cost/Benefit*
1	1000.00	2500.00	0.1239
2	1500.00	4500.00	0.7437
3	5000.00	22000.00	0.2349

The lower portions of the output for each location contain the cost and benefit information. As detailed in the CBM documentation in Appendix 2, the benefit is determined from the National Safety Council estimates of $37,000/fatality accident, $2,200/nonfatal injury accident, and $360/property-damage-only accident. To eliminate the chance effects of fatalities, the upper two severity classes were pooled and prorated at 67 injuries/ fatality for urban areas and 22 injuries/fatality for rural areas.

It was imperative to communicate several facts to the decision makers regarding the use of the cost/benefit information. These involved the limitations of applying this information. First, lowest cost/benefit did not mean "best." Quite often a better overall benefit could be obtained by implementing the alternative without the lowest cost/benefit at a location. Another problem arose in making judgments with respect to the "intangibles." For example, major future modifications that would alter traffic flows could make the current information obsolete. The life of certain alternatives would change in such cases. Also, no present worth of benefits was used. The discussion of this subject is beyond the scope of this text; however, it should be obvious that *immediate* gains should be given some priority over those accrued over many years.

An obvious dilemma developed. The shortcomings listed above coupled with all the usual shortcomings of any mathematical model must be communicated to the users, the decision makers. At the same time, it was necessary to "sell" them on the use of these new methods. It is difficult to sell a product by stressing its limitations. Thus a decision was made to stress the *usefulness* of the results of CBM until a decision was made by the upper levels to *use* the results. At that point, once the results were being used, the decision makers would be thoroughly versed as to the limitations of the model. This decision was justified since the cost/benefit results per se were far superior to any other method available for allocating funds.

Note that this decision was not intended to withhold the limitations at this point. They were disclosed. However, the emphasis was put on the strengths of the model rather than its weaknesses. At the appropriate time the stress would change to compensate for possible decision-maker *overconfidence* in the model.

A direct frontal attack was made upon two of the weaknesses of the system developed to this point. It was immediately announced that an optimization system was under development which would in fact determine the optimal decisons for each location. Thus this system would identify if marginal returns indicated that the minimum cost/benefit was not the "best" alternative to implement. The second weakness, that of estimating the effects on causes, was attacked by the implementation of as many before–after studies as were possible under the data-base and monetary-resource constraints.

In the meantime an effort was made to sell the use of the CBM output. It was stressed that these results were the best guides to effective safety project implementation that had ever existed in the state. However, rather than documenting the use of the raw cost/benefit information, resources were directed into the development of DPM.

9.6 Implementation of the Optimization Technique

Once the cost/benefit module (CBM) was operational, all the resources of the HLIC group were transferred into the development and application of an optimization technique to determine the policy that would yield maximum return. Several techniques, including total enumeration, were tried unsucessfully. Either the assumptions of the optimization algorithms were not satisfied, or else the computer requirements were overly restrictive.

The HLIC group had taken a risk by departing from their modular concept. That is, rather than documenting the results of CBM, thoroughly selling the results, and explaining the limitations to upper management, they decided to forge ahead with the development of an optimization program. They were motivated by the possible abuse that could be made of the cost/ benefit information and also by the delay that the above effort would have afforded. Also, at this particular time they were under no pressure to get a product out.

Fortunately, with the aid of a resident specialist in operations research, it was determined that dynamic programming could be applied to obtain the required optimal solution. Since this specialist had some past experience in programming dynamic programming solutions, his services were borrowed for a few days to lay out the program. The HLIC group took over from that point, programming the algorithm and modifying other programs to form the necessary interface. The optimization program was called the Dynamic Programming Module (DPM).

During this time, field investigation reports, HALIForms, continued to come in. They were checked for consistency. Some of the lives of certain projects seemed to be overestimated, and a decision was made to consider a maximum life of 20 years. Thus any life over 20 years was reduced to 20. The rationale was that the benefits not obtained in 20 years should not be considered because of the uncertainty in transportation systems over that period of time.

The output of CBM was augmented so that card output could be placed directly into DPM. This greatly facilitated the processing such that the first 80 projects could be quickly run. Before announcing the completion of DPM, the results were given very stringent tests to ensure their validity. Attempts were made to find sets of alternatives that could "beat" the DPM system. When all these proved fruitless, it was determined that confidence could be placed in DPM.

Care had to be exercised at this point, since two extreme dangers existed. The DPM specified *one* policy, that is, one alternative at each location. The decision makers might react to this rather narrow specification in two ways.

They might reject it as being too binding or restrictive to be useful. This could easily be done by seizing upon an assumption or the specification of an alternative that was impractical. On the other hand, they might react in veneration of the computer, which many people believe is capable of just about anything. Either of these extremes would prove detrimental to the long-term rightful use of the system.

Since the staff group was not working directly with the decision makers in designing and implementing the system, care had to be taken to neither oversell nor undersell the system. The original policy in this regard was thus implemented. That is, every attempt would be made to sell the system and not stress its shortcomings until upper management realized its worth. At that time the limitations of the system would be thoroughly impressed upon the decision makers. These two aspects of communication are discussed in the next two sections.

9.7 Selling the Total System

The first step in selling a working system is to communicate an under-standing of the total structure of the system. This precedes an explanation of the *use* of the system, since the user's confidence in the system is a prerequisite to its effective usage. Figure 9.3 presents a system overview that described the field–central office interface of the system. This figure was used to demonstrate the method by which routine accident data are translated into a knowledge of critical locations and their respective histories. Traffic engineers then investi-gate these locations, fill out the HALIForms, and submit the results. Finally, the CBM and DPM routines are run to determine the optimal policy.

A clear understanding of what constitutes optimality was crucial at this point. Here two limitations of obtaining optimal policies were stressed. The first is inaccurate data and the second is computer roundoff. If both cases the emphasis was placed upon the fact that the most accurate techniques that were known were being applied. If innovations for improvement were made known, these would be incorporated into the system as soon as possible. Constructive criticism was welcomed and even solicited. In the meantime, the most expert investigators in the state and the most sophisticated computerized optimiza-tion techniques were being utilized. An optimal policy was therefore defined, and it is used here to mean the best policy in terms of total expected return given that the input data are correct and that computer roundoff errors are negligible.

With this definition in mind, the difficulty in finding the optimal solution was stressed. The example was presented that one alternative (two real alternatives) at 30 locations would lead to over 1 billion possible ways to allocate funds. Of course, the pilot project consisted of 80 locations with an

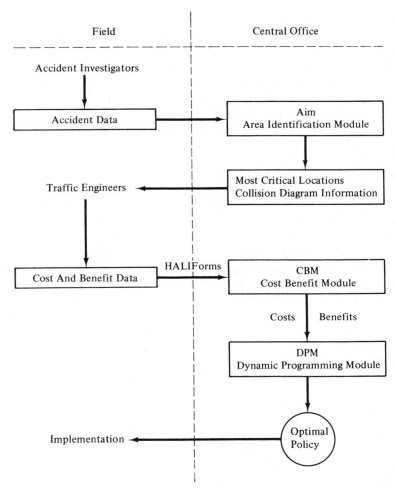

Figure 9.3. System overview

average of nearly three alternatives to be considered for each. Some examples of the complexity involved in simple problems were also demonstrated.

With this foundation it was stressed that the computer portion of the system did nothing magical. Machines for thousands of years have been doing things that man by himself could not. In this case the machine was doing nothing more than performing literally billions of calculations as well as making many comparisons. It was merely calculating the total return for a given policy and comparing it with others, sequentially eliminating bad policies. Thus no control could be lost. Rather, the decision maker had now gained a previously unknown dimension of control, owing to his improved knowledge of the situation.

Once it was established that a system was available to determine the optimal policy, the emphasis shifted to the harm that would be done if the "optimal" policy were not implemented. Some discussion within the group of past safety improvements was put forward. Analyses showed clearly that, even assuming 100% accident reduction, these were very much inferior in terms of cost/benefit to those determined by DPM. However, a decision was made to use this type of ammunition sparingly, for the following reasons: (1) it was unfair and thus damaging in morale since previously neither the decision makers nor their advisors had access to the type of information now being produced; (2) decisions made in the past had been combinational in maximizing returns of both safety and public demand for improved traffic flow; and (3) politically, it would have meant disaster. Thus, although it was made known that this type of analysis was available through confidential memoranda, this was not used as an overt selling tool for the system.

Rather, to illustrate the gains that could be obtained from implementing the optimal policy, it was compared with another "reasonable" policy—that of least cost/benefit first. This served two purposes. First, it demonstrated the value of using the optimal policy over *any other* policy. Minimum cost/benefit was viewed as just an example of another policy—and it was stressed that it was not a particularly bad policy at that. A second purpose was seen to be necessitated by the early exposure of the decision makers to CBM. Many had adopted an attitude that minimum cost/benefit first was a good-enough policy. This attitude had to be countered if the results of DPM were to be given credibility.

Figure 9.4 shows the comparison between these two policies. In presenting this comparison it was first emphasized that minimum cost/benefft first was a policy that selected improvements from the top of a list arranged in order of minimum cost/benefit first. Once an improvement was selected at a location, no other improvement at that location was considered. This policy guarantees a high return on investment, and therefore the results are reasonably good, especially at the lower budget limitations. A strong inference was given that minimum cost/benefit was in all probability superior to the policies that had been adopted in the past.

With this in mind, the $.5 million budget (COST) in Figure 9.4 was discussed. At this level of funding the 80 locations, the policy of putting the minimum cost/benefit first yielded a return of nearly $3 million, whereas the optimal policy, based upon DPM, produced a return of over $1 million more. On the right-hand scale, it could be seen that over the life of these projects, an additional 16 lives* could be expected to be saved. All this could be obtained without paying out one additional penny.

---

*See Appendix 2.4.2 for the method of obtaining this figure.

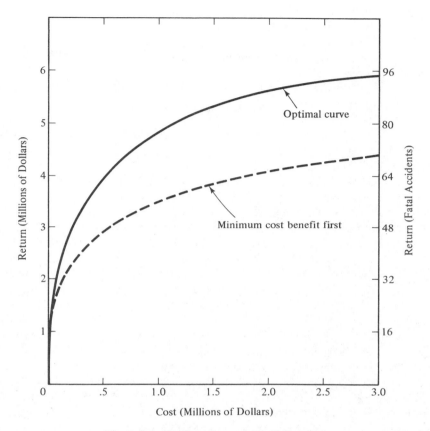

Figure 9.4. Cost/benefit curve for 80 locations

Two things were stressed at this point. First, these were not just theoretical figures. The results presented were based upon real estimates from the field. If anything, they were conservative. Minor errors could not possibly negate the validity of the conclusions presented. A second point was that these additional returns, in this case 16 lives and $1 million, could be obtained by *nothing other than a change in policy*. These were fairly conclusive selling points that could not be ignored. In turn, however, the total system itself was subjected to a higher degree of scrutiny than before. This was both expected and welcomed for the sake of total system improvement.

At the same time that the value of using the system was communicated to higher management, their responsibilities under the system were clarified. It was made clear that the optimization system in no way was designed to relieve them of their prerogatives. Rather, it was designed to sharpen focus on the problems they were most qualified to solve. One of the major global problems was the total budget requirement. A determination had to be made of how much to spend on the "Top 80," as opposed to the amount to be withheld pending a future analysis. The optimization system did not specify this value, nor could it, because global trade-offs were outside the scope of the model. Here the very top level of deicsion making was required to trade-off safety improvements with other social needs.

Figure 9.4 was also very helpful in this regard. Looking at the $.5 million figure, an 8-to-1 return could be expected. The $1 million figure offered only about $4.8 million return, so the *marginal* return of the last $.5 million was only about 1.6 to 1. Although this return is not bad, considerable thought should be given to holding off and evaluating more locations before spending above the $1 million figure. Thus the cutoff should be somewhere between $.5 and $1 million dollars, at the discretion of the decision maker.

Once the value of the optimal policy was emphasized, the use of the computer output was explained. The DPM computer outputs still contained too much information for the decision makers to comprehend and evaluate. Therefore, a final reduction was made to abstract the necessary information by hand. Table 9.2 gives the format in which this was presented in terms of the first 24 locations. The possible budget constraints, in increments of $.2 million, are given at the top. Each location is coded by its reference number in the first column. The policy is given in the body of the table for each budget constraint.

To use the table, the decision maker must have a cross-referenced list of locations in order of reference number. In fact, there may be a file containing all the information with regard to the location. In this case, however, a list was abstracted from the detailed information that contained for each reference number the complete location description, as well as a brief description of each alternative and its cost. The decision maker could use Table 9.2 by looking up the reference number and finding the alternative number given in the table. If more detailed information were required, the CBM output, similar to that given in Table 9.1, could be easily referenced. Note that this is easily located by reference number also.

Thus the decision maker was presented with all the information that he needed to initiate the safety improvement program. He did not, however, have the policy spelled out for him. As discussed above, he had to decide upon a total budget constraint. Also, marginal decisions must be clarified before proceeding. For example, suppose that the $1 million budget were adopted; consider location (reference number) 006. Here the policy is changing from alternative 2 to 1 and back again. Because of the ordering, of lowest-cost alternatives first, alternative 2 will cost more than 1. Generally, as the budget constraints increase, the alternative at a location will increase. Exceptions occur when the particular alternative "fits" a leftover sum of money quite well. In this case that alternative will be selected to maximize total return without exceeding the budget.

Table 9.2 PRESENTATION OF OPTIMAL ALTERNATIVES

Reference Number	Budget Constraints (Millions)														
	.2	.4	.6	.8	1.0	1.2	1.4	1.6	1.8	2.0	2.2	2.4	2.6	2.8	3.0
001	5	5	5	5	5	5	5	5	5	5	5	5	5	5	5
002	—	3	3	3	3	3	3	3	3	3	3	3	3	3	3
003	1	—	—	—	3	—	3	3	3	3	3	3	3	3	3
004	3	3	3	3	3	3	3	3	3	3	3	3	3	3	3
005	—	—	—	—	2	—	2	2	2	2	2	2	2	2	2
006	1	1	1	1	2	1	2	2	2	2	2	2	2	2	2
007	3	3	3	3	3	3	3	3	3	3	3	3	3	3	3
008	2	3	3	3	3	3	3	3	3	3	3	3	3	3	3
009	1	—	—	—	1	—	1	3	1	1	3	1	3	3	3
010	2	2	2	2	2	2	2	2	3	3	3	3	3	3	3
011	—	1	1	1	1	1	1	1	1	1	1	1	1	1	1
012	2	2	2	2	2	2	2	2	2	2	2	2	2	2	2
013	1	1	2	2	2	2	2	2	2	2	2	2	2	2	2
014	1	1	1	1	1	1	1	1	1	1	1	1	1	1	1
015	3	3	3	3	3	3	3	3	3	3	3	3	3	3	3
016	1	1	1	1	1	1	1	1	1	1	1	1	1	1	1
017	1	1	1	1	1	1	1	1	1	1	1	1	1	1	1
018	—	1	1	1	1	1	1	1	1	1	1	1	1	1	1
019	—	—	—	—	1	—	1	1	1	1	1	1	1	1	1
020	—	—	—	—	—	—	—	—	—	—	1	1	1	1	1
021	—	2	2	2	2	2	2	2	2	2	2	2	2	2	2
022	—	1	2	2	2	2	2	2	2	2	2	2	2	2	2
023	—	2	2	2	2	2	2	2	2	2	2	2	2	2	2
024	1	1	1	1	1	1	1	1	1	1	1	1	1	1	1
.															
.					remaining locations excluded										
080															

This is exactly the case in location 006 and a few of the other locations. In these cases, judgment should be exercised to refine the final policy. The CBM output should be consulted. The decision makers were cautioned against making adjustments arbitrarily, however. Although some flexibility exists within the budget, a large increase in the cost at one location will generally mean that some sacrifice will be required at another. All ramifications of a change should thus be thoroughly explored since the stated policy is generally the maximum return policy.

In the above discussion an ongoing rapport between the staff group and upper management was assumed. In actuality, the communication took on a variety of modes. Once the results of the Top 80 project were generated, a meeting with the highest levels was instigated through the normal chain of command. Simultaneously, parallel channels were sought and established to influential individuals who advise the top-level decision makers. Full advantage was taken of the informal and unstructured methods of communication.

At the first indication of top-level approval of the methods and results, the formal report on the Top 80 was widely distributed. This, of course, was intended to build the momentum of the system. The importance of lower-level communication for establishing esprit de corps was not minimized. Investigators who were previously hesitant to take time from their other duties now saw that their investigations were bearing fruit. When called upon to expand the number of locations, they were less complacent. All of this, of course, would be for nought if there were not some clear moves in the near future to implement the now-well-published results.

In the case of state agencies, other methods are available for building system momentum. It must be recognized that our system of government is ultimately responsive to the citizens. Whereas this often leads to short-term sacrifices for political expediency, we must respect the system for its obvious long-term beneficial effects.

With the knowledge of everyone concerned, newspaper contacts were used to publicize the new techniques being used. Obviously, specific improvements could not be disclosed without arousing unnecessary controversy. However, the public knowledge that advanced computerized techniques were being used to identify and correct high-accident locations was an asset to the momentum of the system.

At the same time the broadest coverage that could be obtained was sought within the federal government and with other states. This thorough publicity would guarantee two things. First, those in other states working within this area would have access to the innovations made locally. More important, however, public scrutiny would ensure the type of constructive criticism that could be used for constant improvement.

9.8 Limitations of the System

Of course, the system was not adopted immediately—thus the reason for the selling effort described above. No one can expect a method that has been developed in a few months to be implemented overnight. Procedures that have been in effect for many years as well as established viewpoints are not quickly obliterated. Rather, an evolutionary implementation of the new system should be sought. The minimum of disruption must be coupled with the resolve to complete conversion within a given time limit.

In this case, however, the response by upper management was affirmative. They clearly recognized the total system as being as asset to the citizens of the state. Since this was consistent with their own personal goals, they had no reservation in making the necessary decisions and implementing the results.

With this milestone past, it was necessary to emphasize the limitations of the system so that it would not be abused. Nothing could be more harmful at this point than an overreaction of upper management, thinking that this new system was going to eliminate all the problems. Overfunding of the first 80 locations, for example, could lead to a lack of funds for other locations. At the same time, if too much were expected from these 80 locations, the entire system might be considered a failure.

Thus an immediate effort was launched to get the decision makers to use the system as a tool. The first limitation was explained to be the limitation in the size of the Top 80 data base. The system could not guarantee a global optimal unless all locations, or at least all "accident potential" locations, were included. Of course, this was impossible. Therefore, every effort was made to get the most dangerous locations included in the Top 80. However, a large number of locations are also close in potential accident reduction to the Top 80. Therefore, it was essential to broaden the base of the study.

This was not to say that the Top 80 projects should not be funded while waiting for additional funding. On the contrary, many obvious projects, such as location 001 in Table 9.2, were in desperate need of action. A decision was required by the decision maker—how much to spend now and how much to hold off for more data. Models have not been applied to solving this problem, and therefore the responsibility falls upon the one who is paid (or elected) to make this decision.

Another limitation involved the possible inaccuracy of estimated accident reduction. The importance of this factor is difficult to determine. If before–after studies were completed, the validity of the results could be greatly improved. But decisions were required without this knowledge. Thus, in this case, the decision maker had a choice. He could trust his lower-level personnel, the investigators, and proceed with the results of the DPM. Or, he could question the validity of the estimates and change the results. If the latter course of action were chosen, the best way to proceed would be to vary the input to CBM and DPM and examine the results. This is not very difficult, nor is it time consuming. The alternative, which is adopted by all too many decision makers who do not wish to be bothered with details, is to arbitrarily change the results. This is not as bad as it sounds, provided that the "ideal" (i.e., the optimal policy) is used as the guide from which these modifications are made. Needless to say, if too many modifications are made, the decision maker would be better off to fire his staff and his investigators, and save the money that was spent on computer time. All this extra money could serve quite well in arbitrarily installing a few additional traffic signals each year.

The above remarks, although somewhat sarcastically stated, are intended to demonstrate the inherent contradiction in straying too far from the policy specified by DPM. Now, some deviation from the strict computer output is essential to take into account the many "intangible" factors that could not be considered within the model. To discuss all of these would be to review the assumptions of the models applied. This is best done by learning the basic principles and their application. It is to this end that this entire book has been dedicated.

9.9 Conclusion

This chapter has presented a case study of general nontechnical problems encountered by the systems engineer in developing and implementing a new system. That many of the events did not actually occur as described is irrelevant. No two situations are exactly alike and thus the recurrence of this particular chain of events is just as likely as another actual history. But this is not important. The principles exposed by this case study are difficult to generalize without great oversimplification. Teaching these principles is even harder since the inexperienced cannot understand the problem, and the experienced have already formulated their approach to these matters.

Hopefully, this case study has demonstrated that the safety systems engineer is more than a technically trained individual in all aspects of safety and systems analysis. Especially in the field of safety, the selling of a sound method, both up and down the chain of command, is as essential as the creation of a sound method.

Some of the basic tools of management, mathematics, and psychology have been presented. The development of a system has been documented both in technical and descriptive terms. Chapter 10 presents a more global look at safety from the legal and moral points of view.

QUESTIONS AND PROBLEMS

1. Give an example of the adage "power once used is power lost."
2. Give examples of political problems that may affect a company.
3. What are the advantages of a staff group being located (a) high in the organization and (b) low in the organization? By "high" it is meant that the group reports to someone at a high level, the president, for example.
4. State the problems involved in providing the decision maker with (a) too much information and (b) not enough information. In your opinion, which is worse?
5. State the advantages in the "modular" approach to system development.
6. State the advantages and disadvantages of publicizing a system development. Is it possible to oversell a system?

7. What are the advantages and disadvantages in stressing the limitations of a system?

8. What purpose is served by giving a system or a subsystem a "name."

9. Discuss the concept of "momentum."

10. Why do Roadway Environmental Causes precede Alternative Descriptions on the HALIForm (see Figure 9.1)?

11. What two extreme dangers are associated with higher management acceptance of a new computerized system?

12. Can a policy ever be stated to be optimal in the absolute sense? Explain.

13. Once a policy is spelled out by DPM, what decisions are still required at higher levels?

14. List the advantages of publicizing a system once higher-level acceptance has been attained.

THE RIGHT TO BE UNSAFE

10.1 Introduction

No book on systems analysis for safety is complete without a discussion of safety as a component of a larger system. To this point the safety system itself has been the only system under consideration, and an attempt was made to maximize the returns of this system subject to the constraints imposed. A broader viewpoint is necessary to determine exactly what constraints, and specifically what budget constraint, should be imposed. An understanding of trade-offs and total system optimization is essential to the top decision maker. The responsibility to present these techniques to the top decision maker may fall squarely upon the safety engineer. The purpose of this chapter is to provide the reader with the background upon which to base decisions after an objective review of the facts.

To clarify the problem, it would help to identify the faceless "top decision maker" mentioned above. He is not the safety director, even though some companies claim to give their safety director a blank check. Whether through formal edict or informal guidance, the safety director is generally under some form of control from his superiors.

The logical conclusion is that the president of the company (or Chairman of the Board) is ultimately the one who decides the extent of the safety effort.

10

However, although he may have the final say, his decision will be based heavily upon recommendations of both line and staff personnel. But even here the "top decision maker" is really not identified. Generally there are legal requirements that must be considered by all who influence the decision. Thus government strongly influences the extent and content of the safety program. And, although the process is often slow, generally government will respond to the collective will of the majority.

Being addressed to the top decision maker and those who influence him, this chapter is addressed to virtually everyone. It is philosophical in nature, as its title implies. No new quantitative techniques will be presented. However, the concepts of optimization presented in Chapter 8 will be used extensively.

10.2 Illustrative Example

Sometimes it is necessary in order to illustrate a point to use an extreme example. The most extreme example in the occupational spectrum is that of the daredevil performers—the circus acrobats, automotive stunt performers, and film-studio stunt men. To a lesser extent, other sports activities, such as football and hockey, also illustrate the point. Summed up in a hopefully facetious sort of way: How do you provide a man who wants to jump the Grand Canyon on a motorcycle with a "safe and healthful" workplace?

Obviously, performers such as these will continue to be given the right to be unsafe. However, this is not as significant as the apparent exception to the "rule of safety" and the precedent set thereby. For if performers are given the right to be unsafe purely for entertainment purposes, why should not others be given the right to be unsafe for productive purposes? Of course, we are assuming that the individual who takes the risk is rewarded commensurately with the performer. Further, philosophically speaking, it should not matter whether he is working for a large corporation or is self-employed. The individual is taking the risk voluntarily, and society in general is paying for the service.

There are innumerable analogies apart from the workplace to illustrate the point further. The government long ago gave up the notion that it could protect the common man from the evils of alcoholic beverages. Men have the right in most areas of the country to risk becoming alcoholics. Similarly, many law-enforcement officers inform us that, if the public so desired, traffic deaths could be virtually eliminated by stricter laws coupled with more rigid enforcement. The recent drop in deaths due to the national 55-mph speed limit proved this point. But the public insists upon their right to injure and possibly kill themselves. Obviously they (at least the majority) feel that the risks are justified by the gains obtained, whatever these gains might be.

The old cliche expresses the generally accepted concept best: "One man's right to move his fist ends where another man's nose begins." With notable exceptions, most of our laws today reflect the individual's right to take a risk, emotionally, spiritually, and, of importance here, physically. Put this individual into the workplace, however, and his right is diminished, since under the Williams–Steiger Occupational Safety and Health Act of 1970 the *employer* is responsible for the worker's actions. Now if he takes risks he may lose his job, despite the fact that injury would be incurred by no one other than himself (e.g., the individual who refuses to wear a hard hat).

Please do not misunderstand these introductory remarks. The OSH 1970 law if properly administered will continue to be a tremendous benefit to all American citizens. However, those who fail to put safety into its proper philosophical perspective do a great disservice to those who are trying to accomplish the spirit of the OSH 1970 Act. There is no excuse for the employer who willfully fails to bring the proper safe and healthful environment to the workplace. But, on the other hand, what about the employer who *forces* his employee to do something that the employee does not want to do? True, if the employee's attitude will cause harm to others then his employment should be terminated. But of concern here is the employee who believes it is in his best interest (for safety reasons or otherwise) to violate the letter of the law, knowing that he could not possibly injure another person in the process.

A final truism before stating the problem more specifically: life itself is a gamble. It has long been recognized that total safety involves the avoidance of all risks, a concept that is repugnant to all but a reclusive few. Excellent work has been done in the area of risk acceptance by Rockwell and Burner (3), and the interested reader should be familiar with this work as well as its references. It goes without saying that to wipe out risk is to wipe out the essence of life for many. For risk taking constitutes their life blood, their very satisfaction in life. To remove the possibility of the loss also removes the capability to win. And, in general, life itself would be boring were it not made significant by the specter of death.

But the extreme risk takers are not the primary concern here. Indeed OSHA 1970 is flexible enough to accommodate performers as well as those occupations which of necessity require risks. In the future employees might be able to petition for exceptions to the letter of the law when they feel that it is in the best interest of everyone. The primary concern here is with the right of *society in general* to incur risks for the satisfaction of goals that it deems beneficial. This is a far more global problem than defining exceptions for high risk takers. The extremists merely provide an excellent forum from which a more general discussion can be launched.

10.3 Statement of the Problem

In discussions and writings precipitated by OSHA 1970, two prevalent opinions are heard. One group claims that a given degree of safety as well as a specific set of working conditions (standards) are necessary for the worker to enjoy freedom from bodily harm. The other, less publicized but quite vocal group claims that the costs that must be incurred in this quest are too great to preserve the consumer's freedom from want. To understand this second argument it is important that we realize who pays for safety. It is not the company, since *uniform application* of the law would require an equal burden upon all companies. Theoretically they can uniformly raise their prices to compensate without losing any market advantage. Similarly, the government cannot pay, since their only source of revenue is through increased taxation or inflation, the net effect being the same. To summarize, the consumer pays for safety in terms of increased prices, taxes, or inflation, the net result of which is decreased buying power. That is, money and resources used for protective devices and other safety equipment cannot be used for the production of consumer goods.

Here we must hasten to add that certain safety investments return much more than their cost. Thus a decision *not* to invest in these could lead to a reduction in consumer purchasing power. The almost perfect analogy exists in the area of capital expenditures. Generally, mechanization and automation reduce the cost of goods, thus increasing consumer purchasing power. It is possible, however, to overmechanize to the point where capital expenditures are not warranted by the volume of production. Such a foolish investment would put a company out of business. However, if mechanization were required by law . . . . Quite possibly the analogy breaks down at this point.

To illustrate with an example. Generally, if it costs twice as much to build a house under "safe" conditions, the consumer will have to pay twice as much for the house. Of course, few (some builders interviewed by the author excepted) anticipate such an increase in cost; but this serves to crystallize the problem. At what point is safety to be sacrificed in the interest of more efficient production? Remember that *perfect* safety would dictate that the house would not be built at all.

It is recognized that safety proponents are not arguing for "perfect" safety, as described immediately above. And, since there are reasonable and well-intentioned men on both sides of this issue, we shall assume that an optimum point probably exists between the two positions. However, given our present ignorance of the cause–effect mechanism that triggers economic gains and losses due to safety, the optimal point may well be outside the range currently under negotiation (i.e., the current standards). Regardless, this is *not* a question of maximizing safety, nor is it a problem of minizing the cost of goods; rather it is an optimization problem. This involves *balancing* benefits associated with safety with the benefits of equally desirable alternative usages of resources.

To illustrate this point, consider some results that were obtained from a research effort conducted to determine the optimal investments in forest harvesting systems.* In this study 146 alternative safety investments were evaluated and the optimal investments were determined by means of fault-tree analysis and dynamic programming. The methods and assumptions applied are not as important here as are the general conclusions of the output. Figure 10.1 demonstrates one of the curves based upon the analysis. This curve can be used to determine the expected number of saved workdays that will be obtained for any budget figure from 0 to $1,200/10,000 cords of production. The shape of the curve depends upon the priorities set on individual investments. Figure 10.1 is an optimal curve based upon the techniques presented in Chapter 8.

---

*Some of the factual data used for illustration was obtained during the course of a research project funded by the U.S. Forest Service, FS-SO-3701-1.6. The views expressed are those of the author and do not necessarily reflect opinions of the U.S. Forest Service or Auburn University.

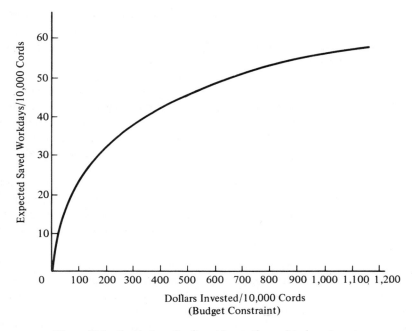

Figure 10.1. Cost/return for forest harvesting safety investments

The costs given in Figure 10.1 are the gross costs of safety. That is, costs do not include the monetary returns that most certainly will be obtained by reducing accidents. Although $1,200/10,000 cords of production for safety must sound like a huge financial burden to the logger, we must hasten to add that it did not exhaust all 146 alternatives. In other words, there were still reasonable safety investments that could have been made! However, we terminated our consideration at $1,200 for obvious practical considerations.

Now it is true that in 1971 logging camps and logging contractors had an injury-frequency rate of 42.4 disabling work injuries per 1 million employee-hours of exposure (reference 1), as compared to the average for all manufacturing industries of 15.2. Therefore, the specific curve that might hold for forest harvesting might be unrealistic for other industries as far as cost/benefit is concerned. However, it is the characteristics of this curve that are important. Comparing this curve with Figure 9.2 reveals that the overall shape is independent of the industry or application. The following characteristics of this curve will be true for all industries:

1. The curve will rise rapidly as the first initial optimal investments are made. This is the $0–100 area of Figure 10.1, which we shall refer to

as the cream of safety investments. These are the obvious investments that require no sophisticated cost/benefit analysis to determine.

2. After the "cream skimming," there is an area of the curve, say from $100 to $300, which rises rapidly but not nearly as fast as the first part. This area requires some thought as to the determination of optimal investments, since there are still many good alternatives available.

3. Above a given point investments become quite questionable. Although there are still many alternatives, the effect of any one is marginal. Hence choosing between alternatives is not so much a problem as choosing the point to refuse purchasing *any* of the remaining alternatives.

4. The curve could be extended ad infinitum and the curve would continue to rise for each additional alternative funded. Obviously, it would never reach that utopian point where all accidents and health problems are eliminated. Hence it would be asymptotic to the line of the maximum expected lost workdays reduced.

Given these facts about safety cost/benefit curves in general, it is possible to restate the optimization problem discussed above in more specific terms. We can assume that there is a curve, such as that given by Figure 10.1, for each industry. The problem is to determine the point for each industry at which additional expenditures for safety are no longer justified. Assuming that the alternative usage of the funds will provide society in general with additional products and services, at what point is the overall well-being of society a maximum?

Before accusing us of supporting a "big brother" attitude, recognize that these decisions *are* currently being made. They may not be described in straightforward terms, but in some way government, management, and labor are determining the point at which the welfare of the worker (or the driver, etc.) is "sacrificed" for the benefit of society. This should not be viewed as a manipulation of individuals for the benefit of the whole if everyone enters into the decision process. By formulating this problem, and presenting a logical solution technique, a greater understanding, if not an optimal solution, can be obtained.

10.4 Solution Technique

Obviously there is a loss which cannot be accurately measured when an accident or health case takes place. But measurement is essential if this problem is to be solved objectively. The alternative is an emotional solution that can only lead to inconsistency and hence inequality. At least when stated, measurements reveal the rationale of the solution technique. Both the measurements and the rationale can be examined and improved as research continues.

One way of proceeding in terms of the benefit of Figure 10.1 (i.e., lost workdays saved) would be to assign a cost to the worker, the company, and society whenever a worker loses one workday. We will call this total cost per lost workday C. On the average, the total cost of n lost workdays will be equal to nC. Figure 10.2 illustrates how this cost will increase with lost workdays.

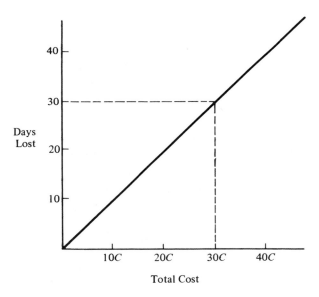

Figure 10.2. Total cost when the cost/lost workday $= C$

Now superimpose upon this curve the particular cost/return curve for the industry in question. For example, the curve developed for the forest products industry is given in Figure 10.1. This would give a representation as indicated in Figure 10.3. Note the point of intersection, called the *break-even point*. At this point the return (in dollars) from the safety investments equals the cost (in dollars) of these investments. Hence the one who is paying the bills, the consumer, is neither gaining nor losing from his investment at this point.

To further understand Figure 10.3, consider the portion to the left of the break-even point. For any investment in this area the return due to the investment will exceed the cost. For example, if $20C$ are invested, 30 days will be saved and hence $30C will be saved, making a net return of $10C. The net return is the difference between the two curves. On the other hand, to the right of the break-even point, the costs exceed the returns. For example, a cost of $50C brings a return of only 45 saved days, which cost out to $45C, a net loss of $5C.

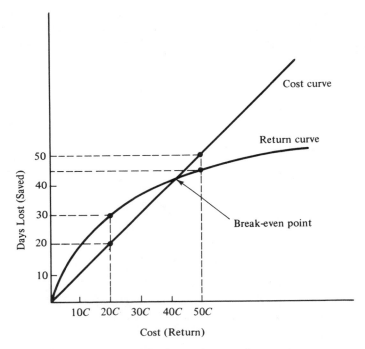

Figure 10.3. Superimposed cost and return curves

Now it should be recognized that no matter what the value of *C* is, the principles still hold. The value of *C* is being, and will continue to be, assigned by decision makers. Whether they realize it or not is not the question. The point is that we face reality when we realize that *C* exists and then proceed to assign a value to it or, from empirical data, attempt to determine its current value.

Once again referring to Figure 10.3, as far as society is concerned, the break-even point is an optimal point with respect to total investments in safety. Additional investments beyond this point will be detrimental to society as far as lost goods and services are concerned. An investment level below this point will be detrimental as far as safety and health is concerned. Obviously, since the curves will be different, the break-even point will be at a different level for each industry. This merely says that some industries are in worse shape and need more safety investment than others. If a consistent value of C, the cost per lost workday to society, could be determined, an optimal point could be determined for each industry. This would yield an optimal for society as a whole, assuming that similar approaches were applied in such areas as public health, environmental pollution, and so on.

Thus the qualitative argument resolves itself to one of determining if current operations are above or below the break-even point. However, there may be an even more crucial problem to be resolved, as discussed in the next section.

10.5 Determination of Standards for an Industry

In this section the effect of applying suboptimal operating decisions will be discussed in light of the model presented above. The cost/benefit curve of Figures 10.1 and 10.3 was assumed to be an optimal curve obtained by applying the concepts of cost/benefit optimization, specifically those employed in Chapter 8. Now, suppose that the "optimal operation decisions" are replaced by a set of arbitrary standards and rules. We do not propose that these standards are overtly harmful in causing accidents. Rather, they are merely assumed to be less than optimal from a cost/benefit standpoint.

Figure 10.4 illustrates the results in this case. Of course, there are an infinite number of "wrong" ways and only one optimal, which in practice is nearly unobtainable. Nevertheless, Figure 10.4 illustrates the effect of poor standards and arbitrary enforcement, as well as bad operating policies. Society would be willing and enthusiastically pay b under the "optimal" plan, whereas they would only be willing to pay a under the suboptimal plan. Further, to obtain the same effect under the suboptimal plan, a cost of c would have to be incurred. To picture c significantly greater than b is not unrealistic, given the shape of Figure 10.1, which is based upon real data.

The importance of developing near-optimal standards is seen from the economic as well as the public-relations point of view. Whereas at the present time it would be quite difficult to evaluate all countermeasures from a cost/benefit point of view before proceding, this is not an impossibility for 5 or 10 years hence. Accident records, possibly augmented as in Chapter 7, can provide all the information required for before–after studies. Industry-wide operation in plowing this information back into the optimization routines can

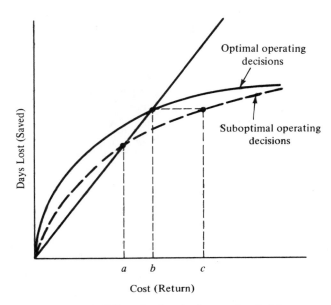

Figure 10.4. Effect of suboptimal operating decisions

lead to the establishment of effective cost/benefit curves. These, in turn, can be used as a basis for the continual improvement of the standards.

10.6 Conclusion

Although the concepts presented in this chapter tend to be philosophical, the chapter is not included merely to provide mental exercise. Methods have been presented by which evaluation of alternative safety investments may be performed. Actually, many of these techniques have been applied by managers since the beginning of the industrial revolution to determine policies that were generally good, if not optimal. These should continue to be applied on both the governmental and industrial levels in the development of both standards and operating policies. Then the optimum level of safety should be determined such that no one individual or group is given undue favoritism over another.

Finally, let us return to the original thought of this chapter. It is man's inherent right to take a risk to accomplish a personal goal as long as neither the risk, nor any detrimental outcome of the risk, will injure another. Further, it is the right of society as a whole to insist upon a balance in its several objectives, such that no one objective is unnecessarily sacrificed for another. Hence society may at times request certain *willing* individuals to assume certain risks and recompense them accordingly so that both the individual and society as a whole will benefit. In so doing the right of the individual and the right of society is being exercised.

The purpose here is not to take sides for or against OSHA '70 and its current effects upon the worker or society in general. Neither its necessity nor the presence of certain deficiencies within it can be disputed. Man's imperfections make both a reality. Rather, the continuous improvement of our national safety system is dependent upon an understanding of the principles presented above. Once labor, management, and government understand and apply these principles, a greater degree of order in resolving safety problems will be forthcoming.

SELECTED REFERENCES

1. *BLS Report* 406, U.S. Department of Labor, Washington, D.C., 1972.

2. ROCKWELL, THOMAS H., "A Systems Approach to Maximizing Safety Effectiveness," *ASSE Monograph* 1, p. 14, June 1971 (System Safety Techniques . . .).

3. ROCKWELL, T. H.; and BURNER, L. R., "Information Seeking in Risk Acceptance," *ASSE Journal*, Feb. 1968.

4. RUBINSKY, STANLEY; and SMITH, N. F., "Process Simulators for Accidental Research," *ASSE Journal*, Nov. 1971.

QUESTIONS AND PROBLEMS

1. Using Figure 10.1, determine the marginal return of spending $100 according to the optimal policy *after* the following has already been spent: $0, $100, $200, . . . , $600. Calculate the cost/benefit of each of these $100 expenditures.

2. In Problem 1, if there were an alternative in the area of public safety that would save 5 workdays for a cost of $100 in the same time frame as in Figure 10.1, what would be the point at which money should be shifted to this area, assuming that it were possible? Ignore total budget limitations.

3. Why is Figure 10.2 linear as opposed to Figure 10.1? Is the assumption of linearity warranted? Explain.

4. In Figure 10.3 assume that $C = \$200/\text{day}$. From Figure 10.3 determine (a) the break-even point, (b) the amount of money required to bring about a $2,000 gross saving, and (c) the range between which at least a $1,000 net savings will be realized.

5. Draw the cost and return curves and find the break-even point where the cost per lost workday is $200 and the following expenditures are expected to bring the corresponding results:

Expenditure	Workdays Saved/Year
$ 500	10
1,000	14
2,000	20
3,000	24
4,000	27
5,000	29
6,000	32
7,000	33
8,000	34
9,000	35

6. Use the diagram that you drew for Problem 5 to determine the following:
   (a) The gross and net return for expenditures of $2,000, $4,000, and $8,000.
   (b) The marginal return for a $500 expenditure after $4,500 has been spent.

7. Suppose that another policy currently in effect for Problem 5 yielded the following suboptimal results:

Expenditure	Workdays Saved/Year
$ 500	7
1,000	10
2,000	16.5
3,000	20
4,000	23
5,000	26.5
6,000	28
7,000	30
8,000	32
9,000	33

Superimpose this curve upon the one drawn in Problem 5 and determine the break-even point. How much less would a manager be willing to spend under this policy if his objective was to break even?

8. In Problem 7, determine the amount that would have to be spent under the suboptimal policy to obtain the same benefits as a $5,000 expenditure would obtain under the optimal policy.

SELECTED CODES FROM ANSI
Z16.2-1962(R1969)

This material is reproduced with permission from American National Standard Method of Recording Basic Facts Relating to the Nature and Occurrence of Work Injuries (Appendix), ANSI Z16.2-1962(R1969), copyright 1963 by the American National Standards Institute, copies of which may be purchased from American National Standards Institute, 1430 Broadway, New York, New York 10018. This is intended to serve as a reference list for the forms suggested in Chapter 7. It is not intended to replace the methodology described in ANSI Z16.2, in that no narrative descriptions explaining code usage have been reproduced. The reader is encouraged to obtain and use the entire standard if it applies to his specific recordkeeping needs.

A1.1 Nature of Injury Classification

Code

100   Amputation or enucleation

110   Asphyxia, strangulation, drowning

120   Burn or scald (heat)—The effect of contact with hot substances. Includes electric

burns, but not electric shock. Does not include chemical burns, effects of radiation, sunburn, systemic disability such as heat stroke, friction burns, etc

130   Burn (chemical)—Tissue damage resulting from the corrosive action of chemicals, chemical compounds, fumes, etc (acids, alkalies)

140   Concussion—Brain, cerebral

150   Contagious or infectious disease—Anthrax, brucellosis, tuberculosis, etc

160   Contusion, crushing, bruise—Intact skin surface

170   Cut, laceration, puncture—Open wound

180   Dermatitis—Rash, skin or tissue inflammation, including boils, etc. Generally resulting from direct contact with irritants or sensitizing chemicals such as drugs, oils, biologic agents, plants, woods, or metals, which may be in the form of solids, pastes, liquids, or vapors and which may be contacted in the pure state or in compounds or in combination with other materials. Does not include skin or tissue damage resulting from corrosive action of chemicals, burns from contact with hot substances, effects of exposure to radiation, effects of exposure to low temperatures, or inflammation or irritation resulting from friction or impact

190   Dislocation

200   Electric shock, electrocution

210   Fracture

220   Freezing, frostbite, and other effects of exposure to low temperature

230   Hearing loss, or impairment (a separate injury, not the sequelae of another injury)

240   Heat stroke, sunstroke, heat cramps, heat exhaustion, and other effects of environmental heat. Does not include sunburn or other effects of radiation

250   Hernia, rupture—Includes both inguinal and noninguinal hernias

260   Inflammation or irritation of joints, tendons, or muscles—Includes bursitis, synovitis, tenosynovitis, etc. Does not include strains, sprains, or dislocation of muscles or tendons, or their aftereffects

270   Poisoning, systemic—A systemic morbid condition resulting from the inhalation, ingestion, or skin absorption of a toxic substance affecting the functioning of the metabolic system, the nervous system, the circulatory system, the digestive system, the respiratory system, the excretory system, the musculo-skeletal system, etc. Includes chemical or drug poisoning, metal poisoning, organic diseases, and venomous reptile and insect bites. Does not include effects of radiation, pneumoconiosis, corrosive effects of chemicals; skinsurface irritations; septicemia or infected wounds

280   Pneumoconiosis—Includes anthrocosis, asbestosis, silicosis, etc

290   Radiation effects—Sunburn and all forms of damage to tissue, bones, or body fluids produced by exposure to radiations

300   Scratches, abrasions (superficial wounds)

310   Sprains, strains

400   Multiple injuries

990   Occupational disease, NEC*

995   Other injury, NEC

999   Unclassified, not determined

A1.2 Part of Body Affected Classification

Code

100   *Head*
    110   Brain
    120   Ear(s)
          121   Ear(s) external
          124   Ear(s) internal (include hearing)
    130   Eye(s) (include optic nerves and vision)
    140   Face
          141   Jaw (include chin)
          144   Mouth (include lips, teeth, tongue, throat, and taste)
          146   Nose (include nasal passages, sinus, and sense of smell)

---

*NEC = Not elsewhere identified

                    148   Face, multiple parts (any combination of above parts)
                    149   Face, NEC
             150   Scalp
             160   Skull
             198   Head, multiple (any combination of above parts)
             199   Head, NEC

200   *Neck*

300   *Upper extremities*
       310   Arm(s) (above wrist)
                    311   Upper arm
                    313   Elbow
                    315   Forearm
                    318   Arm, multiple (any combination of above parts)
                    319   Arm, NEC
             320   Wrist
             330   Hand (not wrist or fingers)
             340   Finger(s)
             398   Upper extremities, multiple (any combination of above parts)
             399   Upper extremities, NEC

400   *Trunk*
       410   Abdomen (include internal organs)
       420   Back (include back muscles, spine, and spinal cord)
       430   Chest (include ribs, breast bone, and internal organs of the chest)
       440   Hips (include pelvis, pelvic organs, and buttocks)
       450   Shoulder(s)
       498   Trunk, multiple (any combination of above parts)
       499   Trunk, NEC

500   *Lower extremities*
       510   Leg(s) (above ankle)
                    511   Thigh
                    513   Knee
                    515   Lower leg
                    518   Leg, multiple (any combination of above parts)
                    519   Leg, NEC
             520   Ankle
             530   Foot (not ankle or toes)
             540   Toe(s)
             598   Lower extremities, multiple (any combination of above parts)
             599   Lower extremities, NEC

700   *Multiple parts* (Applies when more than one major body part has been affected, such as an arm and a leg)

800   *Body system* (Applies when the functioning of an entire body system has been affected without specific injury to any other part, as in the case of poisoning, corrosive action affecting internal organs, damage to nerve centers, etc. Does not apply when the systemic damage results from an external injury affecting an external part such as a back injury which includes damage to the nerves of the spinal cord)
       801   Circulatory system (heart, blood, arteries, veins, etc)
       810   Digestive system

820 Excretory system (kidneys, bladder, intestines, etc)
830 Musculo-skeletal system (bones, joints, tendons, muscles, etc)
840 Nervous system
850 Respiratory system (lungs, etc)
880 Other body systems

900 *Body parts, NEC*

999 *Unclassified* (insufficient information to identify part affected)

A1.3 Source of Injury Classification

A1.3.1 The Basic Classifications

Code

0100 Air pressure (abnormal environmental)

0200 Animals, insects, birds, reptiles (live)

0300 Animal products (not food)

0400 Bodily motion (no lifting, pulling, pushing, etc. See rule 3.3.2.3)

0500 Boilers, pressure vessels

0600 Boxes, barrels, containers, packages (empty or full)

0700 Buildings and structures (not floors, working surfaces, or walkways. See working surfaces)

0800 Ceramic items, NEC

0900 Chemicals, chemical compounds (solids, liquids, gases)

1000 Clothing, apparel, shoes

1100 Coal and petroleum products

1200 Cold (atmospheric, environmental)

1300 Conveyors

1400 Drugs and medicines

1500 Electric apparatus

1700 Flame, fire, smoke

1800 Food products (including animal foods)

1900 Furniture, fixtures, furnishings (not fixed parts of buildings or structures)

2000 Glass items, NEC (glassware, glass fibers, sheets, etc, not bottles, jars, flasks, or glass cloth)

2200 Hand tools, not powered

2300 Hand tools, powered

2400 Heat, atmospheric, environmental (not hot objects or substances)

2500   Heating equipment, NEC (furnaces, retorts, space heaters, stoves, ranges, etc—not electric)

2600   Hoisting apparatus

2700   Infectious and parasitic agents, NEC (bacteria, fungi, parasitic organisms, viruses, etc—not chemicals, drugs, prepared medicinal items, or food infestations)

2800   Ladders (fixed or portable)

2900   Liquids, NEC

3000   Machines

4000   Mechanical power transmission apparatus
         NOTE: Transmission equipment includes all mechanical means of transmitting power from a prime mover up to but not including a shaft, or any pulleys or gears on the shaft, the bearings of which form an integral part of a machine. Directly connected prime movers are defined as having no transmission apparatus

4100   Metal items, NEC (plates, rods, wire, shapes, nuts, bolts, nails, etc—includes molten metal, ingots, and melting scrap, but not ores or other raw materials)

4200   Mineral items, metallic, NEC (products of mining; raw or semi-processed, such as ores and ore concentrates)

4300   Mineral items, nonmetallic, NEC (products of mining, excavating, landslides, etc, such as dirt, clay, sand, gravel, stone, etc)

4400   Noise

4500   Paper and pulp items, NEC

4600   Particles (unidentified)

4700   Plants, trees, vegetation (in natural or unprocessed condition—does not include threshed grains, harvested fruits, limbed logs, etc)

4800   Plastic items, NEC (powders, sheets, rods, shapes, etc, but not uncombined chemicals or components used in plastic manufacturing)

4900   Pumps and prime movers

5000   Radiating substances and equipment (use this code only in cases of radiation injuries)

5100   Soaps, detergents, cleaning compounds

5200   Silica

5300   Scrap, debris, waste materials, NEC

5400   Steam

5500   Textile items, NEC [includes animal fibers after first scouring and cleaning, vegetable and synthetic fibers (except glass), yarn, thread, yard goods, felts, and textile products]

5600   Vehicles (see rule 3.3.2.4 regarding injuries experienced while occupying a vehicle)

5700   Wood items, NEC (logs, lumber, slabs, poles, chips, and wood products)

5800   Working surfaces (surfaces in use as supports for people)

8800   Miscellaneous, NEC

9800   Unknown, unidentified (other than particles)

A1.4 Accident Type Classification

000-199   *Accidents other than motor-vehicle or public transportation accidents*

Code

010   *Struck against*
    011   Stationary object
    012   Moving object

020   *Struck by*
    021   Falling object
    022   Flying object
    029   Struck by, NEC

030   *Fall from elevation*
    031   From scaffolds, walkways, platforms, etc
    032   From ladders
    033   From piled materials
    034   From vehicles
    035   On stairs
    036   Into shafts, excavations, floor openings, etc (from edge of opening)
    039   Fall to lower level, NEC

050   *Fall on same level*
    051   Fall to the walkway or working surface
    052   Fall onto or against objects
    059   Fall on same level, NEC

060   *Caught in, under, or between*
    061   Inrunning or meshing objects
    062   A moving and a stationary object
    063   Two or more moving (not meshing) objects
    064   Collapsing materials (slides of earth, collapse of buildings, etc)
    069   Caught in, under, or between, NEC

080   *Rubbed or abraded*
    081   By leaning, kneeling, or sitting on objects (not vibrating)
    082   By objects being handled (not vibrating)
    083   By vibrating objects
    084   By foreign matter in eyes
    085   By repetition of pressure
    089   Rubbed or abraded, NEC

100   *Bodily reaction*
    101   From involuntary motions
    102   From voluntary motions

120   *Overexertion*
    121   In lifting objects
    122   In pulling or pushing objects
    123   In wielding or throwing objects
    129   Overexertion, NEC

130   *Contact with electric current*

150   *Contact with temperature extremes*
     151   General heat—atmosphere or environment
     152   General cold—atmosphere or environment
     153   Hot objects or substances
     154   Cold objects or substances

180   *Contact with radiations, caustics, toxic and noxious substances*
     181   By inhalation
     182   By ingestion
     183   By absorption
     189   NEC

200   *Public transportation accidents* (Code for type of vehicle in which injured was a passenger)
     201   Aircraft accident
     203   Bus accident
     205   Ship or boat accident
     207   Streetcar or subway accident
     209   Taxi accident
     211   Train accident
     298   Public vehicle accident, NEC

300   *Motor-vehicle accidents* (Code in terms of the event affecting or involving the vehicle in which the injured was an occupant. If more than one of the listed events occurred, code for the first event in the sequence)
     310   Collision or sideswipe with another vehicle—both vehicles in motion
         311   With an oncoming vehicle on same road, street, or trafficway
         312   With a vehicle moving in same direction on same road, street, or trafficway
         313   With a vehicle moving in an intersecting trafficway
     320   Collision or sideswipe with a standing vehicle or stationary object
         321   Running into or sideswiping a standing vehicle or object in the roadway
         322   Running into or sideswiping a standing vehicle or object at side of road (not in trafficway)
         323   Struck by another vehicle while standing in roadway
         324   Struck by another vehicle while standing off the roadway
     330   Noncollision accidents
         331   Overturned
         332   Ran off roadway (out of control)
         333   Sudden stop or start (throwing occupants out of, or against interior parts of the vehicle; or throwing contents of vehicle against occupants)
         338   Other noncollision accidents

899   *Accident type, NEC*

999   *Unclassified, insufficient data*

A1.5 Hazardous Condition Classification

NOTE: Selection of the hazardous condition which caused or permitted the occurrence of the accident does not involve a determination of the feasibility of correcting or eliminating the named condition.

Code

000   *Defects of agencies* (i.e. undesired and unintended characteristics, generally the opposite of the desirable and proper characteristic, such as being dull when it should be sharp. Do not classify an intended and necessary characteristic of an agency as a defect. For example: A knife is expected to be sharp and is not defective because it has this characteristic)

       001   Composed of unsuitable materials
       005   Dull
       010   Improperly compounded, constructed, or assembled
       015   Improperly designed
       020   Rough
       025   Sharp
       030   Slippery
       035   Worn, cracked, frayed, broken, etc
       099   Other defects, NEC

100   *Dress or apparel hazards*

       NOTE: Name this hazardous condition if it, in fact, contributed to the occurrence of the accident even though the condition was created by the injured employee's own choice or unsafe act.

       110   *Lack of necessary personal protective equipment*

          NOTE: Name this hazard only when the personal protective equipment constitutes an essential element in the safe performance of the activity. Does not apply when the use of the protective equipment would merely have minimized the injury without preventing the accident.

       113   Improper or inadequate clothing
       199   Dress or apparel hazards, NEC

200   *Environmental hazards, NEC*

       NOTE: These are general hazards of the workplace which commonly affect everyone in the area regardless of his assignment. *They should be named as the accident cause only when none of the other more specific hazardous condition designations apply.*

       205   Excessive noise
       210   Inadequate aisle space, exits, etc
       220   Inadequate clearance (for moving objects or persons)
       230   Inadequate traffic control (on employers' premises only—refers to maintenance of traffic lanes: elimination of blind corners, etc; control of speeding; direction of traffic away from danger points, etc)
       240   Inadequate ventilation (general—not due to defective equipment)
       250   Insufficient workspace
       260   Improper illumination (Insufficient light for the operation, glare, etc)
       299   Environmental hazards, NEC

300   *Hazardous methods or procedures*

(Caution should be observed in the application of this classification, particularly to avoid its becoming a "catch-all" for cases which cannot be assigned to other specific classifications because of inadequate information. It is not intended that an activity should be classified as a hazardous procedure simply because an injury or injuries occurred in the course of that activity. A hazardous method or procedure in this context is usually a deviation from the normal and generally accepted safe procedures commonly applied in industrial operations. In some respects, this classification parallels the unsafe act classification. The distinguishing characteristic is that the procedures classified here were planned, directed, or condoned by supervision.)

310   Use of inherently hazardous (not defective) material or equipment

320   Use of inherently hazardous methods or procedures

330   Use of inadequate (not defective) or improper tools or equipment

340   Inadequate help for heavy lifting, etc

350   Improper assignment of personnel (i.e. disregard of physical limitations, skill, etc)

399   Hazardous methods or procedures, NEC

400   *Placement hazards* (materials, equipment, etc—not persons)

410   Improperly piled (refers to manner of piling)

420   Improperly placed (refers to position occupied)

430   Inadequately secured against undesired motion (not unstable piling)

500   *Inadequately guarded*

510   Unguarded (mechanical or physical hazards—not electrical or radiation hazards)

520   Inadequately guarded (mechanical or physical hazards—not electrical or radiation hazards)

530   Lack of or inadequate shoring in mining, excavating, construction, etc

540   Ungrounded (electrical)

550   Uninsulated (electrical)

560   Uncovered connections, switches, etc (electrical)

570   Unshielded (radiation)

580   Inadequately shielded (radiation)

590   Unlabeled or inadequately labeled materials

599   Inadequately guarded, NEC

600   *Hazards of outside work environments—other than public hazards* (encountered while working in or on premises not controlled by the employer and not arising from the activities of the injured or his co-employees or from the tools, materials, or equipment used in those activities)

610   Defective premises of others

620   Defective materials or equipment of others

630   Other hazards associated with the property or operations of others

640   Natural hazards (i.e. hazards of irregular and unstable terrain; exposure to the elements, wild animals, etc; encountered in open country operations but not in cleared or regularly designated work areas)

700   *Public hazards* (encountered in public places away from employers' premises)

710   Public transportation hazards (encountered while a passenger is on a public carrier)

720   Traffic hazards (encountered on public streets, roads, or highways)

780   Other public hazards (other hazards of public places to which the general public is also exposed)

980   *Hazardous conditions, NEC*

990   *Undetermined—insufficient information*

999   *No hazardous condition*

A1.8 Unsafe Act Classification

Code

050   *Cleaning, oiling, adjusting, or repairing of moving, electrically energized, or pressurized equipment* (Do not include actions directed by supervision)
     051   Caulking, packing, etc, of equipment under pressure (pressure vessels, valves, joints, pipes, fittings, etc)
     052   Cleaning, oiling, adjusting, etc, of moving equipment
     056   Welding, repairing, etc, of tanks, containers, or equipment without supervisory clearance in respect to the presence of dangerous vapors, chemicals, etc
     057   Working on electrically charged equipment (motors, generators, lines, etc)
     059   NEC

100   *Failure to use available personal protective equipment* (goggles, gloves, masks, aprons, hats, lifelines, shoes, etc)

150   *Failure to wear safe personal attire* (wearing high heels, loose hair, long sleeves, loose clothing, etc)

200   *Failure to secure or warn*
     201   Failure to lock, block, or secure vehicles, switches, valves, press rams, other tools, materials, and equipment against unexpected motion, flow of electric current, steam, etc
     202   Failure to shut off equipment not in use
     203   Failure to place warning signs, signals, tags, etc
     205   Releasing or moving loads, etc, without giving adequate warning
     207   Starting or stopping plant vehicles or equipment without giving adequate warning
     209   NEC

250   *Horseplay* (distracting, teasing, abusing, startling, quarreling, practical joking, throwing material, showing off, etc)

300   *Improper use of equipment*
     301   Use of material or equipment in a manner for which it was not intended
     305   Overloading (vehicles, scaffolds, etc)
     309   NEC

350   *Improper use of hands or body parts*
     353   Gripping objects insecurely
     355   Taking wrong hold of objects
     356   Using hands instead of hand tools (to feed, clean, adjust, repair, etc)
     359   NEC

400   *Inattention to footing or surroundings*

450   *Making safety devices inoperative*
     452   Blocking, plugging, tying, etc, of safety devices
     453   Disconnecting or removing safety devices
     454   Misadjusting safety devices
     456   Replacing safety devices with those of improper capacity (e.g. higher amperage electric fuses, low capacity safety valves, etc)
     459   NEC

500 *Operating or working at unsafe speed*
    502 Feeding or supplying too rapidly
    503 Jumping from elevations (vehicles, platforms, etc)
    505 Operating plant vehicles at unsafe speed
    506 Running
    508 Throwing material instead of carrying or passing it
    509 NEC

550 *Taking unsafe position or posture*
    552 Entering tanks, bins, or other enclosed spaces without proper supervisory clearance
    555 Riding in unsafe position (e.g. on platforms, tailboards, on running boards of vehicles; on forks of lift truck; on hook of crane; etc)
    556 Unnecessary exposure undersuspended loads
    557 Unnecessary exposure to swinging loads
    558 Unnecessary exposure to moving materials or equipment
    559 NEC

600 *Driving errors* (by vehicle operator on public roadways)
    601 Driving too fast or too slowly
    602 Entering or leaving vehicle on traffic side
    603 Failure to signal when stopping, turning, backing
    604 Failure to yield right of way
    605 Failure to obey traffic control signs or signals
    606 Following too closely
    607 Improper passing
    608 Turn from wrong lane
    609 NEC

650 *Unsafe placing, mixing, combining, etc*
    653 Injecting, mixing, or combining one substance with another so that explosion, fire, or other hazard is created (e.g. injecting cold water into hot boiler, pouring water into acid, etc)
    655 Unsafe placing of vehicles or material moving equipment (i.e. parking, placing, stopping, or leaving vehicles, elevators, or conveying apparatus in unsafe position for loading or unloading)
    657 Unsafe placement of materials, tools, scrap, etc (i.e. so as to create tripping, bumping, slipping hazards, etc)
    659 NEC

750 *Using unsafe equipment* (e.g. equipment tagged as defective or obviously defective. Do not include the use of inherently hazardous material for its intended purpose unless it was obviously defective. Do not include use of defective material or equipment when the defect was hidden and not obvious to the user)

900 *Unsafe act, NEC*

993 *No unsafe act*

999 *Unclassified—inadequate data*

COMPUTER PROGRAM DOCUMENTATION

A2.1 Introduction

This appendix presents all the information necessary to use three programs: (1) Fault-Tree Analysis (FTA), (2) Cost/Benefit Module (CBM), and (3) Dynamic Programming Module (DPM). The FTA program performs the probabilistic calculations necessary for the cost/benefit techniques discussed in Chapter 5. The use of CBM requires less detailed input data. This was described in Chapters 7 and 8. The Dynamic Programming Module (DPM) uses the cost and benefit produced by either FTA or CBM in order to determine optimal policies. This was discussed in Chapters 8 and 9.

The same format will be used to document each of the programs. An introductory section will be followed by a specification of data requirements. The processing details will then be given and a sample output will be discussed. The program listings are together in a separate section at the end of the chapter. All programs were written in standard FORTRAN IV. They were tested on the IBM 360 and 370 systems. Processing times are not significant for normal-sized problems. Each depends upon the size of the problem and the machine upon which it is run.

2

A2.2 Fault-Tree Analysis

A2.2.1 Introduction

The FTA program obtains a cost/benefit analysis using fault-tree analysis and the techniques given in Chapter 5. For each alternative specified for a basic event, the probability of the head event, the benefit, and the cost/benefit are output. Thus this program relieves the analyst of the necessity of performing the detailed calculations given in Equations (5.1)–(5.7).

A2.2.2 Data Requirements

In order to supply the input data for this program, the Boolean expression for the fault tree under consideration must be obtained. Section 5.6 demonstrated the methodology of translating the basic events and the Boolean relationships of a fault tree into a Boolean expression. Once this is accomplished, it is necessary to "code" the Boolean expression so that it can be read by the computer. Section 4.11 illustrated the methodology for translating a Boolean function into binary codes. Only the first two steps of the five-step procedure need to be applied to use this program.

First, each term in the Boolean representation should be coded by a binary number. Then these numbers must be transformed into base 10 for input into FTA. Having performed these preliminary steps, the data should be placed on data cards, as given below. An example follows the data format specifications.

Record 1 (one data card per record). Cols. 1 and 2: NV, the number of Boolean variables in the Boolean expression for the fault tree. *Note:* All variables should be right-justified, with leading zeros or blanks, unless otherwise noted.

Cols. 4 and 5: NT, the number of terms in the Boolean expression for the fault tree. There should be one binary representation, subsequently translated to base 10, for each term.

Cols. 6–15: ISEV, the severity of the head event in lost workdays, dollars, or some other loss unit. If a decimal figure is required, it may be punched and the number may appear anywhere in the field. If the decimal is omitted, the number will be read as a whole number and it must be right-justified.

Cols. 16–47: IHEAD, a description of the head event. Any combinations of letters, numbers, or symbols may be used. This will be used to label the output.

Record 2 (one data card per record). This record consists of the IT(I)'s, the base 10 representations for the terms of the Boolean expression for the fault tree. The ordering of the terms is unimportant. The following format is used:

Cols. 1–3: first term
Cols. 4–6: second term
Cols. (3N-2 through 3N): Nth term
There will be NT terms read (see record 1).

Record 3 (as many cards as there are alternatives +2). Generally, each card layout is as follows:

Cols. 1–5: COST, the cost of the safety alternative under consideration. If the decimal is not punched, the last two digits in the field will be assumed to be to the right of the decimal. Placing a decimal in the field overrides this specification.

Cols. 6–5(N + 1): PNY(I)'s, the probabilities of the basic events. These must be in the same order as they were coded (see the example below). The following format applies:

PNY(1)   (Cols. 6–10)

PNY(2)   (Cols. 11–15)

PNY(N)   [Cols. 5N + 1 through 5(N + 1)]

There will be NV probabilities read (see record 1). If a decimal is not placed in the field, three places to the right of the decimal are assumed.

Note the two exceptions to the general case for record 3 presented above. The first card in record 3 will represent the original system. Therefore, the cost placed in this card will be left blank or zero. The original basic event probabilities will be placed in their appropriate positions. The second exception occurs when all alternative cards have been assembled. Following this set of cards should be placed a card with a 2.0 punched in columns 5–7. This will signal the completion of the analysis for that fault tree.

To save computer time, FTA has the option of processing as many alternatives within each fault tree as are required. Also, as many fault trees may be processed during one compilation as is desired. Record 1 for the second fault tree may be placed immediately behind record 3 for the first, and so forth. A blank card in place of a subsequent record 1 will indicate that processing should terminate.

A2.2.3 Example

Consider the fault tree and the corresponding data given in Figure A2.1. The probabilities given in the table below the figure are the new basic event probabilities brought about by the alternative. The original system (alternative 0) lists the probabilities that exist without modification. The objective, of course, is to determine the cost/benefit of an alternative, given the information in Figure A2.1.

The methods explained in Chapters 4 and 5 are used. First, the Boolean expression for the fault tree is determined. It is: $T = A + B + C + D + E$, which translates to the following binary representations: 10000, 01000, 00100, 00010, and 00001. These are translated into base 10 as 16, 8, 4, 2, and 1, respectively. Table A2.1 demonstrates the format in which these input data are keypunched for FTA.

The FTA program performs all the calculations of probabilities for cost/benefit. Table A2.2 demonstrates the output of FTA. The upper portion contains the input parameters. The cost/benefits, labeled COST/EFF, are measured with respect to alternative 0. For example, with no improvements the probability of the head event is .0760 with a severity of 15.15 lost work-

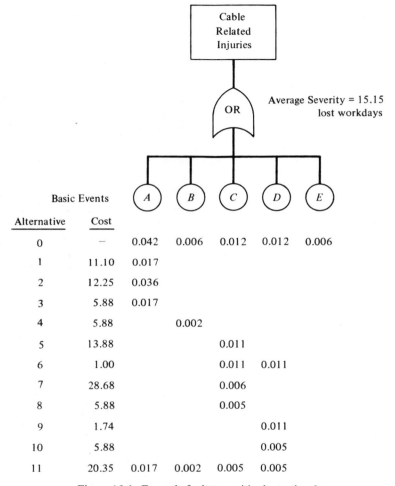

Figure A2.1. Example fault tree with alternative data

days, resulting in an expected loss of 1.151 lost workdays. With alternative 1, the probability is reduced to .0510, so the expected loss is (.0510)(15.15) = .772. The effectiveness of alternative 1 is therefore 1.151 − .772 = .379. Minor errors in the fourth decimal place may accrue from computer rounding.

The program listing for FTA is given in Section A2.5.2.

Table A2.1 EXAMPLE INPUT FOR FAULT-TREE ANALYSIS

	Columns												
	1—5—	1 0	1 5	2 0	2 5	3 0	3 5	4 0	4 5	5 0	5 5	6 0
Record 1	5 5		15.15		2.1	Cable-Related Injuries						
Record 2	16	8 4	2 1									
Record 3		.042	.006	.012	.012	.006						
	1110	.017	.006	.011	.012	.006						
	1225	.036	.006	.012	.012	.006						
	588	.017	.006	.012	.012	.006						
	588	.042	.002	.012	.012	.006						
	1388	.042	.006	.011	.012	.006						
	100	.042	.006	.011	.011	.006						
	2868	.042	.006	.006	.012	.006						
	588	.042	.006	.005	.012	.006						
	174	.042	.006	.012	.011	.006						
	588	.042	.006	.012	.005	.006						
	2035	.017	.002	.005	.005	.006						
		2.0										

Table A2.2 EXAMPLE OUTPUT OF FTA

2.1 Cable-Related Injuries
Severity of Head Event = 15.15
Number of Variables = 5
Number of Terms = 5
Transmission Function = 16   8   4   2   1

Alt	Probability	Cost	Effectiveness	Cost/Eff	Dp Stg-Alt
0	.0760	—	—	—	—
1	.0510	11.10	.3798	29.22	
2	.0703	12.25	.0877	139.73	
3	.0519	5.88	.3653	16.10	
4	.0723	5.88	.0563	104.39	
5	.0751	13.88	.0142	979.62	
6	.0742	1.00	.0283	35.27	
7	.0704	28.68	.0850	337.39	
8	.0695	5.88	.0992	59.29	
9	.0751	1.74	.0142	122.81	
10	.0695	5.88	.0992	59.29	
11	.0346	20.35	.6282	32.40	

A2.3 Cost Benefit Module

A2.3.1 Introduction

The high-accident-location Cost/Benefit Module (CBM) is designed to determine cost/benefit for each alternative proposed project at each high-accident location. Although the purpose of this module is not to determine optimal safety investments, the information provided by it fulfills a twofold objective in this regard. First, it provides the input into the Dynamic Programming Module (DPM), which will obtain as output those projects which will yield the maximum benefit in terms of reduced accidents for any given budget restriction. Second, it provides a tool for decision making to supplement the DPM output. In many cases intangibles not revealed by pure cost/benefit figures must be observed in making final budget allocations among projects. CBM will provide the supplementary information to ensure that such decisions are not far from optimal from a cost/benefit point of view.

A2.3.2 Data Requirements

The most essential part of this system in performing its intended purpose consists of accurate and complete data. Care should be taken that those supplying input data are trained in the meaning of the data itself as well as its ultimate use. To aid the data gatherers, a specific form to collect data for input to CBM has been designed. This form will be explained first followed by the detailed specifications for data entry into the program.

In Section 9.4 the investigation techniques in conjunction with the HALIForm were specified. Figure A2.2 shows how this information is translated into card input for CBM. Several things should be noted from Figure A2.2. First, the names of the investigators as well as the descriptions of causes and alternatives are not keypunched for CBM. Second, there are three records for each location. The first record covers item 1 through the "number of causes." The second record covers item 5 through the "number of alternatives." The third record, which is multiple-card, covers items 10–14, one card for each alternative. Since alternatives will be referenced by number only, it is important that a cross-reference list for the locations and alternatives be maintained.

The following input is required by the CBM:

Record 1 (one data card per job). Col. 1: place a 1 in the first column if punched output for subsequent processing is desired. In this case, the costs and returns for the Dynamic Programming Module (DPM) will be output in the required format.

Col. 2: place a 1 in the second column if only a listing of the locations (record 3, cols. 1–68) is desired.

Col. 3: place a 1 in the third column if two locations per page without maintenance information are desired.

Note: If the options above are not to be exercised, the respective column should be left blank. Option 2 cannot be implemented simultaneously with any other option.

Page _____ of _____

HIGH-ACCIDENT-LOCATION INVESTIGATION FORM

$\overline{1}\quad \overline{2}\quad \overline{3}\quad \overline{4}$

(1) Location: _____   Urban ( )   Rural ( )

(2) Time Period of Accident History (Years): _____
    5

(3) Date: Month   Year
         76 77   78 79   72

(4) Investigators: _____

                                                          68   71   75

Record 3

Number of Causes
_____ 80

(5) Roadway Environment Causes

	Severity During Accident History			
(6) Number of Accidents	(7) Number of Fatality Accidents	(8) Number of Injury Accidents	(9) Number of Cases of Property-Damage Only Accidents
5	6	7	8	9	10	11	12
1.	13	14	15	16	17	18	19	20
2.	21	22	23	24	25	26	27	28
3.	29	30	31	32	33	34	35	36
4.								

Record 4

Number of Alternatives _____

(10) Alternatives (Description)

	Next Column	11. Cost	12. Life	13. Maintenance Cost/Year	14. Effect on Cause			
					1	2	3	4
1.	5	11	12	13	14	18	19	21	22	24	25	27	28	30
2.	5	11	12	13	14	18	19	21	22	24	25	27	28	30
3.	5	11	12	13	14	18	19	21	22	24	25	27	28	30

Record 5

Figure A2.2. Data entry requirements using HALIForm

369

Cols. 5–80: title for the job, alphanumerically.

Record 2 (one data card per job; numbers can be placed anywhere in field and decimal points must be punched).

Cols. 1–10: average cost or negative utility of a fatal accident.

Cols. 11–20: average cost or negative utility of a nonfatal-injury accident.

Cols. 21–30: average cost or negative utility of a property-damage-only accident.

Cols. 31–40: ratio of nonfatal-injury accidents to fatal accidents for urban areas (always greater than 1). (See the example below.)

Cols. 41–50: ratio of nonfatal-injury accidents to fatal accidents for rural areas (always greater than 1). (See the example below.)

Record 3 (one data card per *location*). See Figure A2.2 for the source of the following:

Cols. 1–4: reference number, assigned sequentially by the central office.

Cols. 5–68: alphanumeric description of the location.

Col. 71: if the injury/fatality ratio given in record 2, cols. 31–40, is effective in this location, place a 1 in col. 71; if the ratio given in cols. 41–50 is effective, leave col. 71 blank.

Cols. 72–75: number of years over which the accident history in record 4 has been accumulated. If this is a fractional number of years, the decimal must be punched and the number should be expressed to the nearest .01 year.

Cols. 76 and 77: number of the month in which the investigation took place.

Cols. 78 and 79: number of the year (last two digits) in which the investigation took place.

Col. 80: number of roadway environment causes used for analysis in item 5.

Record 4 (one data card per *location*). See Figure A2.2 for the source of the following:

Cols. 1–4: identical reference number, as given in record 3, cols. 1–4.

Cols. 5 and 6: total number of accidents due to cause 1.

Cols. 7 and 8: number of fatal accidents due to cause 1.

Cols. 9 and 10: number of nonfatal-injury accidents due to cause 1.

Cols. 11 and 12: number of property-damage-only accidents due to cause 1.

Cols. 13 and 14: same as cols. 5 and 6, but for cause 2.

Cols. 15 and 16: same as cols. 7 and 8, but for cause 2.

Cols. 17 and 18: same as cols. 9 and 10, but for cause 2.

Cols. 19 and 20: same as cols. 11 and 12, but for cause 2.

Cols. 21–28: same as cols. 5–12 and cols. 13–20, but for cause 3.

Cols. 29–36: same as cols. 21–28, but for cause 4.

Etc., up to a maximum of eight causes.

Col. $8 \times$ (no. of causes) $+ 5$: this is the very next column after all the numbers have been inserted for all causes. In this column is placed the number of alternatives to be considered in item 10 of the HALIForm. The maximum is nine alternatives.

Record 5 (one card for each *alternative*). See Figure A2.2 for the source of the following:

Cols. 1–4: identical reference number as given in record 3, cols. 1–4.

Cols. 5–11: cost of the alternative.

Cols. 12 and 13: life over which the benefits of the alternative will be obtained without repeating the cost given in cols. 5–11.

Cols. 14–18: yearly maintenance cost required for the alternative to produce its benefits throughout its life.

Cols. 19–21: effect that the alternative will have on cause 1 in terms of percent accident reduction. This number will be expressed as a percentage (i.e., 0–100). Example: If cause 1 had 10 accidents/year and 20 was placed in cols. 19–21, future estimates of accidents due to that cause would be 8 accidents/year.

Cols. 22–24: same as cols. 19–21, but for cause 2.

Cols. 25–27: same as cols. 19–21, but for cause 3.

Etc., 3 columns per cause, up to eight causes.

There will be one card in record 5 for each alternative up to a maximum of nine alternatives. All record 5 cards for a given location will be placed together in the deck and in order of increasing cost (lowest cost first).

The data deck will be arranged as follows. Records 1 and 2 will be followed by records 3 and 4 for the first location. Record 5, containing as many cards as alternatives for the first location, will follow. Then records 3 and 4 will be repeated for the second location, followed by that location's record 5 cards. Any number of locations may be run together in one job in this way. When one of the record 3 cards is left blank, the program will terminate.

A2.3.3 Example

Consider the following input data for an example:

Card 1: no pertinent calculational data.

Card 2: Cols. 1–10: cost of fatality, $C_f = \$37,000$.

        Cols. 11–20: cost of injury, $C_i = \$2,200$.

Cols. 21–30: cost of property damage, $C_p = \$360$.
Cols. 31–40: urban injury/fatality $= 67.00$.
Cols. 41–50: rural injury/fatality $= 22.00$.

Card 3: Cols. 1–68: no pertinence to calculations.
Col. 71: zero, signifying rural area.
Cols. 72–75: time period of data history, $T = 1.5$ years.
Cols. 76–79: not pertinent to calculations.

Cards 4–7: as indicated in Figure A2.3.

This presents all the data required to process the one location. The following steps are followed by the program in determining cost/benefit.

Step 1. Calculate the weighted average of the cost of an injury/fatality accident. Call this value Q, where

$$Q = \frac{C_f + R(C_i)}{1 + R}$$

where C_f is the fatality cost, C_i the injury cost, and R the injury/fatality ratio for both urban and rural.

Using information from card 2,

$$Q = \frac{37,000 + 22(2,200)}{23} = = \$3,713/\text{accident}$$

Step 2. For alternative 1, calculate the savings in injury/fatality accidents as follows:

	Cause		
	1	2	3
Number of injury accidents + fatal accidents	6	6	0
Accident reduction of alternative 1	30%	20%	20%
Number of injury/fatalities saved	1.8	1.2	0

This yields a total savings of 3.0 injury/fatality accidents. The savings in dollars or negative utility is

(3.0 accidents) (Q)

(3 accidents) ($3,713/accident) $= \$11,139$

HIGH-ACCIDENT-LOCATION INVESTIGATION FORM

(1) Location: _W. Lincoln & Washington Ind. No. 33_   Urban (X)   Rural ( )

(2) Time Period of Accident History (Years): _1.5_

(3) Date: _7/73_   (4) Investigators: _John Doe, R. Smith_

Number of Causes __3__

(5) Roadway Environment Causes

	(6) Number of Accidents	(7) Number of Fatality Accidents	(8) Number of Injury Accidents	(9) Number of Cases of Property-Damage Only Accidents
			Severity During Accident History	
1. Narrowing Roadway (E. & N. Approach)	10	2	4	4
2. Inadequate Signing and Markings	12	1	5	6
3. No Transition on North Approach	1	0	0	1
4.				

Number of Alternatives __3__

(10) Alternatives (Description)

		11. Cost	12. Life	13. Maintenance Cost/Year	14. Effect on Cause			
					1	2	3	4
Card 5	1. Transition and Warning Signs	500	12	20	30	20	20	
Card 6	2. Alt. 1 Plus Upgrade Markings	1,000	10	200	40	60	40	
Card 7	3. Alt. 2 plus 4 phase Signals	10,000	5	2,000	50	80	90	

__1__ __2__ __3__ __4__

Card 3

Card 4

Figure A2.3. Example CBM input

373

Step 3.   For alternative 1, calculate the savings in property damage as follows:

	Cause		
	1	2	3
Number of property-damage-only accidents	4	6	1
Accident reduction of alternative 1	30%	20%	20%
Number of property-damage-only accidents saved	1.2	1.2	.2

This yields a total of 2.6 property-damage-only accidents saved. The savings in dollars or negative utility is thus

$$(2.6 \text{ accidents}) \qquad (C_p)$$
$$(2.6 \text{ accidents})(\$360/\text{accident}) = \$936$$

Step 4.   Calculate the total benefit (savings in dollars or reduction in negative utility) from alternative 1. This is the sum of the benefits calculated in steps 3 and 4. This is $11,139 + $936 = $12,075.

Step 5.   Calculate the benefit per year. Since the accident history is 1.5 years, the benefit is 1.5 years. Hence the benefit per year is

$$\frac{\$12,075}{1.5 \text{ years}} = \$8,050/\text{year}$$

Step 6.   Calculate the total benefit from alternative 1. Since alternative 1 has a life of 12 years, the total benefit is

$$(\$8,050/\text{year})(12 \text{ years}) = \$96,600$$

Step 7.   Calculate the total maintenance cost by multiplying the maintenance cost per year by the life in years. For alternative 1, this is

$$(\$20/\text{year})(12 \text{ years}) = \$240$$

Step 8.   Calculate the cost/benefit without maintenance and the cost/benefit with maintenance. Since the cost is 500 for alternative 1, the cost/benefit without maintenance is

$$\frac{\$500}{\$96,600} = .00518 = .0052$$

With maintenance the cost/benefit is

$$\frac{(\$500 + \$240)}{96,900} = .00766$$

Step 9.   Repeat steps 2–8 for all other alternatives.

Step 10.   Repeat steps 2–9 for all other locations.

An example output, which includes the example given above, is given in Table A2.3. The upper portion will be printed out once for each execution of the program. The lower portion (location description and below) will be output for each location.

Table A2.3 SAMPLE OF CBM OUTPUT

	Test Data CBM Cost/Benefit Module					
Option 1 2 3 4						
0 0 0 0						
Inj/Fat Ratio Urban = 22.00, Rural = 67.00						
Neg Utility Fatality = 37000. Injury = 2200. PRP DM = 360.						
1234 W. Lincoln & E. Washington Ind. No. 33						
Accident History 1.50 Years. Month 2, Year 73, 3 Causes.						
Area Considered Urban.						

Roadway Cause	*Tacc*		*Nfat*	*Ninj*		*Npro*
1	10.		2.	4.		4.
2	12.		1.	5.		6.
3	1.		0.	0.		1.
Totals	23.		3.	9.		11.

				1	*Effect On* 2	3
Alternative	*Cost*	*Life*	*Main Cost*	1	2	3
1	500.00	12.	20.00	0.30	0.20	0.20
2	1000.00	10.	200.00	0.40	0.60	0.40
3	10000.00	5.	2000.00	0.50	0.80	0.90

Cost/Benefit Analysis

Alternative	*Cost*	*Benefit*	*Cost/Benefit*
1	500.00	96600.88	0.0052
2	1000.00	161961.56	0.0062
3	10000.00	105779.00	0.0945

Cost/Benefit Analysis, Maintenance Included

Alternative	*Maintenance*	*Total Cost*	*Cost/Benefit*
1	240.00	740.00	.0077
2	2000.00	3000.00	.0185
3	10000.00	20000.00	.1891

Although it is not part of the CBM, the problem of retranslating dollar figures to estimate fatalities will be considered here. This was done for Figure 9.4 to show the expected number of lives that would be saved by each of the budget figures. The values given in card 2 and the procedure used in step 1 can be combined to obtain this estimate.

Consider the return or benefit of $1 million, and assume for simplicity that all areas are rural, so that the injury/fatality ratio is 22 to 1. Let X be the number of fatality accidents and Y the number of injury accidents. Thus from the ratio above,

$$22X = Y$$

Ignoring property-damage-only accidents, the total breakdown of benefit is prorated such that for $1 million benefit there is $37,000/fatality accident and $2,200 per injury accident. Thus it is correct to set

$$37,000X + 2,200Y = 1,000,000$$

Solving these two equations simultaneously yields

$$37,000X + 2,200(22X) = 1,000,000$$
$$85,400X = 1,000,000$$
$$X = 11.7 \text{ fatal accidents/\$1 million benefit}$$

Now this is the number of fatal accidents, not the number of fatalities. From accident records the number of fatalities per fatal accident can be obtained. Assuming 1.37 fatalities per fatal accident yields $(11.7)(1.37) = 16.03$ fatalities per $1 million of benefit. This was the figure used in Figure 9.4.

A2.4 Dynamic Programming Module

A2.4.1 Introduction

The purpose of this program is to determine those projects which should be funded to obtain a maximum return in terms of reduced accidents, injuries, and fatalities. It uses for input the output of the Cost/Benefit Module (CBM). This cost/benefit information is evaluated in terms of the benefits to be derived from all combinations of alternatives at the various locations. Those combinations that are determined to be nonoptimal are sequentially eliminated by the program. The final output consists of a listing of the alternatives, one for each location, which will yield optimal results according to the input. Since generally a project at every location will not fit within the budget limitation, the least desirable locations are also eliminated (alternative 0 is chosen), such that the total budget constraint is not violated. A range of total budget constraints are evaluated such that the optimal set of alternatives that satisfies each is available.

A2.4.2 Data Requirments

Aside from basic system parameters, DPM accepts as input a cost and a benefit from each alternative at each location. Since the objective of DPM is budget allocation, the cost entered should be the cost to be charged to the budget under consideration. Hence maintenance or other yearly costs should be excluded. Costs not to be allocated in the current budget will be considered in the benefit estimate.

The benefit to be input is the benefit to be derived from the corresponding investment. Hence this benefit should be cumulated in time from the point at which the alternative is implemented to the point at which the investment for the alternative must be repeated for replacement or upgrading. If the yearly cost for maintenance is not significantly different (per dollar of initial investment) among alternatives, it can be ignored, because it will not affect the cost/benefit comparisons.

DPM has been designed as an independently operating module. Hence it can be applied wherever budget allocation is required and cost and benefit information is available. However, the Cost/Benefit Module (CBM) produces output that has been specifically designed to be used in DPM for budget allocation. On the printed output (see Table A2.3), the figures labeled "COST" and "BENEFIT" provide the necessary input. An option has also been designed within CBM to punch data cards for DPM. Thus the interface between CBM and DPM consists of cost/benefit information on punched cards. The arrangement of these cards is discussed below.

DPM uses standard dynamic programming procedures in processing. As a result, the total continuous allocation of the budget is approximated by a series of discrete allocations. The size of these discrete trials will be called one "increment." The number of increments to be used is fixed by the program at 300. Hence the size of the increment will be determined by the size of the maximum budget to be considered. The maximum budget is equal to the increment times 300. The above is explained since the increment size is specified by the user on the first data card.

The following are the data specifications for DPM (all numbers right-justified in field):

Record 1 (1 card per record). Cols. 1–4: number of locations to be evaluated.

Cols. 5–8: objective indicator. This value will be 1 (in column 8) for all maximization problems. In the event that DPM is to be used to minimize the returns input, a−1 would be input.

Cols. 9–14: value of the increment, as discussed above. The maximum budget considered will be equal to 300 times this value. Hence to determine the increment, consider the maximum *reasonable* budget and divide by 300. Round appropriately to the nearest whole number.

It is not recommended that the increment be too large, owing to the inaccuracy that might accrue, since the program rounds at each stage to the nearest increment. Accuracy can be checked by adding the costs output and comparing to the total budget output.

On the other hand, the increment must be sufficiently large to reach the maximum amount to be spent. A recommended procedure is to run the program with several increments, to determine if the accuracy is being diminished. This is discussed further below.

Cols. 15–18: this value will determine the starting point for output. To determine the value of this parameter, determine the minimum budget that is to be output. Divide this minimum budget figure by the increment and add 1. This value will be entered in cols. 15–18. For maximum output, enter 1 in col. 18.

Cols. 19–22: this value determines an index between budgets output. Generally, it would be wasteful of paper and printer time to output all 300 budgets for which an optimal allocation has been performed. Therefore, this value enables a cross section of all these budgets to be output. The value entered will be multiplied by the increment to form the index. Example: If there is a 1 in col. 18, and the increment is 1,000, then a 10 in cols. 21 and 22 would produce the results for budgets 10,000, 20,000, 30,000, . . . , 300,000.

Cols. 29–32: value of a four-digit number called the DPM RUN CODE number. This number is strictly for cross-reference purposes. It will be printed at the top of each page of output, and it will be punched for output in DPM if the punch option is in effect.

Col. 34: will contain a 1 if the punch option is in effect. Otherwise, it will be left blank.

Record 2 (several cards per record). This set of cards contains the costs for each alternative at each location. These will be ordered as follows:

First Card:
    Cols. 1–3: number of alternatives for Location 1
    Cols. 4–13: cost, alternative 1, location 1
    Cols. 14–23: cost, alternative 2, location 1
    Cols. 24–33: cost, alternative 3, location 1
    Etc. to cols. 64–73 (alternative 7)

Second Card: (if less than seven alternatives in location 1)
    Cols. 1–3: number of alternatives for location 2
    Cols 4–13: cost, alternative 1, location 2
    Cols 14–23: cost, alternative 2, location 2
    Cols. 24–33: cost, alternative 3, location 2
    Etc. to cols. 64–73 (alternative 7)

Subsequent cards will follow for each location. For any location, X, with more than seven alternatives, another card will follow the first cost card for that location in the following format:

Cols. 1–10: cost, alternative 8, location X
Cols. 11–20: cost, alternative 9, location X
Etc. to cols. 71–80 for alternative 15

Similarly, a card in the same format for alternatives 16–23 would follow if required. This may be repeated up to 30 alternatives.

Note: The option exists that after the cost of the last alternative of each location, a reference number for the location may be inserted. If left blank, this reference number will be read as zero. In either case, this will cause an additional *card* to be read for that location in the event of 6, 7, 14, 15, 22, or 23 alternatives. Because of this, a blank card will be required after the cost card for locations with 6, 7, 14, 15, 22, or 23 alternatives. If the reference-number option is to be exercised, place the reference number followed by a decimal in the next field after the last cost (i.e., the next 10 columns, as indicated for costs above).

Record 3 (same number of cards as record 2). This set of cards contains the benefits for each alternative at each location. These will be ordered as follows:

First Card:

Cols. 1–3: number of alternatives for location 1
Cols. 4–13: benefit, alternative 1, location 1
Cols. 14–23: benefit, alternative 2, location 1
Cols. 24–33: benefit, alternative 3, location 1
Etc. to cols. 64–73 (alternative 7)

Second Card: (if less than seven alternatives in location 1)

Cols. 1–3: number of alternatives for location 2
Cols. 4–13: benefit, alternative 1, location 2
Cols. 14–23: benefit, alternative 2, location 2
Cols. 24–33: benefit, alternative 3, location 2
Etc. to cols. 64–73 (alternative 7)

Subsequent cards will follow for each location. For any location, X, with more than seven alternatives, another card will follow the first cost card for that location in the following format:

Cols. 1–10: benefit, alternative 8, location X
Cols. 11–20: benefit, alternative 9, location X
Etc. to cols. 71–80 for alternative 15

Similarly, a card in the same format for alternatives 16–23 would follow if required. This may be repeated up to 30 alternatives.

Note: The option exists that after the benefit of the last alternative of each location, a reference number for the location may be inserted. If left blank, this reference number will be read as zero. In either case, this will cause an additional *card* to be read for that location in the event of 6, 7, 14, 15, 22, or 23 alternatives. Because of this, a blank card will be required after the benefit card for locations with 6, 7, 14, 15, 22, or 23 alternatives. If the reference-number option is to be exercised, place the reference number followed by a decimal in the next field after the last benefit (i.e., the next 10 columns, as indicated for benefits above).

If another job is to be run immediately following the first, records 1, 2, and 3 for the first job can be followed by records 1, 2, and 3 for the second job, and so on. After the final job, a blank card should be placed in the record 1 position to signify the end of processing.

A2.4.3 Example

The following test set of data is for a seven-location problem in which the maximum budget is 300,000. Hence the increment is 300,000/300 = 1,000. Output is desired to start at 10,000 and index in jumps of 10,000 each. Hence the first data card is

$$\flat\flat\flat7\flat\flat\flat1\flat\flat1000\flat\flat11\flat\flat10\flat\flat\flat\flat\flat\flat9006$$

where $\flat$ = blank. The 11 is the starting point for output, which will start the output at a \$10,000 budget. The 10 is the output step, which in this case will cause a jump of \$10,000 between maximum budgets. The 9006 is the run code, which is purely for purposes of cross reference. Record 2 data cards, which contain the cost information, are as follows:

$$\flat\flat1\flat\flat\flat\flat\flat17587.\flat\flat\flat\flat\flat\flat\flat75.$$
$$\flat\flat1\flat\flat\flat\flat\flat1500.\flat\flat\flat\flat\flat\flat\flat76.$$
$$\flat\flat1\flat\flat\flat\flat\flat\flat720.\flat\flat\flat\flat\flat\flat\flat77.$$
$$\flat\flat1\flat\flat\flat\flat50000.\flat\flat\flat\flat\flat\flat\flat79.$$
$$\flat\flat1\flat\flat\flat\flat10000.\flat\flat\flat\flat\flat\flat\flat82.$$
$$\flat\flat2\flat\flat\flat\flat\flat1000.\flat\flat\flat100000.\flat\flat\flat\flat\flat\flat\flat84.$$
$$\flat\flat2\flat\flat\flat\flat20000.\flat\flat\flat\flat40000.\flat\flat\flat\flat\flat\flat\flat86.$$

This means, for example, that alternative 1 at location 3 costs \$720, alternative 2 at location 6 costs \$100,000, etc. Each card is a location. Locations 1–5 have only one alternative, whereas locations 6 and 7 have two. Note that the reference numbers 75, 76, 77, 79, 82, 84, and 86 have been used for these locations. Their use will become obvious when subsequent processing is discussed.

The record 3 data cards for this job, which contain the benefit information, are as follows:

ƀƀ1ƀƀƀ135360.ƀƀƀƀƀƀƀ75.

ƀƀ1ƀƀƀƀ35400.ƀƀƀƀƀƀƀ76.

ƀƀ1ƀƀƀƀƀ5280.ƀƀƀƀƀƀƀ77.

ƀƀ1ƀƀƀ162540.ƀƀƀƀƀƀƀ79.

ƀƀ1ƀƀƀƀƀ3960.ƀƀƀƀƀƀƀ82.

ƀƀƀƀƀƀƀ21134.ƀƀƀƀƀ9180.ƀƀƀƀƀƀƀ84.

ƀƀ2ƀƀƀƀƀ3960.ƀƀƀƀƀ4320.ƀƀƀƀƀƀƀ86.

This indicates, for example, that a benefit of $162,540 is to be obtained from location 4, alternative 1; or that a benefit of $4,320 is to be obtained from location 7, alternative 2; and so on. Note that the same reference numbers are used, and they must be placed in the same order in record 3 as they were in record 2.

This data can be processed followed by a blank card or another set of locations. In either case a three-part output will be obtained. Since the first two parts only repeat the input for verification, they will not be discussed. The output for some of the maximum budgets is given in Table A2.4. Actually, all the budgets from 10,000 through 300,000 would be output by the computer.

With this output in hand, the decision maker requires only a knowledge of his budget constraint, and he can immediately determine the alternatives to implement at each location. Since the location numbers are given on the output, a cross-reference list of locations could be easily maintained so that all the details of the improvement would be right at the hands of the decision maker.

Also, given the output in Table A2.4, a curve for determining the "correct" budget figure (such as that given in Figure 9.4) can be constructed very easily. This is done by using the maximum budget, labeled "BUDGET" in Table A2.4, as the cost. The benefit figure is the top benefit figure, labeled "RETURN" in Table A2.4, for each budget.

Notice that the costs of the alternatives do not add exactly to the maximum budget figure. Obviously, they cannot, because of the discrete nature of the expenditures. Generally, these will add to something slightly below the maximum budget.

A final word is required to apply this program to large budgets and many locations. When computer storage space is exceeded by the number of locations, it is possible to divide these locations into subsets. A subset should not bᵉ so large that the sum of the feasible investments is greater than the maximum budget limitation (i.e., $300,000 for this example).

Table A2.4 EXAMPLE OF DPM OUTPUT

DPM Run Code 9006					
Budget	Location		Alt-Num	Cost	Return
10000.00					
	7	86	0	0.	41814.
	6	84	1	1000.	41814.
	5	82	0	0.	40680.
	4	79	0	0.	40680.
	3	77	1	720.	40680.
	2	76	1	1500.	35400.
	1	75	0	0.	0.
20000.00					
	7	86	0	0.	176040.
	6	84	0	0.	176040.
	5	82	0	0.	176040.
	4	79	0	0.	176040.
	3	77	1	720.	176040.
	2	76	1	1500.	170760.
	1	75	1	17587.	135360.
.					
.					
.					
130000.00					
	7	86	2	40000.	347994.
	6	84	1	1000.	343674.
	5	82	1	1000.0	342540.
	4	79	1	50000.	338580.
	3	77	1	720.	176040.
	2	76	1	1500.	170760.
	1	75	1	17587.	135360.
.					
.					
.					
220000.00					
	7	86	2	40000.	356040.
	6	84	2	100000.	351720.
	5	82	1	10000.	342540.
	4	79	1	50000.	338580.
	3	77	1	720.	176040.
	2	76	1	1500.	170760.
	1	75	1	17587.	135360.
.					
.					
.					
to 300000.00					

Now each of the subsets can be run for the, say, $10,000–$300,000 budgets. Notice that the output for each subset is a set of cost and return figures. Thus we can view each *budget* as an alternative and each subset as a location. This telescopes the capability of DPM, enabling it to handle any number of locations.

The punch-output option for DPM is designed for this purpose. The punch output provides, in the proper format for DPM reading, all 30 costs and all 30 returns for each subset. To take advantage of this feature, the user need only adjust record 1 and appropriately arrange all the cost cards together and all the return cards together for all subsets. Notice that the DPM RUN CODE number, input in record 1 for each subset, will be output on the punched cards for the location reference number. This provides for complete cross referencing throughout the system. The program listing for DPM is given in Section A2.5.4.

A2.5 Program Listings

A2.5.1 Introduction

The program listings are presented together in this section, for ease of reference. Programs are written in standard FORTRAN IV. Control cards, of course, have been removed from the programs. To activate these programs, consult the systems consultant at your computer center. Keypunch the program cards according to standard FORTRAN rules. Use the Job Control Language (JCL) control cards recommended by your computer center.

A2.5.2 Fault-Tree Analysis

```
C       THIS PROGRAM, BY THE METHOD OF FAULT TREE ANALYSIS,
C       CALCULATES THE PROBABILITY OF A HEAD EVENT BASED ON THE
C       INPUT OF THE BASIC EVENT PROBABILITIES.  ALSO BY APPLY-
C       ING DOLLAR ALLOCATIONS TO THE BASIC EVENTS TO REDUCE
C       THE PROBABILITIES OF THEIR OCCURRENCE, A NEW PROBABILITY
C       OF THE HEAD EVENT IS OUTPUT ALONG WITH ITS ASSOCIATED
C       EFFECTIVENESS (BASED ON THE HEAD EVENT SEVERITY).
C
        DOUBLE PRECISION P
        DIMENSION IT(20),N(10),NNT(20,10),NBY(10),NAN(1024),
       *PNY(10),NNY(10),IHEAD(8)
        REAL ISEV
        WRITE(6,34)
     34 FORMAT('1')
C
C       INPUT THE NUMBER OF VARIABLES, THE NUMBER OF TERMS,
C       AND THE SEVERITY OF THE HEAD EVENT.
C
    205 READ(5,52) NV,NT,ISEV,IHEAD
     52 FORMAT(I2,1X,I2,F1C.0,8A4)
        KKK=C
        IF(NV.EQ.0) GC TO 201
C
C       INPUT THE BASE 10 CONVERSION OF THE TRANSMISSION
C       FUNCTION.
C
```

```
      READ(5,42)(IT(I),I=1,NT)
   42 FORMAT(2CI3)
      MAX=(2**NV)-1
C
C     SET ALL BASE 10 NUMBERS THAT MIGHT BE GENERATED IN THE
C     TRANSMISSION FUNCTION TO ZERO.
C
      DO 103 I=1,MAX
  103 NAN(I)=0
      DO 1 J=1,NT
C
C     CONVERT THE FIRST TERM CF THE TRANSMISSION FUNCTION TO A
C     BINARY NUMBER.
C
      NA=IT(J)
      NAN(NA)=1
      DO 2 I=1,NV
      NBN=NA/2**(NV-I)
      NNT(J,I)=NBN
      NA=NA-2**(NV-I)*NBN
    2 CONTINUE
      I=C
C
C     COUNT THE NUMBER CF ONES IN THE BINARY NUMBER.
      DO 3 K=1,NV
      IF(NNT(J,K).EC.1) I=I+1
    3 CONTINUE
C
  5C0 FORMAT('0',14X,'SEVERITY CF HEAD EVENT = ',F7.2,/'0',
     *14X,'NUMBER OF VARIABLES = ',I2,/'0',14X,
     *'NUMBER CF TERMS = ',I2,/'0',14X
     *,'TRANSMISSION FUNCTION = ',20(I3,1X))
      WRITE(6,600)
  600 FORMAT('0',//////15X,'ALT',2X,'PROBABILITY',3X,'COST',
     *3X,'EFFECTIVENESS',3X,'CCST/EFF',2X,'CP STG-ALT')
   41 SUM=C.0
      DO 32 NUM=1,NX
      IF(NAN(NUM).EQ.C) GO TO 32
      NY=NUM
C
C     CONVERT THE GENERATED BASE 10 TRANSMISSION NUMBERS INTO
C     BINARY NUMBERS AND STORE THEM IN AN ARRAY NNY.
C
      DO 22 I=1,NV
      NN=NY/2**(NV-I)
      NNY(I)=NN
      NY=NY-2**(NV-I)*NN
   22 CONTINUE
      P=1
C
C     EACH NUMBER CF THE ARRAY IS CHECKED TC SEE IF ITS VALUE
C     IS 1 CR 0.
C
      DO 23 K=1,NV
      IF(NNY(K).EQ.1) GC TO 24
C
C     MULTIPLY THE VARIABLE PROBABILITIES TOGETHER TO FIND
C     THE ASSOCIATED PROBABILITY FOR EACH TERM.
C
      P=P*(1-PNY(K))
```

```
         GO TC 23
     24  P=P*PNY(K)
     23  CONTINUE
C
C        SUM THE PROBABILITIES FOR ALL TERMS.
C
         SUM=SUM+P
     32  CONTINUE
         IF(KKK.NE.0) GO TC 76
C
C        CALCULATE THE INITIAL EFFECTIVENESS OF THE HEAD EVENT.
C
         SA=ISEV*SUM
         WRITE(6,700) KKK,SUM
    7C0  FORMAT('C',14X,I2,5X,F7.4,4X,'------',6X,'------',
        *7X,'------')
         KKK=KKK+1
         GO TO 200
C
C        CALCULATE THE NEW EFFECTIVENESS BASED ON THE ALLOCATION
C        OF THE SAFETY RESOURCES.
C
C        SET NZ EQUAL TO THE NUMBER OF ZEROS IN THE TRANSMISSION
C        FUNCTION.
C
         NZ=NV-I
         NTMS=(2**NZ)-1
C
C        FIND THE COMBINATIONS OF NZ BINARY DIGITS.
C
         DO 20 L=1,NTMS
         NM=NTMS-1
         KA=L
         DO 21 K=1,NZ
         M=KA/2**(NZ-K)
         NBY(K)=M
         KA=KA-(2**(NZ-K))*M
     21  CONTINUE
         NUM=0
         M=1
C
C        CHECK THE ORIGINAL TERM OF THE TRANSMISSION FUNCTION
C        AND NOTE THE COLUMN WITH A VALUE OF ONE.
C
         DO 10 K=1,NV
         IF(NNT(J,K).EQ.1) GO TO 9
         VX=NBY(M)
         M=M+1
         GO TO 11
      9  CONTINUE
         VX=NNT(J,K)
     11  NUM=NUM+VX*2**(NV-K)
     10  CONTINUE
C
C        STORE THE GENERATED BASE 10 NUMBERS BY SETTING THEIR
C        ARRAY VALUES EQUAL TO ONE.
C
         NAN(NUM)=1
```

```
      20 CONTINUE
       1 CONTINUE
         NX=(2**NV)-1
C
C        INPUT THE COST OF A PARTICULAR  SAFETY MEASURE   AND THE
C        PROBABILITIES OF THE VARIABLES.
C
    200 READ(5,30)  COST,(PNY(I),I=1,NV)
     30 FORMAT(F5.2,15F5.3)
        IF(PNY(1).EQ.2.) GO TO 205
        IF(KKK.EQ.0) GO TO 93
        GO TO 41
C
C        WRITE THE APPROPRIATE HEADINGS.
C
     93 WRITE(6,78) IHEAD
     78 FORMAT('1',////////' ',23X,8A4///)
        WRITE(6,500) ISEV,NV,NT,(IT(I),I=1,NT)
     76 S=ISEV*SUM
C
C        OBTAIN THE NET RESULT OF THE SAFETY MEASURES.
C
        DS=SA-S
        CE=COST/DS
C
C        OUTPUT THE PROBABILITY OF THE HEAD EVENT, THE COST OF
C        THE SAFETY MEASURES, AND THE EFFECTIVENESS OF THE
C        PARTICULAR SET OF MEASURES.
C
        IF(KKK.NE.16) GO TO 652
        WRITE(6,653)
    653 FORMAT('1',////////15X,'ALT',2X,'PROBABILITY',3X,'COST',3X,
       *'EFFECTIVENESS',3X,'COST/EFF',2X,'DP STG-ALT')
    652 WRITE(6,800) KKK,SUM,COST,DS,CE
    800 FORMAT('0',14X,I2,5X,F7.4,3X,F7.2,5X,F7.4,5X,F8.2)
        KKK=KKK + 1
        GO TO 200
    201 CONTINUE
        WRITE(6,35)
     35 FORMAT('1')
        STOP
        END
```

A2.5.3 *Cost/Benefit Module*

```
        DIMENSION TITL(19),XLOC(17),SEV(9,8),CSEF(10,11),B(8)
        READ(5,10)IOP1,IOP2,IOP3,IOP4,(TITL(I),I=1,19)
     10 FORMAT(4I1,19A4)
        READ(5,11) CFAT,CINJ,CPDO,RURB,RRUR
     11 FORMAT (8F10.0)
        WRITE(6,15) (TITL(I),I=1,19)
     15 FORMAT (20X,19A4)
        WRITE(6,16) IOP1,IOP2,IOP3,IOP4
     16 FORMAT (/'   OPTION   1    2    3    4'/ 9X,4I4)
        WRITE(6,17) RURB,RRUR
     17 FORMAT(/'   INJ/FAT RATIO   URBAN=',F5.2,',   RURAL=',F5.2)
        WRITE(6,18) CFAT,CINJ,CPDO
```

```
   18 FORMAT(' NEG UTILITY FATALITY=',F7.0,'  INJURY=',F6.0,
     1'  PRP DM=',F5.0/////)
C
C     THE ABOVE READS AND PRINTS THE BASIC PARAMETERS CONSTANT
C     FOR THE ENTIRE PROGRAM.  BELOW IS THE INPUT WHICH IS
C     EXECUTED FOR EACH ACCIDENT LOCATICN.
C
      N=0
    1 READ(5,20) NO1,(XLOC(I),I=1,16),ILOC,TIME,NMO,NYR,NCAU
      N=N+1
   20 FORMAT(I4,16A4,2X,I1,F4.0,I2,I2,I1)
      IF(NC1)98,99,98
   98 CONTINUE
      IF(IOP2) 122,123,122
  122 WRITE(6,121) NO1,(XLOC(I),I=1,16)
  121 FORMAT(3X,I4,2X,16A4)
      READ(5,30)NO2,((SEV(I,J),J=1,4),I=1,NCAU),ALT
      NALT=ALT/10.+.1
      NJ=3+NCAU
      DO145  I=1,NALT
      READ (5,50)NO3,(CSEF(I,J), J=1,NJ)
  145 CONTINUE
      GO TO 1
  123 CONTINUE
      IF(IOP3.EQ.0) GO TO 1123
      NCK=N/2
      NCK=2*NCK
      IF(N.EQ.NCK) GO TO 1125
 1123 WRITE(6,22)
   22 FORMAT(1H1)
      WRITE(6,1127)
 1127 FORMAT(//' REF NO')
      GO TO 1128
 1125 WRITE(6,125)
  125 FORMAT(/////' REF NO')
 1128 CONTINUE
      IF(NCAU.EQ.1) GO TO 129
      WRITE(6,21) NO1,(XLOC(I),I=1,16),TIME,NMO,NYR,NCAU
   21 FORMAT(3X,I4,8X,16A4,//9X,'ACCICENT HISTORY ',F4.2,
     2' YEARS.  MONTH ',I
     12,',YEAR ',I2,', ',I1,' CAUSES.')
      GO TO 130
  129 WRITE(6,29)  NO1,(XLOC(I),I=1,16),TIME,NMO,NYR,NCAU
   29 FORMAT(3X,I4,8X,16A4,//9X,'ACCICENT HISTORY ',F4.2,
     2' YEARS.  MONTH ',I
     12,',YEAR ',I2,', ',I1,' CAUSE. ')
  130 CONTINUE
C
C     ROUTINE FOR ASSIGNING RURAL OR URBAN RATIOS BELOW.
C
      IF(ILOC-1) 24,23,24
   23 WRITE(6,25)
   25 FORMAT('     AREA CONSICERED URBAN.')
      R=RURB
      GO TO 28
   24 WRITE(6,26)
   26 FORMAT('     AREA CONSICERED RURAL.')
      R=RRUR
   28 CONTINUE
```

```
C
C      SECCND CARD INPUT FOR EACH CRITICAL LOCATION (SEVERITIES).
C
       READ(5,30)NO2,((SEV(I,J),J=1,4),I=1,NCAU),ALT
   30 FORMAT(I4,38F2.0)
       NALT=ALT/10.+.1
C
C      ROUTINE TO CHECK CARD SEQUENCE COCE.
C
       IF(NO1-NC2) 35,36,35
   35 WRITE(6,37)NC1,NO2
   37 FORMAT('    SEQUENCE/CODE NC. ERROR. CHECK ',I5,' AND',I5,
      2' **PROCESSING CONTINUES')
   36 CONTINUE
C
C      OUTPUT CF SEVERITIES AND TOTALS.
C
       WRITE (6,42)
   42 FORMAT(/ ' ROADWAY CAUSE    TACC    NFAT    NINJ    NPRC')
       TOT1=C
       TOT2=C
       TOT3=C
       TOT4=C
       DO 41 I=1,NCAU
       WRITE (6,40) I,(SEV(I,J),J=1,4)
   40 FORMAT (1X,I7,F12.0,3F6.0)
       CHECK=SEV(I,2)+ SEV(I,3) + SEV(I,4)
       IF(ABS(SEV(I,1)-CHECK)-.1) 72,69,69
   69 WRITE (6,68) I
   68 FORMAT ('    ERROR IN NUMBER COUNT IN SEVERITY, CAUSE',I3/,
      2'    ***PROCESSING CCNTINUES, TOTAL IS IGNORED.')
   72 CONTINUE
       TOT1=TOT1+ SEV(I,1)
       TOT2=TOT2+ SEV(I,2)
       TOT3=TOT3+ SEV(I,3)
       TOT4=TOT4+ SEV(I,4)
   41 CONTINUE

       WRITE(6,43) TOT1,TCT2,TCT3,TOT4
   43  FORMAT('    TOTALS',F12.C,3F6.0)
C
C      INPUT NEXT SET CF NALT CARCS, CNE FOR EACH ALTERNATIVE.
C
       NJ=3+NCAU
       CO 45  I=1,NALT
       READ (5,50)NO3,(CSEF(I,J), J=1,NJ)
   50 FORMAT(I4,F7.C,F2.0,F5.0,8F3.2)
       IF(NC3-NC1)51,52,51
   51 WRITE(6,37) NO1, NO3
   52 CONTINUE
   45 CONTINUE
C
C      OUTPUT CF ALTERNATIVE INFORMATICN.
C
       WRITE(6,58)(I,I=1,NCAU)
   58 FORMAT(/' ALTERNATIVE    COST    LIFE    MAIN COST    ',
      1'EFFECT CN...',8I5)
```

```
C
C      NUMBER COUNT CHECK OF SEVERITIES.
C
       DO 70 I=1,NALT
       WRITE(6,71) I,(CSEF(I,J),J=1,NJ)
    71 FORMAT(I7,F13.2,F8.0,F9.2,F24.2,7F5.2)
    70 CONTINUE

C
C      COMPUTATION OF B(I), THE ITH ALTERNATIVE BENEFIT.
C
       Q=(CFAT + R * CINJ) /(1.+R)
       DO 75   I=1,NALT
       B(I) = 0.
       DO 76   J=1,NCAU
       JEFT = J +3
       B(I) = B(I)+ (Q*(SEV(J,2)+SEV(J,3))*CSEF(I,JEFT))+CPDC*
      1SEV(J,4)*CSEF(I,JEFT)
    76 CONTINUE
    75 CONTINUE

C
C      CALCULATION OF COST/BENEFITS AND OUTPUT.
C
       WRITE (6,80)
    80 FORMAT(  /5X,' COST/BENEFIT ANALYSIS'//' ALTERNATIVE        ',
      1'COST        BENEFIT        COST/BENEFIT')
       DO 90 I=1,NALT
       B(I)=B(I)*CSEF(I,2)/TIME
       CSBN=CSEF(I,1)/B(I)
       WRITE(6,91)I,CSEF(I,1),B(I),CSBN
    91 FORMAT(I7,F14.2,F14.2,F18.4)
    90 CONTINUE
       IF(ICP3.EQ.1)GO TO 980
       WRITE(6,94)
    94 FORMAT(/'   COST/BENEFIT ANALYSIS, MAINTENANCE INCLUDED'/)
       WRITE(6,95)
    95 FORMAT(' ALTERNATIVE         MAINTENANCE      TOTAL COST    ',
      1'COST/BENEFIT')
       DO 97 I=1,NALT
       XMAIN=CSEF(I,2)*CSEF(I,3)
       CBNM=(CSEF(I,1)+XMAIN)/B(I)
       TMCST=CSEF(I,1)+XMAIN
       WRITE(6,96) I,XMAIN,TMCST,CBNM
    96 FORMAT(I7,F23.2,F14.2,F16.4)
    97 CONTINUE
   980 CONTINUE
       IF (ICP1) 105,106,105
   105 XNO1=NO1
       WRITE(7,107) NALT,(CSEF(I,1),I=1,NALT),XNO1
   107 FORMAT(I3,7F10.0/8F10.0,/,8F10.0)
       WRITE(7,107)NALT,(B(I),I=1,NALT),XNO1
   106 CONTINUE
       GO TO 1
    99 CALL EXIT
       STOP
       END
```

A2.5.4 Dynamic Programming Module

```
C       THIS MAIN PROGRAM IS THE CPTIMIZATION ROUTINE FOR THE
C       DYNAMIC PROGRAMMING ALGCRITHM.  IT CALCULATES AND OUT-
C       PUTS THE OPTIMAL CECISICN AND OPTIMAL RETURN FOR ALL
C       INPUT STATES CR BUDGETS.
C
        DIMENSICN ORET(50,301),CDEC(50,301),NIN(50),NDE(51)
        DIMENSICN NOD(50,301),C(50,31),R(31),CO(30),RC(30)
        DIMENSICN CX(50,31),RX(31),XNC1(50)
      1 CONTINUE
        IST=C
        VRET=C.C
        READ(5,4) NSTG,NBJ,XINC,K1,K2,XNC2,IPUNCH
      4 FORMAT(2I4,F6.0,2I4,F10.0,I2)
        IF(NSTG)2,999,2
      2 CONTINUE
        DO 5 I=1,NSTG
      5 NIN(I)=3C1
      3 FORMAT(2CI4)
        WRITE(6,101)
        DO 15 I=1,NSTG
        C(I,1)=C.
        READ(5,1CO) NDEC,(CX(I,IC),IC=1,NCEC),XNO1(I)
    1CC FORMAT(I3,7F1C.C/8F10.0/8F10.0/8F10.0)
        NDE(I)=NCEC
        DO 301 IC=1,NCEC
        ICP1=IC+1
        C(I,ICP1)=CX(I,IC)
    3C1 CONTINUE
     15 CONTINUE
        WRITE(6,200) XNC2
    2CO FORMAT(8X,'DPM RUN CODE',F6.0//)
        WRITE(6,201)NSTG,NBJ,XINC,K1,K2
    2C1 FORMAT(5X,'STAGES---MAXIMUM---INCREMENT---LLIMIT---ULIMIT'
       1,/,2I9,F13.2,2I9,//,'-----STAGE---INPUTS---CECISICNS')
        DO 2C3 I=1,NSTG
        WRITE(6,202)I,NIN(I),NDE(I)
    2C2 FORMAT(I9,I8,I10)
    2C3 CONTINUE
        WRITE(6,101)
        WRITE(6,200) XNC2
        WRITE (6,204)
    204 FORMAT (1H ,'  STAGE    CECISION    CCST    RETURN    REF NC',
       1'   CST/BEN')
        IPAP=C
        NSP1=NSTG+1
        NDE(NSP1)=0
        DO 1C I=1,NSTG
        R(1)=C.
        READ(5,1CO) NDEC,(RX(IC),IC=1,NDEC),XNO1(I)
        DO 302 IC=1,NCEC
        ICP1=IC+1
        R(ICP1)=RX(IC)
    302 CCNTINUE
        NDEC=NDE(I)+1
        DO 206 IC=1,NCEC
        IX=IC-1
```

```
      RECB=R(IC)
      IF(R(IC).EQ.0.0) RECB=1
      CBEN=C(I,IC)/RECB
      WRITE(6,205)I,IX,C(I,IC),R(IC),XNO1(I)     ,CBEN
 205  FORMAT(I6,I9,F11.0,F11.0,F7.0,F10.4)
 206  CONTINUE
      IPAP=IPAP+NDEC
      IP1=I+1
      IPC=IPAP+NDE(IP1)
      IF(IPC.LT.50) GO TO 303
      WRITE(6,101)
      WRITE(6,200) XNO2
      WRITE(6,204)
      IPAP=0
 303  CONTINUE
      NINP=NIN(I)
      DO 20 J=1,NINP
      XIN=(J-1)*XINC
      DUM=-100000000000.*NBJ
      NDEC=NDE(I)+1
      DO 30 K=1,NDEC
      CALL XOUT(I,IST,XIN,K,TDEC,KICK,XINC,C)
      IF(KICK)901,901,902
 902  GO TO 30
 901  CONTINUE
      VRET=R(K)
      IF(I-1)7,7,8
   7  TEST=VRET
      GO TO 11
   8  TEST=VRET+ORET(I-1,IST)
      GO TO 11
  11  IF(NBJ*(DUM-TEST))13,12,12
  13  DUM=TEST
      ODEC(I,J)=TDEC
      ORET(I,J)=DUM
      NOD(I,J)=K
  12  GO TO 29
  29  CONTINUE
  30  CONTINUE
  20  CONTINUE
  10  CONTINUE
      NINP=NIN(NSTG)
      NOPC=40/NSTG
      KKK=NOPC
      ICCRC=0
      DO 40 M=K1,NINP,K2
      ICORC=ICCRC+1
      J=M
      XIN=(J-1)*XINC
      IF(KKK.NE.NOPC) GO TO 67
      WRITE(6,101)
      WRITE(6,200) XNO2
      WRITE(6,16)
  16  FORMAT(' ',           5X,'BUDGET  LOCATION ',4X,'ALT-NUM',
     *5X,'COST',5X,'RETURN')
      KKK=0
  67  KKK=KKK+1
      WRITE(6,18) XIN
```

```
   18 FORMAT('0',1X,F12.2)
      CO(ICORO)=XIN
      DO 45 L=1,NSTG
      I=NSTG+1-L
      XIN=(J-1)*XINC
      TDEC=ODEC(I,J)
      KK=(NCD(I,J)-1)
      NO1=XNO1(I)
      WRITE(6,17) I,NC1,KK,ODEC(I,J),ORET(I,J)
   17 FORMAT(' ',12X,I3,1X,I4,6X,I4,F12.0,F12.0,F10.4)
      K=NCD(I,J)
      CALL XOUT(I,IST,XIN,K,TDEC,KICK,XINC,C)
      J=IST
   45 CONTINUE
      RO(ICORO)=ORET(NSTG,M)
   40 CONTINUE
      IIX=30
      IF(IPUNCH.EQ.0) GO TO 998
      WRITE(7,100) IIX,(CO(I),I=1,30),XNO2
      WRITE(7,100) IIX,(RO(I),I=1,30),XNO2
  998 WRITE(6,101)
  101 FORMAT('1')
      GO TO 1
  999 STOP
      END




C     THIS SUBROUTINE CALCULATES THE OUTPUT STATE NUMBER
C     RESULTING FROM THE INPUT XIN AND SAFETY MEASURE K.  IT
C     ALSO DETERMINES THE COST CF A PARTICULAR SAFETY MEASURE
C     CORRESPONDING TO STAGE I.
C
      SUBROUTINE XOUT(I,IST,XIN,K,TDEC,KICK,XINC,C)
      DIMENSICN C(50,31)
      TDEC=C(I,K)
      OUT=XIN-TDEC
      IF(OUT) 10,20,20
   10 KICK=1
      GO TO 30
   20 KICK=0
      IST=(OUT/XINC) + 1.5
   30 RETURN
      END
```

INDEX

INDEX